Crop Production Science in Horticulture Series

Series Editors: Jeff Atherton, Professor of Tropical Horticulture, University of the West Indies, Barbados, and Alun Rees, Horticultural Consultant and formerly Editor, *Journal of Horticultural Science and Biotechnology*.

This series examines economically important horticultural crops selected from the major production systems in temperate, subtropical and tropical climatic areas. Systems represented range from open field and plantation sites to protected plastic and glass houses, growing rooms and laboratories. Emphasis is placed on the scientific principles underlying crop production practices rather than on providing empirical recipes for uncritical acceptance. Scientific understanding provides the key to both reasoned choice of practice and the solution of future problems.

Students and staff at universities and colleges throughout the world involved in courses in horticulture, as well as in agriculture, plant science, food science and applied biology at degree, diploma or certificate level will welcome this series as a succinct and readable source of information. The books will also be invaluable to progressive growers, advisers and end-product users requiring an authoritative, but brief, scientific introduction to particular crops or systems. Keen gardeners wishing to understand the scientific basis of recommended practices will also find the series very useful.

The authors are all internationally renowned experts with extensive experience of their subjects. Each volume follows a common format covering all aspects of production, from background physiology and breeding, to propagation and planting, through husbandry and crop protection, to harvesting, handling and storage. Selective references are included to direct the reader to further information on specific topics.

Titles available:

1. **Ornamental Bulbs, Corms and Tubers** A.R. Rees
2. **Citrus** F.S. Davies and L.G. Albrigo
3. **Onions and Other Vegetable Alliums** J.L. Brewster
4. **Ornamental Bedding Plants** A.M. Armitage
5. **Bananas and Plantains** J.C. Robinson
6. **Cucurbits** R.W. Robinson and D.S. Decker-Walters
7. **Tropical Fruits** H.Y. Nakasone and R.E. Paull
8. **Coffee, Cocoa and Tea** K.C. Willson
9. **Lettuce, Endive and Chicory** E.J. Ryder
10. **Carrots and Related Vegetable *Umbelliferae*** V.E. Rubatzky, C.F. Quiros and P.W. Simon
11. **Strawberries** J.F. Hancock
12. **Peppers: Vegetable and Spice Capsicums** P.W. Bosland and E.J. Votava
13. **Tomatoes** E. Heuvelink
14. **Vegetable Brassicas and Related Crucifers** G.R. Dixon

*This book is dedicated to the memory of my brothers
Peter Thomas and Michael James Dixon*

Vegetable Brassicas and Related Crucifers

Geoffrey R. Dixon
Centre for Horticulture and Landscape, School of Biological Sciences, University of Reading, UK
(formerly at The Department of Bioscience, The Royal College, University of Strathclyde, Glasgow, UK)

CABI is a trading name of CAB International

CABI Head Office
Nosworthy Way
Wallingford
Oxfordshire OX10 8DE
UK

Tel: +44 (0)1491 832111
Fax: +44 (0)1491 833508
E-mail: cabi@cabi.org
Website: www.cabi.org

CABI North American Office
875 Massachusetts Avenue
7th Floor
Cambridge, MA 02139
USA

Tel: +1 617 395 4056
Fax: +1 617 354 6875
E-mail: cabi-nao@cabi.org

A catalogue record for this book is available from the British Library, London, UK.

Library of Congress Cataloging-in-Publication Data

Dixon, Geoffrey R.
Vegetable brassicas and related crucifers / Geoffrey R Dixon.
p. cm. -- (Crop production science in horticulture series ; 14)
Includes bibliographical references and index.
ISBN 0-85199-395-8 (alk. paper)
1. Brassica. 2. Cole crops. 3. Bok choy. I. Title. II. Series: Crop production science in horticulture ; 14.

SB317.B65D59 2006
635'.34--dc22

2006006061

ISBN: 978 0 85199 395 9

Typeset by Columns Design Ltd, Reading, UK.
Printed and bound in the UK by Biddles Ltd, King's Lynn.

Contents

Preface

Brassica vegetables serve mankind as they have done for millennia as sources of food, fodder and forage. The quiet mundanity of rural cabbage yards belies the botanical miracles taking place within them by these arch-exponents of genotypic and phenotypic diversity and flexibility. Brassicas provide one of the finest examples of convergent evolution in the horticultural forms of the European *Brassica oleracea* and Oriental *B. rapa*. These mirror each other in a diverse spectrum of fresh foodstuffs that have been selected and bred for particular regional preferences over many centuries. Travelling alongside through evolutionary time are numerous other taxonomic variants brought about through natural genetic manipulation and polyploidy that produce oil-rich and condiment crops.

This book *per force* of size concentrates on the vegetable brassicas, but has made great use of the science underlying the biggest world *Brassica* crop on an area basis, oilseeds and canola. It is impossible to study vegetable brassicas without being influenced by *B. napus* and the other oil-rich crops, but to include them in any detail would have required a vastly larger volume. Only very recently has science begun recognizing the nutritional benefits of both vegetables and oilseed brassicas. In comparative terms, only more recently have the biological properties of some of their wild relatives come to the fore. That tiny weed or rock garden speciality, depending on your horticultural predilection, *Arabidopsis thaliana*, has achieved enormous scientific clout as a model system for molecular biological studies. In future decades it could outstrip some of the crop vegetable brassicas in its economic significance.

In considering the horticultural science underpinning *Brassica* crops, I have attempted to think forward into those aspects that will be important for students and others who use this book and whose working careers may have three or more decades still to run. Consequently, there is an emphasis on husbandry as the key tool in pest, pathogen and physiological disorder control. Plant breeding and the potential for genetic manipulation are emphasized. The brassicas are of pivotal significance in providing wholesome and nutritious foodstuffs for our burgeoning world population and are capable of doing even greater service to mankind when tailored to use land with an elevated salt content and minimal water supply. Rich sources of pharmaceuticals could also flow from the brassicas in the next couple of decades.

The view of vegetable brassicas as expressed here is entirely my own, and I fully accept responsibility for all errors, omissions and unconventional thinking. Much tribute is due, however, to my very good friend Dr Michael Dickson of Cornell University, New York State, USA. He has been a constantly supportive mentor during the overlong gestation of the book. Mike's comforting and ever-kindly help is greatly valued. Indeed, he penned parts of Chapter 2 relating to Plant Breeding and Genetics, and sections concerned with pest and pathogen resistance in Chapter 7, for which his help is gratefully acknowledged.

Similarly, substantial gratitude is also due to Professor Paul Williams of Wisconsin University at Madison, USA; he has offered abiding friendship, deep intellectual sustenance and a common view of the convergence of biological science and practice over many decades.

The roots of this book are set deep in my own lifetime of scientific fascination with the biology of *Brassica* allied to their commercial production. These bind together early horticultural experiences in the Thames Valley, the broad fenlands of East Anglia from a Cambridge plant pathologist's perspective, the Scottish Mearns, Fife and the Moray Firth as Head of Horticulture in the Aberdeen School of Agriculture and Scottish Agricultural College. These experiences are topped by all too brief acquaintances with the kale yards of Europe, extensive and intensive efficiencies of the Valley Lands of California, the fertile fields of Wisconsin and the sheer magical diversities of Asian cropping. Founding influences in understanding the true nature of scholarship in the natural sciences based around the concept of '*One Foot in the Furrow*' were my trio of mentors at Wye College (then London University's Faculty of Agriculture and Horticulture). These true teachers, Professor Herbert W. Miles, Dr Edward H. Wilkinson and Mr Alan A. Jackson, entranced young minds in the early 1960s with a rich mixture of ecology, crop physiology, plant breeding and commercial instinct founded firmly in the nascent implications of Rachel Carson's *Silent Spring*.

Committing this text to paper needed the support and encouragement of many other friends and professional colleagues. Particular gratitude goes to Professors John Watson and John Anderson, previous and current Heads of the Department of Bioscience; Mr Stewart Roy, Faculty Officer in the Faculty of Science; and Professor Sir John Arbuthnott, Vice-Chancellor and Principal of the University of Strathclyde in Glasgow. Collectively they provided a welcome haven for scholarship. Recently, this Scottish haven has been exchanged for Professor Paul Hadley's equally warm welcome in Reading University's Centre for Horticulture and Landscape.

My dear wife, Kathy, continues with her unstinting support of my life and interests, aided by Lucy and Richard whose mirthful tolerance of Dad's odd interest in cabbages keeps me sane and smiling. Finally, but not least by any means, I thank my editor, Tim Hardwick, for his continued support and enthusiasm for this book that has helped ensure its completion.

Geoffrey R. Dixon
Sherborne, UK
2006

Acknowledgements (Copyrights)

Fig. 1.1. The United Nations Food and Agriculture Organisation (UNFAO), Rome.
Fig. 1.3. Professor P.H. Williams, University of Wisconsin, and Springer-Verlag GmbH, Heildeberg.
Fig. 1.4. New York Botanic Garden.
Fig. 1.5. Ms Joy Larcom and John Murray, London.
Fig. 1.6. Dr M.H. Dickson, Cornell University.
Fig. 2.1. Dr M.H. Dickson, Cornell University.
Fig. 2.2. Dr M.H. Dickson, Cornell University.
Fig. 2.3. Dr M.H. Dickson, Cornell University.
Fig. 2.4. Dr M.H. Dickson, Cornell University.
Fig. 2.5. Dr M.H. Dickson, Cornell University.
Fig. 2.6. Dr M.H. Dickson, Cornell University.
Fig. 2.7. Dr M.H. Dickson, Cornell University.
Fig. 2.8. Professor P.H. Williams, University of Wisconsin, and *Science* journal.
Fig. 2.9. Professor P.H. Williams, University of Wisconsin.
Table 2.4. Professor P.H. Williams, University of Wisconsin.
Fig. 3.1. Professor C.P. Thiagarajan, Tamil Nadu and *Phyton*, Horn.
Fig. 3.2. Professor C.P. Thiagarajan, Tamil Nadu and *Phyton*, Horn.
Fig. 3.3. Professor C.P. Thiagarajan, Tamil Nadu and *Phyton*, Horn.
Fig. 3.4. Professor C.P. Thiagarajan, Tamil Nadu and *Phyton*, Horn.
Fig. 3.5. Professor C.P. Thiagarajan, Tamil Nadu and *Phyton*, Horn.
Fig. 3.6. Professor C.P. Thiagarajan, Tamil Nadu and *Phyton*, Horn.
Fig. 3.7. Professor C.P. Thiagarajan, Tamil Nadu and *Phyton*, Horn.
Fig. 4.3. Professor Y. Fujima, Kagawa University, and the International Society for Horticultural Science, Leuven.
Fig. 4.4. Professor Y. Fujima, Kagawa University, and the International Society for Horticultural Science, Leuven.
Fig. 4.5. Dr D. Wurr, Warwickshire, and *Journal of Horticultural Science and Biotechnology*.
Table 5.1. Her Majesty's Stationery Office (HMSO), Norwich.

Table 5.2. Her Majesty's Stationery Office (HMSO), Norwich.
Table 5.4. Her Majesty's Stationery Office (HMSO), Norwich.
Table 5.5. Her Majesty's Stationery Office (HMSO), Norwich.
Fig. 5.1. Drs Kristen Stirling and Rachel Lancaster, Department of Primary Industries, Perth, Western Australia.
Table 5.11. Drs Kristen Stirling and Rachel Lancaster, Department of Primary Industries, Perth, Western Australia.
Table 5.12. Drs Kristen Stirling and Rachel Lancaster, Department of Primary Industries, Perth, Western Australia.
Table 6.2. *Australian Journal of Agricultural Research*, Adelaide.
Table 6.4. Collins Publishers, London.
Fig. 6.1. Professor R. Cousens, University of Melbourne and Association of Applied Biologists.
Fig. 6.3. Dr A. Grundy, Warwick-Horticulture Research International, Wellesbourne, and *Journal of Applied Ecology*.
Fig. 6.4. Dr D.A.G. Kurstjens, University of Wageningen, and Elsevier Science bv.
Fig. 7.1. Skye Instruments Ltd, Llandrindod Wells, Wales.
Fig. 7.4. Professor J. Kingsolver, University of North Carolina, and *Physiological and Biochemical Zoology*.
Fig. 7.6. Pearson Education, London.
Fig. 7.7. Dr D.S. Hill, England.
Fig. 7.8. Dr D.S. Hill, England.
Fig. 7.9. Pearson Education, London.
Fig. 7.10. Pearson Education, London.
Fig. 7.11. Dr D.S. Hill, England.
Fig. 7.12. Dr D.S. Hill, England.
Fig. 7.13. Pearson Education, London.
Fig. 7.20. Dr J. Vincente, University of Warwick-Horticulture Research International, Wellesbourne, UK.
Fig. 7.21. Dr M.H. Dickson, Cornell University.

1

Origins and Diversity of *Brassica* and its Relatives

Understanding the brassica[1] vegetables involves a fascinating, biological journey through evolutionary time, witnessing wild plant populations interbreeding and forming stable hybrids. Mankind took both the wild parents and their hybrid progeny, refined them by selection and further combination, and produced over biblical time crops that are, together with the cereals, the mainstay of world food supplies. Genetic diversity and flexibility are characteristic features of all members of the family Brassicaceae (previously the Cruciferae). Possibly, these traits encouraged their domestication by Neolithic man. Records show that the Ancient Greeks, Romans, Indians and Chinese all valued and used them greatly. The etymology of *Brassica* has been contested since Herman Boerhaave suggested in 1727 that it might come from the Greek *αποτουβραξει-υ*, Latin *vorare* ('to devour') (Henslow, 1908). An alternative derivation from *Bresic* or *Bresych*, the Celtic name for cabbage, was suggested by Hegi (1919). This is a contraction of *praesecare* ('to cut off early'), since the leaves were harvested for autumn and early winter food and fodder. Another suggested origin is from the Greek *β-ραοσειν*, crackle, coming from the sound made when the leaves are detached from the stem (Gates, 1953). A further suggestion is a Latin derivation from 'to cut off the head' and was first recorded in a comedy of Plautus in the 3rd century BC. Aristotle (384–322 BC), Theophrastus (371–286 BC), Cato (234–149 BC), Columella (1st century AD) and Pliny (23–79 AD) all mention the importance of brassicas.

Further eastwards, the ancient Sanskrit literature *Upanishads* and *Brahamanas*, originating around 1500 BC, mention brassicas, and the Chinese *Shih Ching*, possibly edited by Confucius (551–479 BC), refers to the turnip (Keng, 1974; Prakash and Hinata, 1980). European herbal and botanical treatises of the Middle Ages clearly illustrate several *Brassica* types, and Dutch paintings of the 16th and 17th centuries show many examples of brassicas. In the 18th century, species of coles, cabbages, rapes and mustards were described in the genera *Brassica* and *Sinapis* in *Institutiones Rei Herbariae* (de Tournefort, 1700) and *Species Plantarum* (Linnaeus, 1735). Probably the most important early, formal classification of *Brassica* was made by Otto Eugen Schultz (1874–1936), published in *Das Pflanzenreich* and *Die Naturlichen Pflanzenfamilien* (Schultz, 1919 and 1936, respectively). These classifications were supported broadly by the great American botanist and horticulturist L. H. Bailey (1922, 1930).

Brassica crops worldwide provide the greatest diversity of products used by man derived from a single genus. Other members of the family Brassicaceae extend this diversity. Collectively, brassicas deliver leaf, flower and root vegetables that are eaten fresh, cooked and processed; used as fodder and forage, contributing especially overwintering supplies for meat- and milk-producing domesticated animals; sources of protein and oil used in low fat edible products, for illumination and industrial lubricants; condiments such as mustard, herbs and other flavourings; and soil conditioners as green manure and composting crops. New avenues for our use of brassicas are becoming apparent. The tiny cruciferous weed *Arabidopsis thaliana* has become a power house for molecular biology, Fast Plants™ are forming important educational and research tools, and more generally brassicas are seen as functional foods with long-term roles in the fight against cancer and coronary diseases.

Wild diploid *Brassica* and their related hybrid amphidiploids (Greek: *amphi* = both; *diploos* = double; possessing the diploid genomes from both parents) evolved naturally in inhospitable places with abilities to withstand drought, heat and salt stresses. The Korean botanist U (1935) deduced that three basic diploid brassica forms were probably the parents of subsequent amphidiploid crops. *Brassica nigra* (black mustard), itself the ancestor of culinary mustards, is found widely distributed as annual herbs growing in shallow soils around most rocky Mediterranean coasts. Natural populations of *B. oleracea* and associated types are seen as potential progenitors of many European cole vegetables. These inhabit rocky cliffs in cool damp coastal habitats. They have slow, steady growth rates and are capable of conserving water and nutrients. The putative ancestor of many Oriental brassica vegetables, *B. rapa*,[2] originates from the fertile crescent in the high plateaux regions in today's Iran–Iraq–Turkey. Here these plants grow rapidly in the hot, dry conditions, forming copious seed. Other family members evolved as semi-xerophytes in the Saharo-Sindian regions in steppe and desert climates. Early human hunter-gatherers and farmers discovered that the leaves and roots of these plants provided food and possessed medicinal and purgative properties when eaten either raw or after boiling. Some types supplied lighting oil extracted from the seed and others were simply very useful as animal feed. These simple herbs have developed into a massive array of essential vegetable crops grown and marketed around the world (see Fig. 1.1).

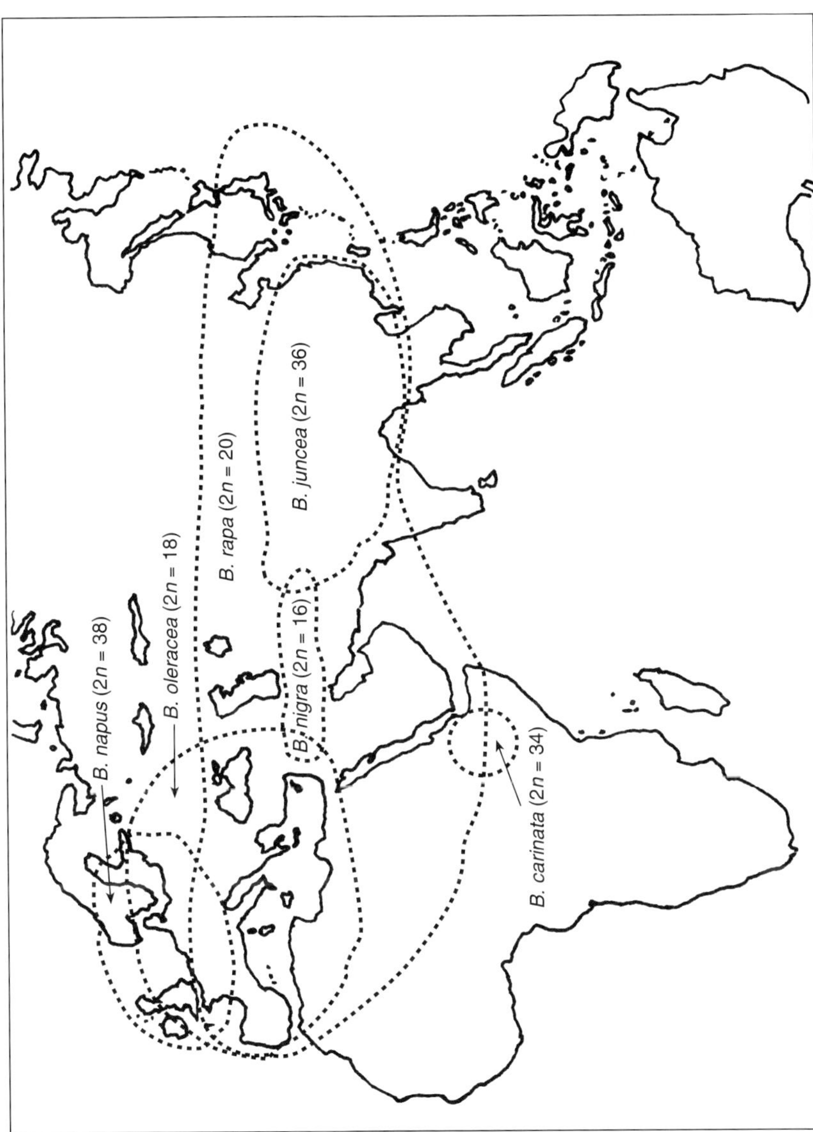

Fig. 1.1. Biogeography of the origins and diversity of the major crop-founding *Brassica* species (United Nations Food and Agriculture Organization, Rome).

BIODIVERSITY

An array of wild diploid *Brassica* species still cling to survival in inhospitable habitats and they are indicative of the natural diversity of this genus (see Table 1.1). This diversity has expanded many fold in domestication and the service of mankind.

Wild hybrids

Wild *Brassica* and its close relatives hybridized naturally, forming polyploids. These amphidiploids together with their parental wild diploids were key building blocks from which our domesticated brassica crops have evolved. Three hybrid species are of particular interest as ancestors of the crop brassicas.

Brassica carinata (n = 17) was formed from *B. oleracea* (n = 9) × *B. nigra* (n = 8). This species is characterized by the slow steady growth of *B. oleracea* and the mustard oil content of *B. nigra.* Wild forms of *B. carinata* are not known, but primitive domesticated types are cultivated in upland areas of Ethiopia and further south into Kenya. This hybrid may have originated from kale land races of *B. oleracea* types hybridizing with wild or semi-domesticated forms of *B. nigra.* Both kale and carinata crops thrive in cool environments typical of the Ethiopian plateau. The local farmers grow them in 'kale gardens'. This term, translated into many languages and dialects, is commonly found throughout those rural areas using vegetables derived from *B. oleracea* types. Carinata crops themselves are alternatively named 'guomin', Abyssinian mustard or Ethiopian cabbage, and provide leafy vegetables and sources of oil. Cabbage itself can be a ubiquitous term used to describe cole brassicas and not necessarily synonymous with the sophisticated heads seen on today's supermarket shelves.

Brassica juncea (n = 18) is a hybrid between *B. rapa* (n = 10) and *B. nigra* (n = 8) (Fransden, 1943) producing large leaves and with the rapid growth of *B. rapa* and the mustard oil of *B. nigra*. Use of *B. juncea* as a source of vegetable oil is gaining importance in India; while throughout Asia, especially in China and Japan, the plant has a great diversity of cultivated forms used as staple vegetables of immense dietary importance. Reputably, wild forms are still found on the Asia Minor plateau and in southern Iran.

The third hybrid, *B. napus* (n = 19), has wild forms in Sweden, Denmark, The Netherlands and the UK. It developed from *B. rapa* (n = 10) × *B. oleracea* (n = 9). This hybrid may have formed as *B. oleracea* types expanded their range along the coasts of northern Europe and *B. rapa* extended from the Irano-Turanian regions. Alternatively, *B. napus* may have Mediterranean origins or, as seems likely, there were several centres of evolution. The wild populations of *B. napus* have acquired major scientific significance recently as they present

Table 1.1. Examples of the diversity and characteristics of selected natural *Brassica* species.

Name	Chromosome count	Geographical distribution	Habitat
B. amplexicaulis	$n = 11$	Intermountain area south-east of Algiers	Small plant, colonizes colluvial slopes, especially medium-sized gravel
B. barrelieri	$n = 9$	Iberian peninsula, extending to Morocco and Algeria	Common on wastelands, especially areas of more dense vegetation
B. elongata	$n = 11$	Plateau steppe lands of southeastern Europe and western Asia as far as Iran	Semi-arid areas
B. fruticulosa subsp. *fruticulosa*	$n = 8$	Found around the Mediterranean coasts especially amongst pine trees; the subspecies *cossoniana* extends beyond the coastal zone and is found on the Saharan side of the Moyen Atlas in Morocco	Biennial or perennial
B. fruticulosa subsp. *cossoniana*	$n = 8$		Annual Both subspecies require well-drained sites; on inland sites characteristically colonizing stony mountain slopes and alluvial areas
B. maurorum	$n = 8$	Endemic to North Africa; inhabits arable land in Morocco and Algeria	Colonizes stony pastures in semi-arid areas from the coast to low mountainous zones; on arable land grows to 2 m high in dense clumps similar to *B. nigra*
B. oxyrrhina	$n = 9$	Southern Portugal, Spain and northwestern Morocco	Coastal sandy habitats
B. repanda and *B. gravinae*	$n = 10$	Inland rocky areas	These species grow together in the lithosol in crevices of rocky outcrops. A polypoid ($n = 20$) form has been found north of Biskra in Algeria
B. spinescens	$n = 8$	Endemic to North Africa	Coastal calcareous or siliceous cliffs. Diminutive growth habit with small, thick glabrous leaves; *B. fruticulosa, B. maurorum* and *B. spinescens* possibly form a single cytodeme
B. tournefortii	$n = 10$	Coastal areas of the Mediterranean extending to western Asia as far as India	Capable of colonizing arid alluvial sand where other vegetation is sparse

After Tsunoda *et al.* (1984).

means of determining the potential for gene flow to and from genetically modified cultivars of oilseed rape (also *B. napus*) (Hailles *et al.*, 1997; Hall *et al.*, 2000; Bond *et al.*, 2004).

The relationships between the hybrid amphidiploids and their parental species are summarized in the gene flow triangle of U (1935) (Fig. 1.2).

Diversity within the amphidiploids

Considerable genetic diversity is present within the three amphidiploid species (Song *et al.*, 1996). Based upon studies of genetic diversity, *B. napus* may be considered as the most ancient amphidiploid, succeeded by *B. juncea* and *B. carinata*. Two major factors are responsible for general diversity within amphidiploids: multiple hybridizations with different diploid parents and genome modifications following polyploidization. A good example of multiple hybridizations was found in *B. napus*. Four cytoplasmic types developed in a group of *B. napus* accessions; these matched with different parental diploid

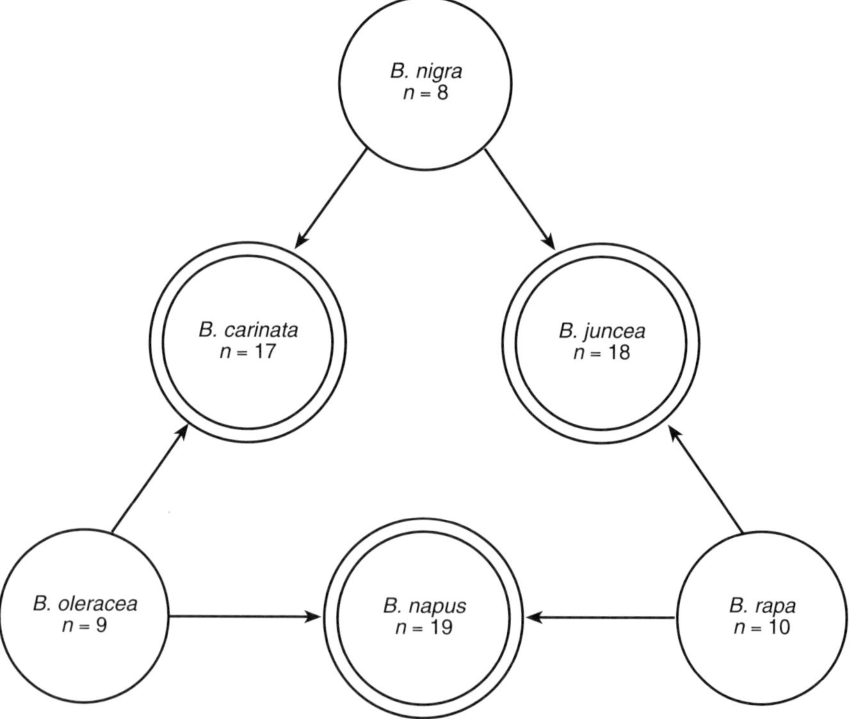

Fig. 1.2. Relationships between diploid and amphidiploid crop-founding *Brassica* species (the 'triangle of U'; U, 1935).

cytoplasms. Direct evidence for genome changes after polyploidization was obtained by studying synthetic amphidiploids (Song *et al.*, 1993).

Artificial synthesis of three analogues of *B. napus*, *B. juncea* and *B. carinata* permitted their analysis for morphological traits, chromosome number and restriction fragment length polymorphisms (RFLPs) in chloroplast, mitochondrial and nuclear DNA clones, and comparison with the natural amphidiploids. This showed that the synthetic hybrids were closer to their diploid parents as compared with the polyploids. Genome changes, mainly involving either loss or gain of parental fragments and novel fragments, were seen in the early generations of the four synthetic amphidiploids. The frequency of genome change and the direction of its evolution in the synthetic hybrids were associated with divergence between the parental diploid species. Rapid genome changes resulted in faster divergence among the derivatives of synthetic amphidiploids which, by providing raw materials for selection, played an important role in the evolutionary success of *Brassica* amphidiploids and many polyploid lineages. Based on RFLP data from *Brassica*, Song *et al.* (1988) generated a phylogenetic tree that permitted the quantitative analysis of the relationships between *Brassica* species. Results from this study suggest the following:

1. *Brassica nigra* originated from one evolutionary pathway with *Sinapis arvensis* or a close relative as the likely progenitor, whereas *B. rapa* and *B. oleracea* came from another pathway with a possible common ancestor in wild *B. oleracea* or a closely related species possessing nine chromosomes.
2. The amphidiploid species *B. napus* and *B. juncea* have evolved through different combinations of the diploid morphotypes and thus polyphyletic origins may be a common mechanism generating the natural occurrence of amphidiploids in *Brassica*.
3. The cytoplasm has played an important role in the nuclear genome evolution of amphidiploid species when the parental diploid species contain highly differentiated cytoplasms.

This research has generated a hypothetical scheme for the possible evolution of *Brassica* species and their subsequent domesticated variants (Fig. 1.3).

Contrasting the physiology and morphology of wild and cultivated brassicas

It will be evident by now that many of the wild *Brassica* spp. and their close allies inhabit dry coastal, arid rocky or desert habitats. These wild plants have very thick leaves containing less chlorophyll and many more cell wall components compared with cultivated plants. Typically, they have well-developed xylem vessels and small leaf areas; these characteristics increase the efficiency of water conservation in plants. The foliage of wild xerophyllous

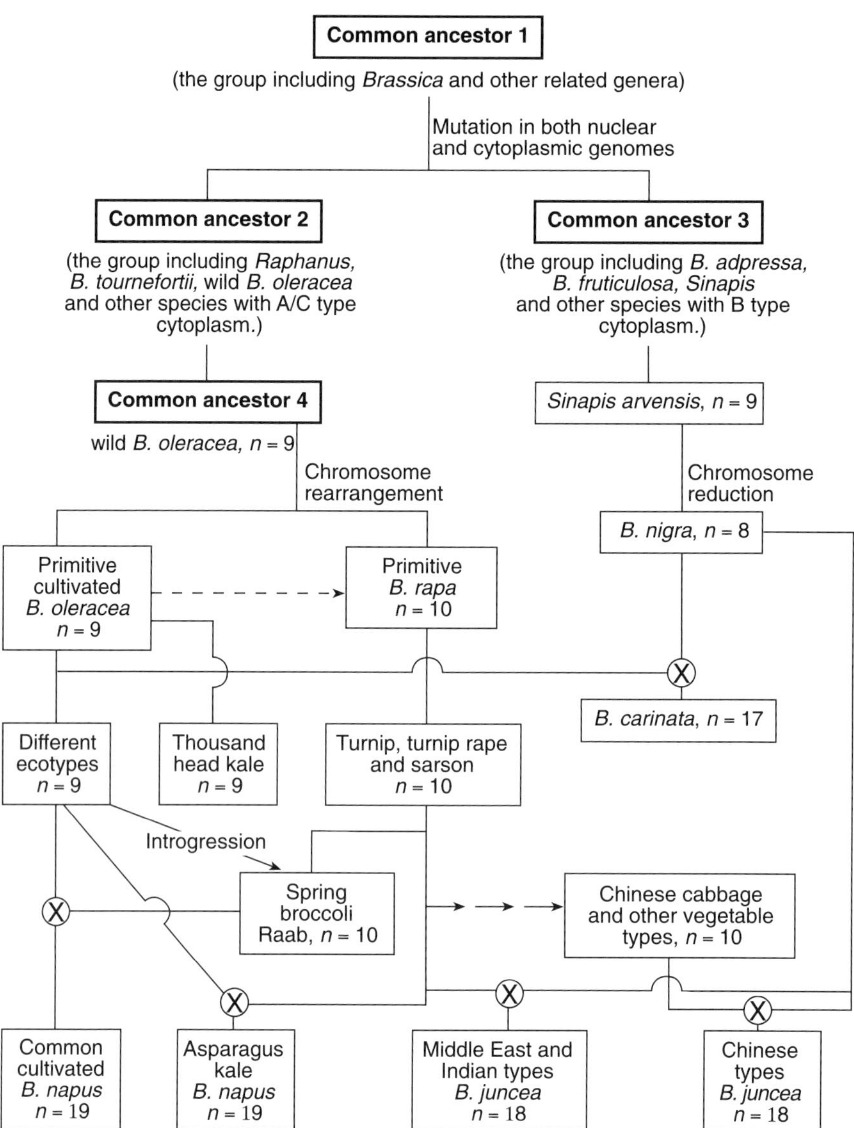

Fig. 1.3. Scheme of genome evolution in crop-founding *Brassica* species and cultivated forms (*Song et al.*, 1988).

plants has evolved high photosynthetic rates per unit leaf area (or per quantum of light received) even in dry air conditions.

Conversely, cultivated brassicas have broadly expanded, thin leaves well supplied with chlorophyll. These characteristics are advantageous for receiving, absorbing and utilizing solar radiation when there are ample supplies of water and nutrients available. Typically these are mesophyllous environments found in fertile, cultivated fields where growers strive to provide optimal nutrition and water.

Similar contrasts between the ecology of wild progenitors of crops and cultivated plants are found between the wild allies of wheat and artificial cultivars. Wild forms possess small, thick leaves, whereas wheat cultivars have large, thin leaves. Both wild *Brassica* and *Hordeum* spp. evolved strategies for successful growth under arid conditions involving the restriction of transpiration, intensification of water movement to sites of photosynthesis, restriction of light absorption and efficient fixation of the absorbed solar radiation. Such traits became redundant in cultivation and consequently were removed by many generations of on-farm selection and, since the start of the 20th century, breeding based on increasing knowledge of the biological components of plant productivity.

Other *Brassica* relatives illustrating the biodiversity of this family

Several other members of the Brassicaceae illustrate the diversity of this family and its evolution in cultivation; examples are summarized in Tables 1.2 and 1.3.

CULTIVATED *BRASSICA* SPECIES AND FORMS

Brassica oleracea group (*n* = 18) – the European group

The European (Occidental) *Brassica* vegetables originate from *B. oleracea* and probably some closely related Mediterranean species. They can be divided into subordinate groups often at the variety (var.), subvariety or cultivar (cv.) levels. Much of the basis of current understanding of diversity within this group comes from the detailed studies of American horticultural botanist Liberty H. Bailey (1922, 1930, 1940).

Multiple origins and parents

One school of thought suggests that the constituent crops within the *B. oleracea* group have multiple origins derived from cross-breeding between closely related *Brassica* species living in geographical proximity to each other. In consequence, the taxonomy of parents and progeny is confused and clouded still further by millennia of horticultural domestication. For example, the progenitors of headed cabbages and kales were postulated by Neutrofal (1927) as *B. montana* and of kohlrabi as *B. rupestris*. Later, Schiemann (1932)

Table 1.2. Characteristics of *Eruca, Raphanus* and *Sinapis* – genera allied to crop-founding *Brassica* (after Tsunoda *et al.*, 1984).

Eruca sativa (n = 11, syn. *E. vescaria* subsp. *sativa*) commonly know as 'garden rocket', which is a salad vegetable in southern Europe and in India forms an oilseed (taramira) and is used for fodder. *E. sativa* has recently become a popular salad vegetable in western Europe. It occurs widely distributed across southern Europe, North Africa, western Asia and India. Variation within the cytodeme *Eruca* is substantial, with ecotypes evolved for several habitats in relation to available water.

Sinapis alba (n = 12) (white mustard) grows wild in Mediterranean areas with abundant moisture and ample soil nutrients, often coexisting with *S. arvensis* (n = 9) (charlock or wild mustard), although the latter prefers lower soil moisture levels. In dry areas, *S. turgida* (n = 9) is common. In wet habitats, *S. alba* reaches 2 m in height with a high mustard oil content; it is grown in Europe as 'white' or 'yellow' mustard. Cultivation is increasing particularly in Canada in combination with 'brown' mustard, *B. juncea.*

Raphanus (radish). There are about 18 genera in the subtribe Raphininae including *Rapistrum, Cakile* and *Crambe.* Possibly all *Raphanus* spp. with n = 9 form a single cytodeme. The genus is found along Mediterranean coasts, forming the dominant plant on coastal areas of the Sea of Marmara and the Bosphorus Straits. Around field margins, *R. raphanistrum* and *R. rapistrum rugosum* tend to coexist. Culinary radish (*R. sativum*) possibly evolved in southern Asia and provides a wide diversity of root forms and flavours. Fodder radish (*R. sativus* var. *oleiferus*) is a source of animal fodder obtained from both roots and foliage. Other varieties *radicula, niger* and *mougri* are identified by Banga (1976). *Crambe* is a source of seed oil and is beginning to be used more extensively in North America.

realized that several of the Mediterranean wild types had formed the origins for locally cultivated land races. Schulz (1936) supported this view, and identified *B. cretica* as a progenitor of cauliflower and broccoli.

Lizgunova (1959) grouped cultivars into five different species and proposed a multiple origin from wild forms. Helm (1963), in devising a triple origin, combined cauliflower, broccoli and sprouting broccoli in one line, thousand-headed kale and Brussels sprouts in another and all other crop forms in a third. Further analysis was completed by Toxopeus (1974), and Toxopeus *et al.* (1984) suggested that for simplicity a horticulturally based taxonomy was preferable to attempted botanical versions.

Potential wild species contributing to the *B. oleracea* group

Populations of wild relatives which cross-fertilize with *B. oleracea* and form interbreeding groups are found on isolated cliffs and rocky islets, forming distinct units often displaying phenotypic differentiation, leading to several layers of variation superimposed on each other (see Table 1.4).

Crops developed within *Brassica oleracea* and allies

The vast array of crop types that have developed within *B. oleracea* (and also *B. rapa*) is probably unique within economic botany (Nieuwhof, 1969). This

Table 1.3. Characteristics of natural relatives of crop-founding *Brassica*.

Name	Chromosome count	Geographical distribution	Habitat
Diplotaxis acris	$n = 7$	Saharo-Sindian area	Dry, desert regions
Diplotaxis harra	$n = 13$		
Diplotaxis tenuisiliqua	$n = 9$	North Africa	Moist areas, sometimes associated with *Sinapis alba*
Erucastrum cardaminoides	$n = 9$	Endemic to the Canary Isles	
Erucastrum laevigatum	$n = 28$	Southern Italy, Sicily and North Africa	Probably an autotetraploid of *E. virginatum* ($n = 7$)
Erucastrum leucanthum	$n = 8$	Endemic to North Africa, particularly the seaward side of the Moyen Atlas	Coexists with *Hirschfeldia incana*
Hirschfeldia incana	$n = 7$	Dominant component of Mediterranean flora	Large roadside colonies, especially favoured by fine textured soils
Hutera spp.	$n = 12$	Iberian peninsula	Rocky outcrops and colluvial sites; *Hutera* and *Rhynchosinapis* form a cytodeme complex found in the Spanish Sierra Montena regions
Sinapidendron spp.	$n = 10$	Madeira	Rocky cliffs; perennial habit

After Tsunoda *et al.* (1984).

has led to acceptance at the subspecies and variety (cultivar)[3] levels of descriptions based around the specialized morphology of the edible parts and habits of growth within the crop types (Wellington and Quartley, 1972).

Brassica oleracea (the cole or cabbage brassicas)

BRUSSELS SPROUTS: *B. OLERACEA* L. VAR. *GEMMIFERA* DC Brussels sprouts may have emerged in the Low Countries in the medieval period and risen to prominence in the 18th century around the city of Brussels. Subsequently, they became established as an important vegetable crop in northeastern Europe, especially the northern Netherlands and parts of the UK. Local open-pollinated land races were developed suited to specific forms of husbandry but usually subdivided into early, midseason and late maturity groups. Often they would be capable of resisting pests and pathogens common within their locality and have morphologies adapted to prevailing climatic conditions. In the 1930s, early maturing types were developed in Japan where many of the original F_1 hybrids (e.g. 'Jade Cross') were produced; these and their derivatives formed the basis for cultivars ideally suited to the emerging 'quick-freeze' vegetable processing industry. The entire worldwide crop of Brussels

Table 1.4. Potential progenitors which interacted to found crop-founding *Brassica oleracea* in Europe.

Brassica cretica: populations occur around the Aegean, including Crete, southern Greece and in southwestern Turkey. The plant has a woody much-branched habit carrying glabrous, fleshy leaves persisting in perennial form for 5–8 years but flowering in the first year under favourable environments. The inflorescence axis elongates between buds prior to opening with light yellow to white flowers. A subspecies *B. cretica nivea* grows on the mountainous cliffs and gorges of the Peleponnese and Crete, while the subspecies *B. cretica cretica* populates mountainous cliffs of the same area. Differentiation of *B. cretica nivea* rests on possession of small white to creamy flowers, strongly reticulated seeds, a tall stem and long-petiolated leaves. Hybrids between both subspecies are found where populations currently overlap or have done so in the recent past.

The *Brassica rupestris–incana* complex: this group has developed a number of distinct regional variants. The complex is characterized by forming a strong central stem, a large apical inflorescence succeeded by further branching. The top of each partial inflorescence forms a tight grouping of buds which may open before the axis between them elongates. Foliage consists of large petiolate leaves that are poorly structured since they shrivel when past maturity, in contrast to *B. cretica* which retains mature leaves with a recognizable morphology. Hairs are characteristically found on both seedlings and adults. Distribution includes Sicily, southern and central Italy and parts of the west of the former Yugoslavia. Species which have been separated but which are included within the complex include: *B. incana* Ten., *B. villosa* Biv., *B. rupestris* Rafin., *B. tinei* Lojac., *B. drepanensis* (Car.) Dam., *B. botteri* Vis., *B. mollis* Vis. and *B. cazzae* Ginzb. & Teyb.

Brassica macrocarpa Guss: a species restricted to Isole Egadi in the west of Sicily with a habit similar to *B. rupestris–incana* with which it forms fertile hybrids, and characterized by possessing smooth-surfaced, thick fruits containing seeds produced in two rows within the loculus. There are questions as to whether this species should be placed in the genus *Brassica* in view of the seed capsule morphology. Conceivably this species is a subordinate of the *B. rupestris–incana* complex.

Brassica insularis Moris: found in Corsica, Sardinia and Tunisia. The plants are similar in low branching habit to *B. cretica* but the stiff glabrous leaves with pointed lobes and large white fragrant flowers and seed glucosinolates differ from other *Brassica* spp. Species rank has been given to the Tunisian form as *B. atlantica* (Coss.) Schulz., but the morphology and intercrossing indicate clear inclusion in *B. insularis* without subdivision.

Brassica montana Pourr. (syn. *B. robertiana* Gay): found growing in northeastern coastal areas of Spain, southern France and northern Italy, these plants are shrubby perennials with lobed green, glabrous or possibly hairy leaves. An intermediate status between the *B. rupestris–incana* complex and *B. oleracea* is suggested by biogeographical and morphological evidence.

Brassica oleracea L. s.str.: found on coasts of northern Spain, western France and southern and southwestern Britain as a stout perennial forming strong vegetative stocks which then flower and branch. The greyish coloration and glabrous nature of leaf surfaces distinguish *B. oleracea*.

Brassica hilarionis Holmb.: located solely in the Kyrenia Mountains of Cyprus, *B. hilarionis* has fruit reminiscent of the size and texture of *B. macrocarpa* and vegetatively a habit and leaf morphology close to *B. cretica*.

Note: Species such as *B. balearica* Pers. and *B. scopulorum* Coss & Dur. have been excluded since their failure to cross with the *B. oleracea* group make it dubious as to whether they should be included with it.
After Tsunoda *et al.* (1984).

sprouts is now dominated by a very limited F_1 germplasm derived by American, Dutch and Japanese breeders originating from the initial crosses. Botanically, the plants are biennial with simple erect stems up to 1 m tall. Axillary buds develop into compact miniature cabbage heads or 'sprouts' that are up to 30 mm in diameter. At the top of the stem is a rosette of leaves; the leaves are generally petiolate and subcircular and the leaf blade is rather small (see Chapter 2 for floral biology).

CAULIFLOWER AND BROCCOLI: *B. OLERACEA* L. VAR. *BOTRYTIS* L. (CAULIFLOWER), *B. OLERACEA* L. VAR. *ITALICA* PLENCK (BROCCOLI) A remarkable diversity of cauliflower- and broccoli-like vegetables developed in Europe, probably emanating from original highly localized crops in Italy and possibly evolved from germplasm introduced in Roman times from the eastern Mediterranean. A classification of the colloquial names used to describe these crops was proposed by Gray (1982) and is shown in Table 1.5.

Over the past 400 years, white-headed cauliflowers (derived from the Latin *caulis* (stem) and *floris* (flower)) have spread from Italy to central and northern Europe, which became important secondary centres of diversity for the annual and biennial cauliflowers now cultivated worldwide in temperate climates. Cauliflowers adapted to hot humid tropical conditions have evolved in India during the past 200 years from biennial cauliflower mainly of British origins following traders and colonizers.

Crisp (1982) proposed a taxonomic basis for grouping the various types of cauliflower found in cultivation; he admitted that this has limitations but at least it gives order where little previously existed (Table 1.6). Molecular biological techniques offer means to understand the relationships between

Table 1.5. *Brassica oleracea* varieties *botrytis* and *italica* with associated colloquial crop names.

Brassica oleracea L. var. *italica* Plenck	Purple-sprouting broccoli Cape broccoli Purple cauliflower Calabrese and other green-sprouting forms (broccoli in North America) White-sprouting broccoli
Brassica oleracea L. var. *botrytis* DC	Cauliflower Heading broccoli Perennial broccoli Bouquet broccoli White-sprouting broccoli[a]

[a]White-sprouting broccolis are thought to have evolved independently in northern Europe. Their close affinity to winter-hardy cauliflower suggests that the late form may be more correctly regarded as a form of *B. oleracea* var. *botrytis*.
After Gray (1982).

vegetable brassicas. Microsatellites, also known as simple sequence repeats (SSRs), are short tandem repeats of nucleotides of 1–6 bp in length that can be repeated up to 100 times. These offer means for discriminating between varieties and cultivars, and measuring levels of genetic diversity within species such as *B. oleracea*. By screening large numbers of DNA sequences from *B. oleracea*, Tonguç and Griffiths (2004a) demonstrated that diversity was least in cauliflower, intermediate in broccoli (calabrese) and greatest in cabbage. This illustrates the highly intensive nature of current cauliflower and broccoli commercial breeding programmes frequently centred on the use of a limited genepool, while until very recently at least cabbage breeding has made use of wider crosses. Several other molecular techniques such as the use of RFLPs, randomly amplified polymorphic DNA sequences (RAPDs) and amplified fragment length polymorphisms (AFLPs) are also useful tools for such studies. The relationships between wild Sicilian populations of *B. oleracea* are discussed by Geraci *et al.* (2001), while Astarini *et al.* (2004) have begun to unravel the descent of cauliflower types used in Lembang in Western Java from material introduced from India in the 19th century and related them to current Australian cultivars.

Table 1.6. Biogeography and characters of cauliflower (*B. oleracea* var. *botrytis*) groups.

Group name	Characteristics	Common types
Italian	Very diverse, includes annuals and biennials and types with peculiar curd conformations and colours	Jezi Naples (= Autumn Giant) Romanesco Flora Blanca
North-west European biennials	Derived within the last 300 years from Italian material	Old English Walcheren Roscoff Angers St Malo
Northern European annuals	Developed in northern Europe for at least 400 years. Origin unknown, perhaps Italian or possibly eastern Mediterranean	Le Cerf Alpha Mechelse Erfurt Danish
Asian	Recombinants of European annuals and biennials, developed within the last 250 years. Adapted to tropical climates	Four maturity groups are recognized by Swarup and Chatterjee (1972)
Australian	Recombinants of European annuals and biennials, and perhaps Italian stocks; developed during the last 200 years	Not yet categorized

From Crisp (1982).

Cauliflower is a biennial or annual herb, 50–80 cm tall at the mature vegetative stage and 90–150 cm when flowering. The root system is strongly ramified, concentrating in the top 30 cm of soil, with thick laterals penetrating to deeper layers. The stem is unbranched, 20–30 cm long and thickened upwards. The leaves are in a rosette (frame) of 15–25 large oblong more or less erect leaves surrounding the compact terminal flower head (curd). Usually lateral buds do not develop in the leaf axils. The glabrous leaves are almost sessile and coated with a layer of wax; the leaf blade is grey to blue-green in colour with whitish main and lateral veins. Leaves vary in shape from short and wide (40–50 cm × 30–40 cm) with curly edges to long and narrow (70–80 × 20–30 cm) with smooth edges. The curd consists of a dome of proliferated floral meristems that are white to cream or yellow colour growing on numerous short and fleshy peduncles. The curd varies from a rather loose to a very solid structure, with a flattened to a deeply globular shape from 100 to 400 mm in diameter. Young leaves may envelop the curd until a very advanced stage of development is reached. Bolting cauliflower plants often have several flower stalks (see Chapter 2 for floral biology).

Broccoli is an Italian word from the Latin *brachium*, meaning an arm or branch. In Italy, the term is used for the edible floral shoots on brassica plants, including cabbages and turnips, and was originally applied to sprouting forms, but now includes heading types which develop a large, single, terminal inflorescence. Broccoli with multiple green, purple or white flower heads (sprouting broccoli) became popular in northern Europe in the 18th century. Broccoli with a single main green head (calabrese; the name has been taken from the Calabria region of Italy) was introduced into the USA by Italian immigrants during the early 20th century (Fig. 1.4). It has become a popular 'convenience' vegetable, and has spread back into Europe from the USA and into Japan and other parts of the Pacific Rim over the past 50 years.

The white-heading forms are also colloquially referred to as cauliflower. Broccoli is often used to describe certain forms of cauliflower, notably in the UK where the term heading or winter broccoli is traditionally reserved for biennial types. The term broccoli without qualification is also generally applied in North America to the annual green-sprouting form known in UK and Italy as calabrese. The term sprouting as used in sprouting broccoli refers to the branching habit of this type, the young edible inflorescences often being referred to as sprouts. The term Cape used in conjunction with broccoli or as a noun is traditionally reserved for certain colour heading forms of *B. oleracea* var. *italica*. A classification of broccoli is given in Tables 1.7 and 1.8.

Broccoli (the single-headed or calabrese type) differs from cauliflower in the following respects: the leaves are more divided and petiolate; and the main head consists of clusters of fully differentiated green or purple flower buds which are less densely arranged with longer peduncles. Axillary shoots with smaller flower heads usually develop after removal of the dormant terminal shoot. The flower head is fully exposed from an early stage of development.

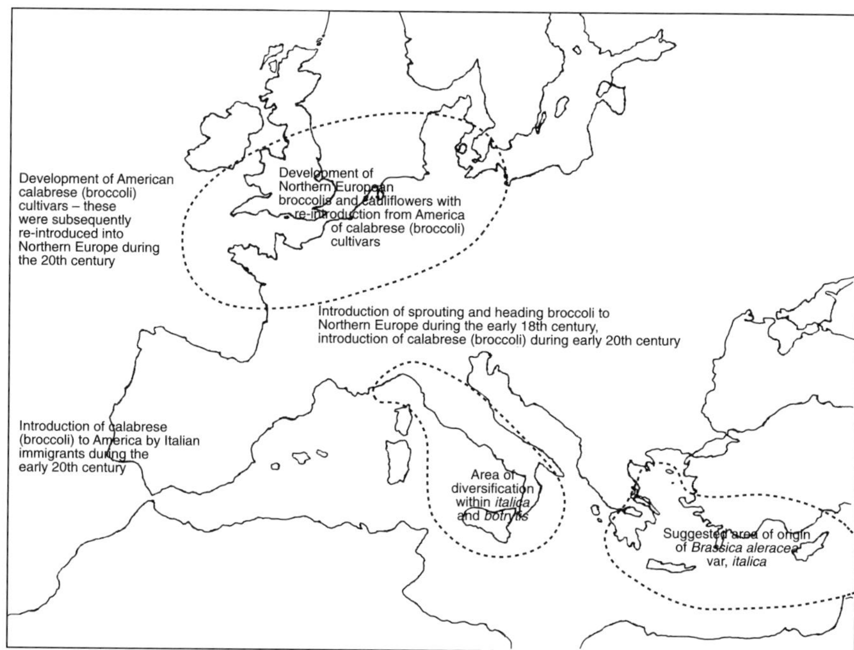

Fig. 1.4. Biogeography of the evolution of broccoli and cauliflower in Europe and north Africa (after Gray, 1982).

Table 1.7. Classification of colour-heading and sprouting broccoli (*B. oleracea* var. *italica*).

Coloured-heading types	
Dark purple heading	Early, intermediate and late maturing cultivars
Copper coloured or purplish-brown	
Green heading	
Sulphur-coloured or yellowish-green heading	
Sprouting types	
Green sprouting	
Purple sprouting	

After Giles (1941).

Broccoli plants also carry inflorescences from the lateral branches. Sprouting forms of broccoli bear many, more or less uniform, relatively small flower heads instead of the single large head of the calabrese type.

CHINESE KALE: *B. OLERACEA* L. VAR. *ALBOGLABRA* (L.H. BAILEY), MUSIL Chinese kale has formed a cultivated stock since ancient times without apparently obvious wild progenitors, but there are possible similarities to *B. cretica* subsp. *nivea*. Following early cultivation in the eastern Mediterranean trade centres, it

Table 1.8. Classification of Italian sprouting broccoli (*B. oleracea* var. *italica*) by morphological types.

Type	Description
Green sprouting of Naples	Small shoots produced on long stems, considered to be a counterpart to the white-sprouting broccoli
Early summer broccoli of Naples	Shows aggregation of smaller shoots with fewer larger shoots. Stems shorter than the Naples green-sprouting type
Calabrese	Further reduction in stem length which gives the plant a heading appearance

After Giles (1944).

could have been taken to China. The lines cultivated in Europe may have lost their identity through uncontrolled hybridization. Recently, much horticultural attention has focused on Chinese kale (*B. oleracea* var. *alboglabra*). Chinese kale is now a cultigen native to southern and central China. It is popular and widely cultivated throughout China and Southeast Asia and is used as leaves in salads and other dishes. It is a new vegetable for Japan, Europe and North America. The flower bud, flower stalk and young leaves are consumed. A classification into five groups, which vary in flower colour from white through yellow and in depth of green coloration in the leaves and their shape, was produced by Okuda and Fujime (1996) using cultivars from Japan, Taiwan, China and Thailand as examples.

It is an annual herb, up to 400 mm tall during the vegetative stage and reaching up to 1–2 m at the end of flowering. All the vegetative organs are glabrous and glaucous. The narrow single stem forks at the top. Leaves are alternate, thick, firm and petiolate, and leaf blades are ovate to orbicular-ovate in shape; the margins are irregularly dentate and often undulate and characteristically auriculate at the base or on the petiole. The basal leaves are smaller and sessile without auricles. The inflorescence is a terminal or axillary raceme 300–400 mm long, with pedicels 10–20 mm long. The taproot is strongly branched (see Chapter 2 for floral biology).

OTHER KALES: *B. OLERACEA* L. VAR. *ACEPHALA* DC Many groups are distinguished: borecole or curly kale, collard, marrow stem kale, palm tree kale, Portuguese kale and thousand-headed kale. Kales are ancient cole crops, closely related to the wild forms of *B. oleracea*, and many distinctive types were developed in Europe. There are residual populations of the original progenitors such as the wild kale of Crimea, variously ascribed to *B. cretica* and *B. sylvestris* but now identified as a hairy form of *B. rupestris–incana*. It is suggested that as a consequence of trade around the Mediterranean, this form was transferred to the Crimea and is evidence of early widespread cultivation of *B. rupestris–incana* types. A similar relic population exists in the wild kale of Lebanon, inhabiting the cliffs near Beirut, which is morphologically similar to *B. cretica* subsp. *nivea*. Both are possible evidence for widespread trade by the earliest Mediterranean civilizations which moved the botanical types around, and this allowed interbreeding, resulting in the widening diversity of horticultural crops which were artificially segregated from the botanical populations.

Brussels sprouts, the kales and kohlrabi are part of a similar group of polymorphous, annual or biennial erect herbs growing up to 1.5 m tall, glabrous and often much branched in the upper parts. Kales, in particular, are extremely variable morphologically, most closely resembling their wild cabbage progenitors. The stem is coarse, neither branched nor markedly thickened, and 300–1000 mm tall. At the apex is a rosette of generally oblong, sometimes red coloured leaves; sometimes the leaves are curled. This is caused by disproportionately rapid growth of leaf tissues along the margins. In borecole or curly kale, the leaves are crinkled and more or less finely divided, often green or brownish-purple, and they are used as vegetables. Collards have smooth leaves, usually green; they are most important as forage in western Europe. Marrow stem kale has a succulent stem up to 2 m tall and is used as animal forage. Palm tree kale is up to 2 m tall with a rosette of leaves at the apex; it is mainly used as an ornamental. Portuguese kale has leaves with succulent midribs that are used widely as a vegetable. Thousand-headed kale carries a whorl of young shoots at some distance above the soil; together they are more or less globular in outline. It is mainly used as forage.

KOHLRABI: *B. OLERACEA* L. VAR. *GONGYLODES* L. Kohlrabi first appeared in the Middle Ages in central and southern Europe. The crop has become well established in parts of Asia over the last two centuries and is economically important in China and Vietnam. Kohlrabi are biennials in which secondary thickening of the short stem produces the spherical edible portion, 50–100 mm in diameter and coloured green or purple. The leaves are glaucous with slender petioles arranged in compressed spirals on a swollen stem.

WHITE-HEADED CABBAGE, RED-HEADED CABBAGE AND SAVOY-HEADED CABBAGE White-headed cabbage, *B. oleracea* L. var. *capitata* L. f. *alba* DC; red-headed cabbage, *B. oleracea* L. var. *capitata* L. f. *rubra* (L) Thell; and savoy-headed cabbage, *B. oleracea* L. var. *sabauda* L. were defined by Nieuwhof (1969). Heading cabbages are the popular definitive image of vegetable brassicas in Europe; indeed the terms 'cabbage garden' and 'vegetable garden' were synonymous in some literature.

Early civilizations used several forms of 'cabbage' and these were probably refined in domestication in the early Middle Ages in northwestern Europe as important parts of the human diet and medicine and as animal fodder. It is suggested that their progenitors were the wild cabbage (*B. oleracea*) that is found on the coastal margins of western Europe, especially England and France, and leafy unbranched and thick-stemmed kales that had been disseminated by the Romans. Pliny described methods for the preservation of cabbage, and sauerkraut was of major importance as a source of vitamins in winter and on long sea journeys. In most cabbages, it is chiefly the leaves that are used. Selection pressure in cultivation has encouraged the development of closely overlapping leaves forming tight compact heads, the heart or centre of which is a central undeveloped shoot surrounded by young leaves. Head shape varies from spherical, to flattened to conical. The leaves are either smooth, curled or savoyed (Milan type). Seed propagation of cabbage is relatively straightforward and, in consequence, large numbers of localized regional varieties, or land races, were selected with traits that suited them to particular climatic and husbandry niches such as: Aubervilliers, Brunswick, de Bonneuil, Saint Denis, Strasbourg, Ulm and York. From the 16th century onward, European colonists introduced cabbages worldwide. Scandinavian and German migrants introduced cabbage to North America, especially the mid-Western states such as Wisconsin. In the tropics, cultivation is usually restricted to highland areas and to cooler seasons. White-heading cabbage is especially important in Asia and India. The majority of cultivars are now F_1 hybrids coming from a circumscribed group of breeders using similar parental genotypes. Forms derived originally from the Dutch White Langedijk dominated the market for storage cabbage, and more recently fresh white cabbage in supermarkets. Refinement of savoy types through breeding of F_1 hybrids has expanded the range of this type now offered to consumers and increased its popularity.

Cabbages are biennial herbs 400–600 mm tall at the mature vegetative stage and 1.5–2.0 m tall when flowering in the second year. Mature plants have a ramified system of thin roots, 90% in the upper 200–300 mm of the soil, but some laterals penetrate down to 1.5–2 m deep. Stems are unbranched, 200–300 mm long, gradually thickening upward. The basal leaves form in a rosette of 7–15 sessile outer leaves each 250–350 × 200–300 mm in size. The upper leaves form in a compact flattened globose to ellipsoidal head, 100–300 mm in diameter, composed of a large number of overlapping fleshy leaves around the single growing point. These leaves are grey to blue-green, glabrous, coated with a layer of wax on the outside of the rosette, and light green to creamy white inside the head, especially with white-headed cabbage. The leaves are red-purple in red-headed cabbage and green to yellow-green and puckered in savoy-headed cabbage. The inflorescence is a 500–1000 mm bractless long raceme on the main stem and on axillary branches of bolted plants. Germination is epigeal and the seedlings have a thin taproot and cordate cotyledons; the first true leaves are ovate with a lobed petiole (see Chapter 2 for floral biology).

Hybridization between taxa

Crossings occur even between distant taxa of the Brassicaceae giving at least semi-fertile hybrids, and this may be analogous to the means by which genetic mixing between wild forms led to the horticultural types grown commercially today. Meiotic pairing is normal and indicates close identity particularly throughout the $2n = 18$ forms. Although pollen fertility and seed set are variable, there is usually enough to provide for the survival of further generations.

These characteristics indicate that where races, varieties or species are cultivated in close proximity, crossings will occur. Self-sterility is found in many of the taxa. It is far from absolute, but sufficiently robust to ensure high proportions of out-breeding. Out-breeding normally results from a high frequency of similar *S* genes between individuals belonging to the same population (see Chapter 2). It is concluded that present-day cultivars include much introgressed genetic material derived from other cultivated or wild forms. Consequently, it is important to understand and use the historical literature that describes crops derived from *B. oleracea* alongside that derived from genetic and taxonomic sources in order to interpret the status of modern forms and hybrids.

Comparative taxonomy using ancient and medieval literature and science

Some syntheses of the literature have been attempted specifically for the Brassicaceae, notably that of Toxopeus (1974). Greek writers, especially Theophrastos (370–285 BC), discussed cole crops. It is evident that 'branching' types were known at that time, and these may have resembled bushy kales that were also found in uncultivated ground. Possibly this indicates the domestication of *B. cretica.* Comments are found describing bushy kales with curled leaves; thus both forms may have been present and undergoing hybrization before spreading to other parts of Europe. The Romans (Cato, 234–149 BC; Plinius, AD 23–79) knew of both stem kales and heading cabbage which were cultivated together. Since seed would be produced locally, hybrids could form and the best selected for further improvement, thereby developing local cultivars. Highly prized types might then have spread further as items of trade. Zeven (1996) suggested that the 'perpetual kale' (*B. oleracea* var. *ramosa*) was the Tritian kale of the Romans (referred to by Pliny in AD 70) which they took throughout their Empire. Some relic populations are still grown in various parts of Europe (Belgium, England, France, Ireland, Portugal, Scotland and The Netherlands) and in Brazil, Ethiopia and Haiti. The crop is known as 'Hungry Gap' in England and 'Cut and Come Again' in Scotland. Plants reach up to 3 m in height, some forms appearing to have lost the capacity to flower, as in those found in the Dutch province of Limburg, resulting from long selection pressure for leafiness and multiple branching habit. The patchwork of dissemination in Europe suggests previously widespread distribution by traders.

The cole and neep crops were grown throughout Europe (Sangers, 1952, 1953). Analysis of archives indicates that cole crops were well recognized in the early 14th century to the extent that the name *Coolman* or *Coelman* (cabbage-man) were common surnames, while the term *coeltwn* (modern Dutch = *kooltuin*) indicated a cabbage garden (Metzger, 1833). Trade was established between the Low Countries and England for the export of cabbages by the 1390s. Dodenaeus had by 1554 (Zeven and Brandenburg, 1986) classified cole crops as white cabbage, savoy cabbage, red cabbage and curly kale, and had recognized the turnip, which in 1608 he had differentiated into flat-rooted and long-rooted forms.

Useful evidence of the forms of brassicas in cultivation comes from studies of the Dutch and Flemish painters of the 15th and 16th centuries where red and white cabbages and cauliflowers figure prominently. Only turnips (*B. rapa*) are seen in these paintings, with an apparent absence of swede (*B. napus*) at this time (Toxopeus, 1974, 1979, 1993). Some evidence is available for the presence of radishes in these paintings, but unfortunately there is also conflict with similarities to turnips; however, it is probable that the French 'icicle' radish can be distinguished. It is likely that all subgroups of *Brassica* (kohlrabi, cauliflower and sprouting broccoli) were developed by medieval times and spread westwards and northwards. The appearance, origins and colloquial naming of rutabaga in Scandinavia, especially Finland and Sweden in the Middle Ages, is described in detail by Ahokas (2004). Further south and south-east, other *Brassica* groups were developed for cultivation, but generally with the exception of *B. alboglabra.*

Diversification of *Brassica* crops is well demonstrated in Portugal where original cole crops were introduced by Celtic tribes several centuries BC and in advance of the Roman conquests. These developed into the Galega kale (*B. oleracea* var. *acephala* or *B. oleracea* var. *viridis*), Tronchuda cabbage (*B. oleracea* var. *costata* (var. *tronchuda*)) and Algarve cabbage (*B. oleracea* var. *capitata*). The Tronchuda types are vigorous growing collard-type plants with a small loose head and large thick leaves, while the Galega types are leafy plants, headless, with large leaves having long petioles and a single indeterminate stem which can attain 2 or 3 m before bolting (Monteiro and Williams, 1989). These crops together with vegetable rape (*B. napus* var. *napus*), Nabo (turnip) (*B. rapa* var. *rapa*), Nabica (turnip-greens) (*B. rapa* var. *rapa*) and Grelos (turnip-tops) (*B. rapa* var. *rapa*) form an essential part of the rural diet in Portugal. There are numerous land races of these crops distributed throughout Portugal which have very low within-population uniformity due to the allogamic pollination mechanism associated with the poor isolation used by farmers for seed production. The high levels of variability in shape, size, colour, taste, earliness, and pest and pathogen resistance in these populations constitute an immense reservoir of diversity for breeding purposes.

Comparative morphology gives further information on the origins of brassicas; an important character is the greyish surface texture of the west

European *B. oleracea* found principally in headed cabbage and Brussels sprouts. The strong dominating central structure of stem kales is found in the *B. rupestris–incana* from which primary origin could be inferred. *B. cretica* is probably the origin of the bushy kales since they share common branching, shrubby habit and fleshy leaves. The presence of white flowers may have been derived from *B. alboglabra* and *B. cretica* subsp. *nivea* or combinations between them.

Local stocks (land races)

Various localities still utilize old cultivars and land races, although there is great economic pressure for these to be supplanted by high yielding standardized often hybrid cultivars. Information concerning the older open-pollinated types is fragmentary but of great value in understanding the history of *Brassica* in cultivation. Around the Aegean, primitive kales closely similar to *B. cretica* are still cultivated, some with branching inflorescences similar to sprouting broccoli. Even wild types may be utilized in some island villages as salad vegetables. In the former Yugoslavia, wild-type kales grow on field margins, waste areas and building sites, and are still used as animal fodder. Two forms are apparent: a tall single stem type similar to marrow stem kale and a more branching type with high anthocyanin content often with a habit similar to cabbage. Neither of these produces heads, but they are possibly early kale types closely similar to the wild progenitors. In the harsh North Sea habitat of the Shetland Islands, kale has provided winter food for man and fodder for sheep for centuries. Virtually every croft maintains an individual selection which is propagated by seed and grown within a walled 'kale yard'. The plants have high anthocyanin content giving a reddish appearance to the foliage, and grow to 1–1.5 m high.

In general, it is inferred that the west European headed cabbage types derived from *B. oleracea* on the grounds of morphology with cross-fertilization with Roman kales. The savoy type is possibly a result of further introgression between other coles. The branching bushy kales possibly originate from *B. cretica* with perhaps 2000 years of hybridization with other forms. Stem kales may well originate from the *B. rupestris–incana* complex in the Adriatic or more southerly parts of Italy. Hybridization to cabbages and perhaps *B. cretica* will have added to variation and type differentiation. Origins for the inflorescence of kales, cauliflower and broccoli are still unresolved. Their rapid growth and morphology may have suggested that *B. cretica* is involved, but the leaf characteristics would also indicate that *B. oleracea* is a progenitor. The rapid flowering *B. alboglabra* possibly segregated from *B. cretica* subsp. *nivea*. Via cultivation and trade, it then spread from ancient Greece into the eastern Mediterranean and further eastwards.

The relationships between regional groups of Italian land race cauliflower (*B. oleracea* var. *botrytis*) and broccoli (*B. oleracea* var. *italica*) have recently been unravelled (Massie *et al.*, 1996). Pools of regional genetic diversity exist within cauliflower and broccoli grown throughout Italy following centuries of selection for local conditions and preferences. Different provinces of Italy have been associated with specific variant types, e.g. Romanesco cauliflower in the Lazio Region; Di Jesi, Macerata and Tardivo di Fano varieties in the Marche Region; and Cavolfiore Violetto di Sicilia (Sicilian purple cauliflower) in Sicily, except for the Palermo region where a green cauliflower is typical. Intermediate forms are also reported as between the Macerata and Sicilian Purples (Gray and Crisp, 1985).

Brassica rapa group (*n* = 18) – the Oriental group

The International Code of Botanical Nomenclature (Gilmour *et al.*, 1969; Greuter *et al.*, 2000) rules that the author who first combines taxa of similar rank bearing epithets of the same date chooses one of them for the combined taxon. Metzger (1833) first united *B. rapa* L. and *B. campestris* L. of 1753 under *B. rapa* L. as used in this text.

Brassica rapa L. (*n* = 10, A) and the amphiodiploids *B. carinata* Braun (*n* = 17, BC), *B. juncea* Coss (*n* = 18, AB), *B. napus* L. (*n* = 19, AC) and *R. sativus* (*n* = 9, R) are grown extensively throughout Asia with a huge number of distinct varieties. Differentiation results from selection both in nature and by the forces of cultivation; this is at least a two-way process with great intermingling and recombination. Several distinct groupings are distinguished especially in the headed Chinese cabbages and Japanese radishes.

Initially the centre of origin of *B. rapa* is postulated as the Mediterranean, from where it spread northwards to Scandinavia and eastwards to Germany and into Central Europe, and eventually towards Asia (Mizushima and Tsunoda, 1967). Along the way, great local variation in cultivation developed. The plant reached China via Mongolia as an agricultural crop and was introduced to Japan either via China to the western part of the country or via Siberia and into the eastern part of the country. In India, *B. rapa* and *B. rapa* var. *sarson* (derived from the former species) are used as oil plants, with no records of wild progenitors having been found.

Seven vegetable groups of *B. rapa* can be differentiated: var. *campestris*, var. *pekinensis*, var. *chinensis*, var. *parachinensis*, var. *narinosa*, var. *japonica* and var. *rapa*. Until recently, these were considered as separate species due to their wide ranges of variation. The leaf vegetables are thought to have developed after they entered China, with the exception of the var. *japonica* group that has a common ancestor in the oil rapes. Since use is for vegetables and oil, possibly other forms have developed from them (Kumazawa, 1965). Parallels are frequently drawn between the development of *B. oleracea* vegetables in Europe (the Occident) and *B. rapa* vegetables in Asia (the Orient). The headed Chinese cabbages tend to dominate the use for cultivation, but there are parts of China itself where *B. juncea* is of particular prominence, and in Central China the forms var. *chinensis* and *narinosa* are of importance largely for climatic reasons.

There are close parallels between the manner of differentiation and selection in cultivation of *B. oleracea* in Europe and of *B. rapa* and *B. juncea* in Asia. Species may have developed along similar lines in a single region as, for example, the trend towards entire glabrous leaves in *B. rapa, B. juncea* and *Raphanus sativus*. It is possible to find atypical characters such as thick stems which are poorly represented in the putative parents but which have formed extensively in the progeny. Thus neither *B. rapa* (A genome) nor *B. nigra* (B genome) possess thick stems, but there are forms of *B. juncea* (AB genome) which are far more developed than the parents and are similar to *B. oleracea* (C genome) (see Table 2.1).

Crop plants of *B. rapa, B. juncea* and *R. sativus* are used as leaf or root vegetables in Japan and China, whereas in India they are developed as oil plants, and *R. sativus* in particular has produced very impressive siliquae. Thus differentiation has arisen over time in the respective directions of cultivation for either vegetative or reproductive organs. In China, differentiation of headed Chinese cabbages has produced types adapted to several climatic zones. In the north, cold-tolerant forms are used in the early summer, with a similar but separate segregation in the south for types suited to winter culture. Distinct types have been developed separately in the two areas capable of accomplishing similar tasks.

Leaf greens types of crop have been developed in the main from *B. rapa* and *B. juncea* to the exclusion of radishes and turnips, which could form the same product only with a lower level of efficacy. Vegetables which are eaten in high volume as fresh forms tend to have been differentiated into the greatest numbers of improved types, whereas those used for preserves and processing have less variation produced by segregation in cultivation. Figure 1.5 summarizes leaf greens developed from both *B. oleracea* var. *alboglabra* and *B. rapa*.

Crops developed within *Brassica rapa* and allies

CAISIN: SYN. VAR. *B. PARACHINENSIS* BAILEY (1922), *B. CHINENSIS* L. VAR. *PARACHINENSIS* (BAILEY) TSEN & LEE (1942) Caisin is generally thought to have differentiated along with the leaf neeps (Chinese cabbage and Pak Choi) from oil-yielding turnip rapes, which were introduced into China from the Mediterranean area through western Asia or Mongolia. Caisin originated in middle China where it was selected and popularized for its inflorescences. It may be seen as parallel variation in *B. rapa* comparable with Chinese kale (*B. oleracea* L. cv. group Chinese kale) in *B. oleracea*.

Where headed Chinese cabbage is grown, caisin and the non-headed leaf neeps (e.g. Pak Choi) are also indispensable vegetables. Caisin is cultivated in southern and central China, in southeastern Asian countries such as Indonesia, Malaysia, Thailand and Vietnam, in other parts of Indo-China and in areas of western India.

The variety *parachinensis* is possibly a derivative of var. *chinensis*, used for the flower stalk and very popular in central China; the plants will bolt readily

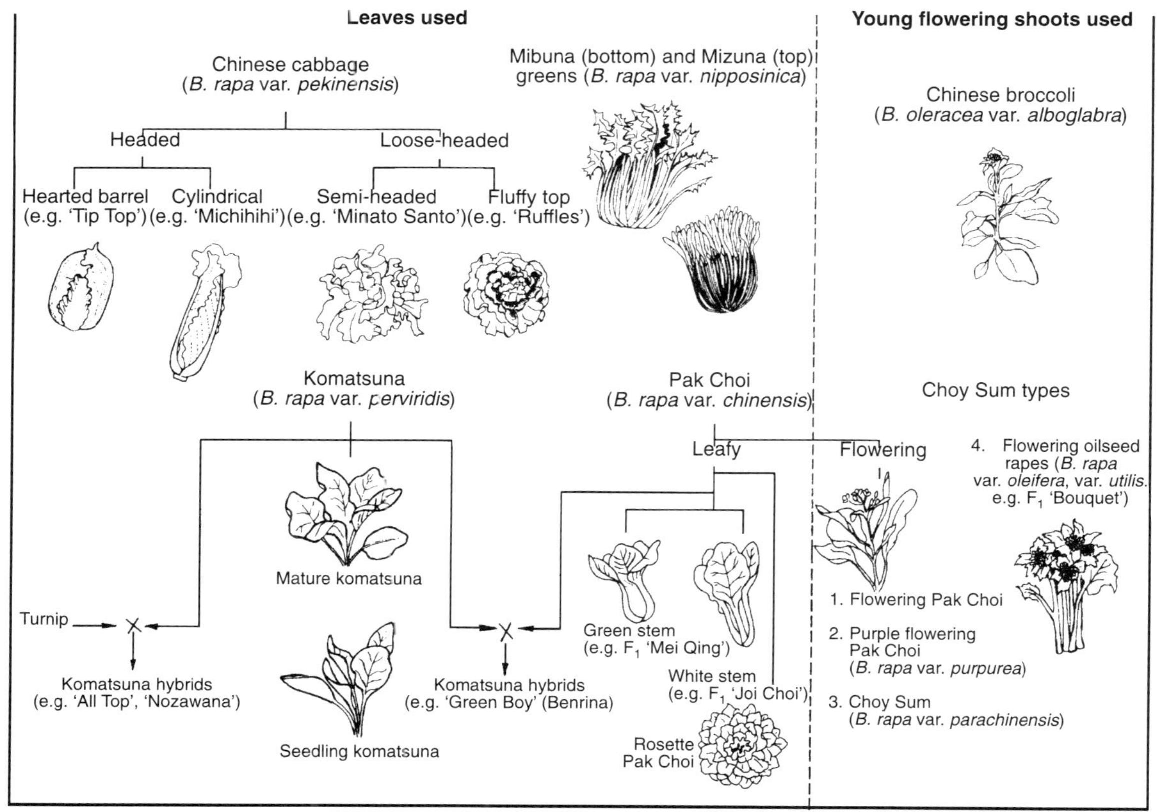

Fig.1.5. Oriental leafy and flowering forms developed from *B. rapa* and *B. oleracea* (after Larkcom, 1991).

and the time from seeding to harvest can be 40–80 days dependent on the cultivar used (Herklots, 1972). Var. *campestris* is the most primitive of vegetables, closely resembling oil rape and used for flower stalks and rosette leaves.

Caisin is an annual, taprooted herb, 200–600 mm tall with usually open, erect or sometimes prostrate growth habit. The stems are usually <10 mm in diameter, small in comparison with other leafy cabbages, and usually profusely branched. There are few leaves in the rosette, usually with only 1–2 leaf layers. There are long petiolate, spathulate or oblong, bright green, stem-leaves. These may be glabrescent to glabrous, green to purple-red and finely toothed when young. Lower stem-leaves are ovate to nearly orbicular, and the central stem leaves ovate to lanceolate to oblong with long and narrow grooved petioles that are sometimes obscurely winged. Upper stem leaves gradually pass into narrow bracts. Inflorescences form as a terminal raceme, elongating when in fruit.

CHINESE CABBAGE: SYN. *B. PEKINENSIS* (LOUR) RUPR. (1860), *B. CAMPESTRIS* L. SUBSP. *PEKINENSIS* (LOUR) OLSSEN (1954), *B. RAPA* L. SUBSP. *PEKINENSIS* (LOUR) HANELT (1986) Chinese cabbage is a native of China; it probably evolved from the natural crossing of Pak Choi (non-headed Chinese cabbage), which was cultivated in southern China for >1600 years, and turnip, which was grown in northern China. Much of its variety differentiation took place in China during the past 600 years. Its derivatives were introduced into Korea in the 13th century, the countries of southeastern Asia in the 15th century and Japan in the 19th century. An illustration of the headed shape of Chinese cabbage with wrapping leaves was first recorded in China in 1753. Chinese cabbage is now grown worldwide.

The origins of var. *pekinensis* may correlate with the oil rapes of northern China, developing first as a headed type where the lower parts of the plant were swollen, and latterly where the entire plant developed a headed form. Several variants exist as to the manner by which the head forms, with the 'wrapped-over' and 'joined-up' forms being distinguished. In the former, the leaves overlap at the top of the conical head, which does not happen in the 'joined-up' or multileaved types. The 'wrapped-over' forms are heavy leaved, early maturing and round headed, with an adaptation to warmer climates.

The 'joined-up' types are late maturing with firm texture and adaptation to cooler climates. It is possible that both forms originated first in Shandung peninsula and the 'joined-up' types spread northwards from there, while the 'wrapped-over' forms developed in a southerly direction. The latter eventually differentiated into a more 'southerly' type, which has very early maturity, small head and heat tolerance (Fig. 1.6).

Shapes of Chinese cabbage are classified on the degree of heading, as non-headed, half-headed and completely headed types, with further refinements to give long, short, tapered, round topped, wrapped-over and joined-up forms. The

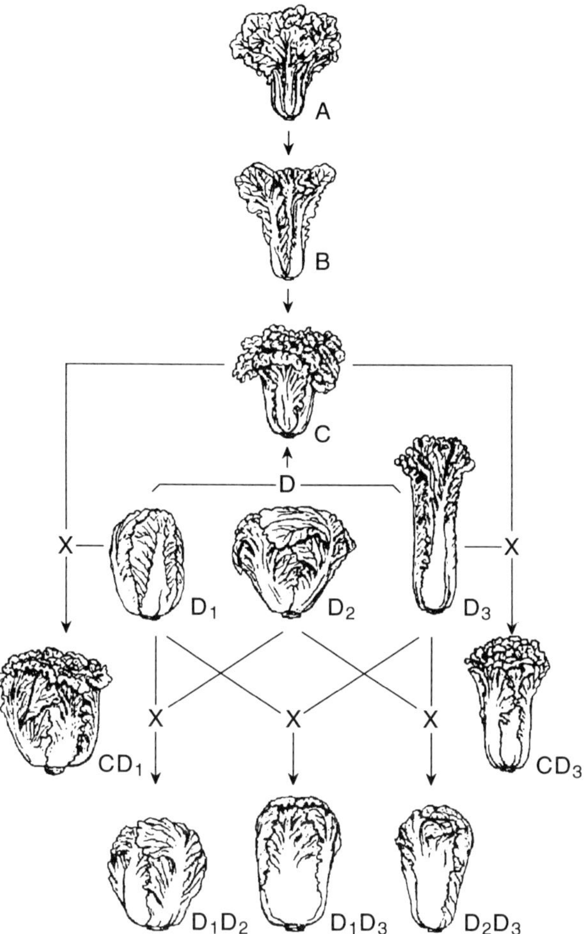

Fig. 1.6. Evolutionary relationships of Chinese cabbage (*Brassica rapa* subsp. *pekinensis*) (M.H. Dickson); A = var. *dissoluta*, B = var. *infarcta*, C = var. *laxa*, D = var. *cephalata*, D_1 = var. f. *ovata*, D_2 = var. f. *depressa*, D_3 = var. f. *cylindrica*, CD_1 = var. *laxa* × f. *ovata*, CD_3 = var. *laxa* × f. *cylindrica*, D_1D_2 = f. *ovata* × f. *depressa*, D_1D_3 = f. *ovata* × f. *cylindrica*, D_2D_3 = f. *depressa* × f. *cylindrica*.

Chinese use them largely in the autumn for preserving and processing. In Japan, they are freshly boiled or salt-pickled, with early and late types, but there is insufficient production to meet consumer demands so consequently crops are distributed between regions in Japan according to seasonal availability. The crop probably entered Japan from Shandung in the Meiji era, being well adapted to the Japanese climate. In South Korea, plant breeders have worked hard in the past 30 years to develop a sequence of cultivars that allows cultivation all year-round. Koreans eat copious quantities of Chinese cabbage fresh, cooked and most of all fermented with garlic and tomato extract called 'kimchi'.

Chinese cabbage is a biennial herb, cultivated as an annual, 200–500 mm tall during the vegetative stage and reaching up to 1.5 m in the reproductive state. The taproot and lateral roots are prominent in older plants, forming an extensive, fibrous, finely branched system. During vegetative stages, the leaves are arranged in an enlarged rosette. This forms a short conical more or less compact head, with ill-defined nodes and internodes and alternate heading and non-heading leaves; the leaves are 200–900 × 150–350 mm in size. Leaves vary in shape with different growth stage; the dark green outer leaves are narrowly ovate with long laminae and winged petioles. Inner heading leaves are broad, subcircular, whitish-green. The flowering stem carries lanceolate leaves, much smaller than heading leaves, with broad compressed petioles and blades, clasping the stem.

MIZUNA: SYN. *B. JAPONICA* MAKINO (1912) NON (THUNB) SIEB (1865/66), *B. CAMPESTRIS* L. SUBSP. *NIPPOSINICA* (BAILEY) OLSSEN (1954), *B. RAPA* L. SUBSP. *NIPPOSINICA* (BAILEY) HANELT (1986) The variety *japonica* is a vegetable unique to Japan, carrying many basal branches and leaves. This plant resembles *B. juncea* since the inflorescence stalk has no leaves, the petioles and siliquae are slender and seeds are small (Matsumura, 1954). Possibly this variety is derived from hybridization from *B. rapa* and *B. juncea*. Grouping of cultivars rests on the level of dissection of the leaves.

NEEP GREENS SYN. *B. PERVIRIDIS* (BAILEY) See Bailey (1940) for certain forms (komatsuna).

TAATSAI: SYN. *B. NARINOSA* BAILEY (1922), *B. CAMPESTRIS* L. SUBSP. *NARINOSA* (BAILEY) OLSSEN (1954), *B. RAPA* L. SUBSP. *NARINOSA* (BAILEY) HANELT (1986) The variety *narinosa* developed in middle China; it is very cold tolerant with thick leaves and crisp petioles that are used as boiled vegetables.

VEGETABLE TURNIP: SYN. *B. CAMPESTRIS* L. SUBSP. *RAPA* (L.) HOOK. F. & ANDERS. (1875), *B. CAMPESTRIS* L. SUBSP. *RAPIFERA* (METZGER) SINSK. (1928), *B. RAPA* L. SUBSP. *RAPA* SENSU AUCT MULT The origin of *B. rapa* is not known; the wide variation of neep crops evolved in different parts of the Eurasian continent. Besides Chinese cabbage, Pak Choi and caisin, the leafy vegetable types or leaf neeps comprise cultivar groups Mizuna, Neep Greens and Taatsai, developed in temperate regions of Asia.

The many forms of vegetable turnip are highly regarded in Japan as well as in Europe, where fodder turnip also used to be a very popular crop (root neeps). Oilseed types (seed neeps), grown for rape oil, are important in India and Canada.

The turnip is the oldest *B. rapa* crop on record. It was described in ancient Greek times of Alexander the Great, whose empire included the Middle East and Persia, from where it may have found its way to eastern Asia. Quite independently of each other in Europe and in Japan, well-defined polymorphic groups of vegetable turnips were created by the 18th century.

Although turnips (var. *rapa*) are grown around the world, Japan has one of the major areas of variety development, while in eastern Asia radishes have predominated. Turnips grow best in cool climates and are mainly used in autumn or winter in southern areas and in spring and summer in more northerly areas. Worldwide, turnips can be classified into Teltou turnips, West European turnips with dissected leaves, Asia Minor and Palestine turnips, Russian turnips of Petrovskij type, Asiatic turnips with dissected leaves of the Afghan type and a subgroup of Asiatic turnips of the Afghan type with entire leaves, Japanese entire-leaved turnips with entire glabrous leaves and European entire-leaved turnips with pubescent leaves (Sinskaia, 1928).

The Afghan types are thought to be close to the original var. *rapa* with glabrous dissected leaves, ascending rosette and small taproots. Japanese forms possibly developed from these with entire leaves, no hairs and well-formed taproots, demonstrating the effects of selection in cultivation. A further classification (Shebalina, 1968) identified Asia Minor, Central Asia, Europe, Iraq and Japan types. The latter include types either derived indigenously in Japan or developed from European types with several intermediates. Characters used to differentiate types include the form of the seed coat which either swells (type A) or forms a think layer of epidermal cells (type B) (Shibutani and Okamura, 1954).

The cultivar groups Mizuna, Neep Greens, Taatsai and Vegetable Turnips are annual or biennial herbs with stout taproots, often becoming fusiform to tuberous (turnips). Stems are erect, branched and up to 1.5 m tall. Leaves are very variable, depending on the cultivated type. They grow in a rosette during the vegetative phase. The basal leaves are more or less petioled, bright green and lyrate-pinnatipartite, dentate, crenate or sinuate, bearing large terminal lobes and up to five pairs of smaller lateral lobes. Lower cauline leaves are sessile, clasping and pinnatifid, and upper cauline leaves are sessile, clasping, undivided, glaucous, entire to dentate. Cultivar group Mizuna consists of spontaneously tillering plants with pinnate leaves (mizuna cultivars) or entire leaves (mibuna cultivars).

The cultivar group Neep Greens comprises essentially non-heading plants including crops such as komatsuna, zairainatane, kabuna and turnip greens. The cultivar group Taatsai typically grows a flat rosette of many small dark green leaves. Cultivar group Vegetable Turnip consists of forms in which the storage organ (swollen hypocotyl and root), i.e. the turnip, is used as a vegetable as well. Turnip roots vary widely in shape, from flat through globose to ellipsoid and cylindrical, blunt or sharply pointed, with the flesh white, pink or yellow, and the apex white, green, red, pink or bronze. All these characteristics may occur in cultivars in any combination imaginable.

PAK CHOI: SYN. *B. CHINENSIS* L. (1759), *B. CAMPESTRIS* L. SUBSP. *CHINENSIS* (L.) MAKINO (1912), *B. RAPA* L. SUBSP. *CHINENSIS* (L.) HANELT (1986) Pak Choi evolved in China, and its cultivation was recorded as far back as the 5th century AD. It is widely grown in southern and central China, and Taiwan. This group is a relatively new introduction in Japan where it is still referred to as 'Chinese vegetable'. It was introduced into southeastern Asia in the Malacca Straits Settlement in the 15th century. It is now widely cultivated in the Philippines and Malaysia, and to a lesser extent in Indonesia and Thailand. In recent years, it has gained popularity in northern America, Australia and Europe where there are concentrations of eastern Asiatic immigrants.

The variety *chinensis* is the 'large white cabbage' of China, while var. *narinosa* and the like are 'small white cabbage' (Wu Geng Min, 1957). These are particularly important where headed forms are not grown and are extensively used in Malaysia and Indonesia. As with var. *pekinensis*, this form differentiated in China from oilseed rape with distinctions between type related to the width and cross-section of the petioles. Those with narrow flat petioles are nearest to the parental origin, forming both branching and non-branching types. Petiole colour varies from white to green, the former being used in autumn while the latter are more cold tolerant and resistant to bolting. Frequently the Chinese forms are larger than those grown in Japan.

Pak Choi is an erect biennial herb, cultivated as an annual. In the vegetative state, it is glabrous, dull green and 150–300 mm tall, and in the generative stage reaches 700 mm. The leaves are arranged spirally, not forming a compact head but spreading in groups of 15–30. Petioles are enlarged, terete or flattened, 15–40 mm wide and 5–10 mm thick, growing in an upright manner forming a subcylindrical bundle. Each white, greenish-white to green leaf blade is orbicular to obovate, 70–200 × 70–200 mm. Stem leaves are entire, tender, smooth or blistering, shiny green to dark green and auriculate-clasping.

BRASSICA JUNCEA (L.) CZERNJAEW

Syn. *Sinapis juncea* L. (1753), *S. timoriana* DC (1821), *B. integrifolia* (West) Rupr. (1860).

Brassica juncea crops are grown worldwide, from India to northern Africa, to central Asia (the southern and southeastern part of the former Soviet Union), to Europe and North America. The exact origin is unknown, but as an amphidiploid it seems logical that it originated in an area where the parental species *B. nigra* (L.) Koch and *B. rapa* L. overlap in their distribution (e.g. central Asia). It is generally agreed that the primary centre of diversity of *B. juncea* is central Asia (northwest India, including the Punjab and Kashmir) with secondary centres in central and western China, Hindustan (east India and Burma) and Asia Minor (through Iran). In *B. juncea*, two types of mustards with varying uses have evolved, i.e. oilseed types and vegetable types. The oilseed types (oilseed mustard) are particularly important in India, Bangladesh and China. The vegetable types include forms with edible leaves (leaf mustard), stems (stem mustard) and roots (root mustard).

The vegetable mustards are widely cultivated in Asian countries. The highest degree of variation occurs in China, which is regarded as the primary centre of varietal differentiation. The early Chinese traders might well have carried the crop into southeastern Asia, whereas the appearance of *B. juncea* near European ports suggests a connection with grain imports. It has also been suggested that Indian contract labourers brought it to the West Indies. In southeastern Asia, it is the leaf mustards which are the most common.

Brassica juncea is an erect annual to biennial herb, 300–1600 mm tall, normally unbranched, sometimes with long ascending branches in the upper parts; in appearance it is subglabrous and subglaucose. The taproot is sometimes enlarged (root mustard). Leaves are very variable in shape and size, either pinnate or entire with petioles pale to dark green, smooth or pubescent, heading or non-heading.

Oil mustards are grown as crops worldwide but are chiefly found in India to western Egypt, central Asia (including southern and southeastern Russia) and Europe. As vegetables, they appear mainly in Asia, especially China, with great diversity of types. In India, they are spice plants used in curries, while the young leaves may be eaten as greens. Most of the *B. juncea* vegetables of Asia have been developed as pickles, hence their use in Europe is limited.

In China, vast amounts of different types of mustard are consumed. Where leaf vegetables of *B. rapa* and *R. sativus* are difficult to grow, such as in southern China, *B. juncea* is also used as a leaf vegetable. Growth is slow, with tolerance to high temperatures and humidity; long days are required to stimulate flowering, with some varieties being day neutral. This is in contrast to *B. rapa* and explains the distribution of crop types. The greatest differentiation of *B. juncea* for greens is found in China, especially in Sichuan province. Sinskaia (1928) differentiated on leaf characters in relation to geographical occurrence with an eastern Asian group with bipinnate leaves; a central Asian group with entire leaves; and a Chinese group with crisp leaves.

Variations of stem characters were used to differentiate forms in Taiwan (Kumazawa and Akiya, 1936); some had compact stems (Kari-t'sai) and others had enlarged stems (Ta-sin-t'sai). Indian types are characterized by their content of the oil 3-butelyl isothiocyanate, which is absent from Chinese types (Narain, 1974), and with the recognition of early bolting northeastern, late bolting northwestern and intermediate types (Singh *et al.*, 1974). In Japan, succulent leafy types were reputedly introduced in the Meiji era.

Seven types may be distinguished by geographical and botanical characters:

1. Hakarashina group (*B. juncea* Coss) with pinnate leaves, distributed in India, central Asia and Europe; these are coarse textured, dissected leaves like those of radishes. The majority are cultivated for oil but may form greens when young.

2. Nekarashina group (*B. juncea* var. *napiformis* Bailey) with enlarged roots, distributed in Mongolia, Manchuria and northern China. The taproot is well enlarged like those of turnips; the crop is absent in Japan. Variation in root character is simple as the root shape is solely conical and root colour is equally upper green and lower white. This group is acclimatized to cold, unlike most *B. juncea* crops, and hence is an important Chinese crop.

3. Hsueh li hung (*B. juncea* var. *foliosa* Bailey) and Nagan sz kaai (*B. juncea* var. *japonica* Bailey) group with glabrous leaves and many branches, distributed in middle and northern China; characterized by bipinnate leaves dissected in thread-like segments like those of carrots, which form a cluster of rosettes. Dissection is deeper in Hseuh li hung and rosettes more vigorous in Nagan sz kaai; cold tolerance is relatively high. The leaves are pubescent, dark green and pungent to taste.

4. Azamina group (*B. juncea* var. *crispifolia* Bailey), ecologically similar to Takana, but the leaves are firmly dissected like parsley with a general appearance of curly kale. They are used for salads and as ornamentals in the USA, possibly as progeny of the introduced cultivar 'Fordhook Fancy'.

5. *Brassica juncea* var. *integrifolia* (Stokes) Kitam with entire succulent leaves, distributed in southern and middle China, and southeastern Asia towards the Himalayas (Kumazawa, 1965). They have either dissected or entire leaves, the latter being more frequent, forming vigorous large plants up to 1 m high, soft in texture with midribs varying from narrow to wide and crescent shaped in transverse section; internodes are extended. The leaves are glabrous or crisp and green to dark green. They are cultivated as specialized regional products in Japan.

6. *Brassica juncea* var. *rugosa* (Roxb) Kitam are large plants with leaves with wide flat entire midribs, and are extremely succulent. The Paan sum of Canton province has a moderately wrapped head and the Chau chiu tai kai tsai forms a completely wrapped head. The Liu tsu chieh of Zheijang province bears protuberances on the midrib.

7. Ta hsin tsai group (*B. juncea* var. *bulbifolia* Mas), plants with succulent stems and elongated internodes. The leaves are not eaten; those used for Za-tsai have enlarged stems up to 6–7 cm with protuberances on the petioles; stem enlargement is encouraged by low temperature.

RAPHANUS SATIVUS L. – RADISH

Four cultivar groups of culinary radish are recognized: Chinese radish (daikon), leaf radish, rat-tailed radish and small European types. The origins of culinary radish are not known since there is no immediately obvious wild progenitor. The zone of maximum diversity runs from the eastern

Mediterranean to the Caspian Sea and eastwards to China and Japan. Radish crops were cultivated around the Mediterranean before 2000 BC and are reported in China in 500 BC and Japan in 700 AD. They are now found worldwide. World radish production is estimated at 7 Mt/year, which is 2% of total vegetable production. They are particularly prominent in Japan, Korea and Taiwan. The mild flavoured Chinese radish produces large roots that can be sliced, diced and cooked. The daikon form is increasing in popularity worldwide especially as it is reputed to be tolerant to clubroot disease caused by *Plasmodiophora brassicae* (see Chapter 7). The small rooted European radish is used as relishes, appetisers and to add variety and colour to green leaf salads. Other forms are used as leaf greens, green manure crops and sources of oil seeds.

NOTES

[1] Throughout this book, '*Brassica*' is used in botanical contexts and 'brassicas' is used in horticultural contexts.
[2] *Brassica rapa* is used is used throughout this book as the preferred name with precedence over *Brassica campestris*.
[3] The term 'cultivar' denotes subspecific variants produced by man-directed breeding, while a 'variety' has arisen in nature.

Breeding, Genetics and Models

The classic triangle of U (1935) (Fig. 1.2) shows the inter-relationships of brassicas based on their chromosome numbers and $2n$ genome descriptors. These species can be inter-crossed using embryo rescue, fusion and other methods. In addition, massive opportunities are emerging from studies of model brassicas such as thale cress (*Arabidopsis thaliana*) and Wisconsin Fast Plants™ that identify genes and their products which can be applied in crop species. Further, some breeders are now working with the less well known species in Brassicaceae, such a *Brassica carinata*, to extract valuable genes for resistance to pathogens, pests and other economic characters. The Brassicaceae is one of the most flexible plant families in terms of interspecifc and intergenomic crosses. Rapid progress is being made in our understanding of their component genes, genomic interactions, protein products and resultant phenotypic characteristics, leading eventually to even wider and more diverse crosses. Concurrently, restriction fragment length polymorphism (RFLP) and linkage maps are being made for most of the major species, and these have shown many common chromosome linkage groups occurring across these species. This is encouraging the research into the potential of single gene transfer into economic crops from the model types.

GENOMIC CHARACTERS AND TAXONOMY

The following six *Brassica* species, plus *Raphanus sativus*, radish $2n = 18$, have been inter-crossed, with varying levels of difficulty requiring embryo culture or fusion to obtain hybrids: *Brassica nigra* Koch, black mustard, $2n = 16$; *Brassica carinata* Braun, Ethiopian mustard, $2n = 34$; *Brassica juncea* L. Coss, brown mustard, $2n = 36$; *Brassica napus*, swede or rutabaga, rape or oilseed rape (canola) $2n = 38$; *Brassica rapa*, turnip and Chinese cabbage, $2n = 20$; and *Brassica oleracea*, cole crops, $2n = 18$. Table 2.1 shows the cytoplasmic and genomic descriptors for these species.

Table 2.1. Designation of cytoplasmic and nuclear genomes of *Brassica* and *Raphanus* species.

Species	Subspecies or variety	Cytoplasm	2*n* genome	Common name
B. nigra		B	*Bb*	Black mustard
B. oleracea		C	*Cc*	Cole vegetables
B. rapa		A	*Aa*	
	Chinensis		*aa.c*	Pak Choi
	Nipposinica		*aa.n*	Mizuna
	Oleifera		*aa.o*	Turnip rape
	Parachinensis		*aa.pa*	Caisin (choy sum)
	Pekinensis		*aa.p*	Chinese cabbage, petsai
	Rapifera		*aa.r*	Vegetable turnip
	Trilocularis		*aa.t*	Sarsons
B. carinata		BC	*Bbcc*	Ethiopian mustard
B. juncea		AB	*Aabb*	Vegetable mustard
B. napus		AC	*Aacc*	Fodder rape, oilseed rape (canola), swede
R. sativus		R	*Rr*	Vegetable radish, daikon

After Williams and Heyn (1980); these authors suggested using the single upper case letter representing the genome descriptor to designate the cytoplasm in which the nuclear genes are functioning.

Hybridization between the seven species in Table 2.1 is easier than to the other cruciferous species. It can still be a difficult and frustrating task, however, to obtain true-breeding new lines with the desired attribute irrespective of whether the species is a diploid or allopolyploid as with *B. carinata*, *B. juncea* and *B. napus*. Using fusion rather than classical hybridization with embryo rescue has improved the chances of success.

FLORAL BIOLOGY AS RELATED TO CONTROLLED POLLINATION

The flower (Fig. 2.1) differentiates by the successive development of four sepals, six stamens, two carpels and four petals. The carpels form a superior ovary with a 'false' septum and two rows of campylotropous ovules. The nucellar tissue is largely displaced by the embryo sac and, when the buds open, the ovules mainly consist of the two integuments and the ripe embryo sac. The buds open under pressure from the rapidly growing petals. Opening starts in the afternoon, and usually the flowers become fully expanding during the following morning. The bright yellow petals grow to 10–25 mm long and 6–10 mm wide. The sepals are erect. Pollination of the flowers is by insects, particularly bees, which collect pollen and nectar.

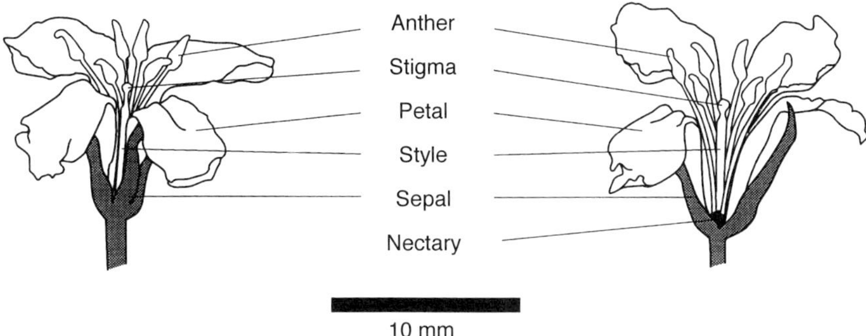

Fig. 2.1. Generalized structure of the flower and half-flower of *Brassica* (M.H. Dickson).

Nectar is secreted by four nectaries situated between bases of the short stamens and the ovary. The flowers are borne in racemes on the main stem and its axillary branches. The inflorescences may attain lengths of 1–2 m. The slender pedicels are 15–20 mm long. After fertilization, the endosperm develops rapidly, while embryo growth does not start for some days. The embryo is generally still small 2 weeks after pollination, and at this stage embryo rescue can be first attempted.

The embryo fills most of the seed coat after 3–5 weeks, by which time the endosperm has been almost completely absorbed. Nutrient reserves for germination are stored in the cotyledons, which are folded together with the embryo radicle lying between them. The fruits of cole crops are glabrous siliquae (pods) (Fig. 2.2), 4–5 mm wide and 40–100 mm long, with two rows of seeds lying along the edges of the replus. One silique contains 10–30 seeds.

Three to 4 weeks after the opening of a flower, the silique reaches its full length and diameter. When it is ripe, the two valves dehisce. Separation begins at the attached base and works towards the unattached distal end, leaving the seeds attached to the placentas. Physical force ultimately separates the seeds, usually by the pushing of the dehisced siliquae against other plant parts either by wind or in threshing operations.

MICROPROPAGATION

Microspore-derived embryogenesis was first completed successfully on broccoli by Keller *et al.* (1975) from isolated anthers. Now embryogenesis of microspores is quite routine in most large-scale crucifer breeding programmes. There are still differences between sources of microspores which affect the number of embryos that are regenerated per anther, but the system is now well developed, and >1000 per bud can be produced. The size of the

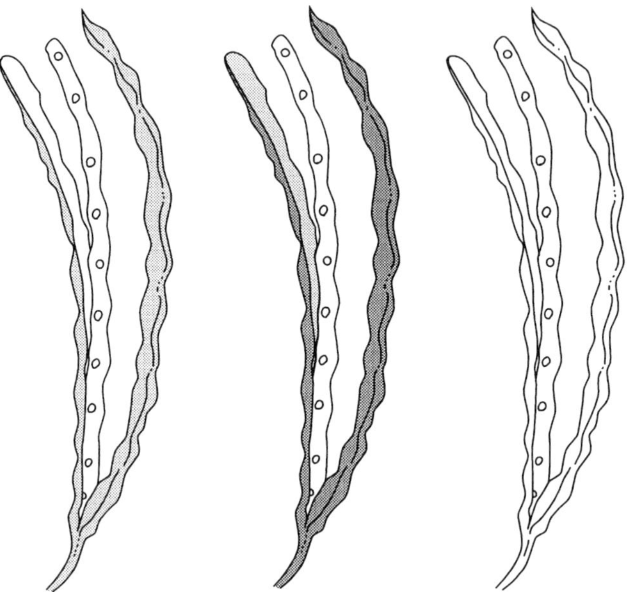

Fig. 2.2. Generalized structure of the dry fruit of *Brassica* (M.H. Dickson).

bud and the environment in which the parent plant has been grown make a considerable difference to success in production of embryos. The breeder must ensure that the numbers of regenerates saved does not become larger than can be handled during evaluation. In most cases, a high percentage of the embryos will be haploid and will double spontaneously providing homozygous plants, saving the necessity of working through several generations of inbreeding. Embryogenesis is an expensive procedure, but the time saved in obtaining homozygous lines compared with using four generations of single seed descent for conventional inbreeding can be well worthwhile.

GENOMICS AND MAPPING

Crucifers enjoy a pivotal role in developing our understanding of plant genomics and mapping, mainly through the extensive study of *Arabidopsis thaliana* (thale cress). *Arabidopsis* has a simple five-chromosome genome with minimal levels of duplication, making it an ideal candidate for the first plant genome project, and a model plant for research, as discussed later in this chapter. This work is of special interest to those working with the *Brassica* crop species, since they share at least 87% coding sequence conservation with *Arabidopsis*.

Molecular markers have been utilized in several *Brassica* species (particularly *B. oleracea*, *B. napus* and *B. rapa*), and simple maps have been created using isozyme, RFLP and randomly amplified polymorphic DNA (RAPD) markers (Quiros, 1999). The developments in the *Arabidopsis* Genome Initiative and expressed sequence tag (EST) databases have subsequently displaced these efforts with more powerful tools for the mapping and understanding of *Brassica* genomes. The uses of comparative genomics have recently accelerated, providing both comparative maps and consensus genetic markers between the species based on ESTs (Brunel *et al.*, 1999; Lan *et al.*, 2000).

The research from *Arabidopsis* is being used to study the functioning of important genes involved in the horticulture and agronomy of *Brassica* crop species, including the genes responsible for head formation in cauliflower and broccoli (Lan and Paterson, 2000). While current efforts are complicated by loci duplication within the *Brassica* crop species (as even the diploid *Brassica* species are derived from smaller ancestral genomes that are thought to be partial amphidiploids), the research should clarify much of the underlying cytogenetics of *Brassica* genomes (Cavell *et al.*, 1998).

Genomic research is currently focused on several *Brassica* species, which will demonstrate the lateral use of genomics from *Arabidopsis* in commercially important crop plants such as oilseed rape (canola), mustard and the vegetable crops. These research foci result in the rapid development of molecular tools that can be utilized by breeders in improving germplasm and cultivars.

CALLUS CULTURE, FUSION AND TRANSFORMATION

Tissue culture techniques have been applied to several *B. oleracea* vegetables, either for clonal propagation or for development of novel and sometimes improved plant types. Plants have been regenerated from diverse multicellular explants, including: immature embryos; seedling parts such as hypocotyls or cotyledons; stem pieces; leaves; roots; and floral tissues such as flowering stalks or caulilfower curds. Plants can be recovered from single wall-free protoplasts, usually isolated from leaves or hypocotyls of plantlets grown *in vitro*. Strong genotype specificity is usually noted in studies comparing regeneration from various lines of a given vegetable, with some performing very well and others showing little or no response. This is particularly true in the case of protoplasts.

Plant regeneration from somatic tissues makes it possible to maintain populations that fail to produce seeds and are difficult to propagate using standard methods. Embryo culture has been utilized to recover progeny from interspecific hybrids produced either by sexual crosses (e.g. *B. napus* × *B. oleracea*) or by protoplast fusion.

Several vegetable lines with new combinations of nuclear and cytoplasmic genomes have been created by protoplast fusion. The usual procedure is to fuse protoplasts of regenerable *B. oleracea* lines with protoplasts from a cytoplasm donor irradiated to eliminate its nuclear DNA. Some plants recovered from such experiments are cybrids containing the nuclear genome of the *B. oleracea* partner and the mitochondria and/or chloroplasts of the other partner.

The initial fusion product contains the mitochondria and chloroplasts of both fusion partners, in contrast to sexually derived zygotes which almost always contain only the chloroplasts of the maternal parent. Sorting out a mixed population of organelles during subsequent cell division and regeneration has made it possible to overcome the chlorosis seen when the Ogura type (Ogura, 1968) of cytoplasmic male sterile (CMS) *Brassica* lines are grown at low temperatures (Walters *et al.*, 1992). This chlorosis results from interaction of radish chloroplasts with a *Brassica* nucleus, so fusion-mediated replacement of the radish chloroplasts with *Brassica* chloroplasts can produce CMS vegetable material suitable for hybrid production. Further fusions allow transfer of a cold-tolerant Ogura CMS from broccoli to cabbage in one step, eliminating the need for many generations of back-crossing (Sigareva and Earle, 1997a). Other types of male-sterile cytoplasms (e.g. Polima and Anand) have also been transferred to *B. oleracea* from other *Brassica* species by protoplast fusion.

Cauliflower and broccoli cybrids with chloroplast-encoded resistance to atrazine herbicides have been created through fusions using *B. rapa* or *B. napus* as the source of the resistant chloroplasts. The broccoli cybrids combined two organelle phenotypes (atrazine resistance and the 'nigra' type of CMS) (Jourdan *et al.*, 1989). Atrazine-resistant *B. oleracea* did not show the negative effects seen in atrazine-resistant rapeseed (*B. napus*).

Attempts to alter the nuclear genome of *B. oleracea* via protoplast fusion have had limited success. *B. oleracea* can readily be fused with *B. rapa* to synthesize new forms of the amphidploid *B. napus* for example (Zhang *et al.*, 2004). This approach increases the genetic diversity in rapeseed and may also provide a way to transfer useful traits from *B. oleracea* to the *B. rapa* vegetables (since sexual crosses of somatic hybrid *B. napus* with *B. rapa* are usually not difficult). Resynthesis of *B. napus* is not, however, an obvious route for the improvement of *B. oleracea* vegetables.

Fusion of *B. oleracea* with other *Brassica* or crucifer species carrying useful resistance genes may produce resistant somatic hybrids. A dominant gene for resistance to *Xanthomonas campestris* pv. *campestris* (blackrot) has been transferred from *B. carinata* to broccoli by fusion, followed by back-crossing to broccoli via embryo rescue and conventional sexual crosses (see Chapter 7). Somatic hybrids combining *B. oleracea* with more distantly related brassicas often suffer, however, from poor fertility; this makes back-crossing more difficult.

Production of transgenic plants is less advanced for *Brassica* vegetables than for rapeseed. Work on transformation of rapeseed is stimulated by its high economic value and the potential to expand the crop's usefulness through modification of its fatty acid content. Transgenic broccoli, cauliflower and cabbage have, however, been created, primarily through procedures involving the vectors *Agrobacterium tumefaciens* or *A. rhizogenes*. This approach has been used in conferring resistance to insects (from *Bacillus thuringiensis*), herbicides and viruses (see Chapter 7). Floral phenotypes have also been altered through transfer of genes preventing pollen development or modifying the self-incompatibility response.

BREEDING PROCEDURES

Male sterility

A considerable number of dominant male sterility genes have been found. Some have been used to a limited extent employing asexual propagation to multiply the sterile lines for hybrid seed production. The prime future method for hybrid production, however, is likely to be via cytoplasmic male sterility. This is rapidly moving to the forefront of methods for production of new hybrids in the vegetable brassicas. Older established hybrids will continue to use self-incompatibility.

The first *B. oleracea* cytosterile was developed by Pearson (1972) when he crossed *B. nigra* with broccoli, and this was developed further in cabbage. Unfortunately, this system was complicated by petaloidy and lack of development of the nectaries, giving a male sterile that was unattractive to pollinating bees. A better male sterile was developed when Bannerot *et al.* (1974) crossed a CMS radish (R cytoplasm) 'Ogura' with cabbage. In the fourth back-cross generation, he obtained normal plants with $2n = 18$ that were totally male sterile, the flowers having only empty pollen grains or vestigial anthers. Male sterile plants had the problem of pale or white cotyledons and also pale yellow leaves during development, accentuated at low temperatures. The cold temperature chlorosis has been overcome, however, by replacing the *Raphanus* chloroplasts with those from *Brassica* via protoplast fusion (Peltier *et al.*, 1983; Yarrow *et al.*, 1986; Jourdan *et al.*, 1989). Initially, however, there was reduced female fertility (Pellan-Delourme and Renard, 1987) which resulted in limited acceptance of this material. Walters *et al.* (1992) transferred the resistance to cauliflower and subsequently into broccoli. This and similar material has resulted in lines with good seed production and no apparent horticultural problems after vigorous selection for seed yield in early generations (Fig. 2.3).

Some plant breeders developing similar male sterility apparently did not select sufficiently rigorously for seed setting in early generations following fusion. This resulted in subsequent problems with a failure of seed production. Apart from the academically produced CMS seeds, various seed companies have patented their own CMS seeds in *Brassica*. This has resulted in confusion

Fig. 2.3. Flowering by cytoplasmic male-sterile and fertile *Brassica* plants (M.H. Dickson).

and inhibited use of the Ogura type of CMS due to high patent charges for the right to use this regardless of its source. More recently, cold-tolerant Ogura male-sterile cytoplasm in Pak Choi and other types of Chinese cabbage has been produced. Plant breeders in China have actively used the genic male-sterile AB line in Chinese cabbage since the 1980s. Characterization of the active male sterility gene *CYP68MF* from the AB line was reported by Wan-Zhi *et al.* (2003).

The 'Anand' cytoplasm was originally discovered in *B. juncea*, but recently its origins were also attributed to the wild species *B. tournefortii*. Cardi and Earle (1995) transferred it from 'rapid cycling' (see later in this chapter) *B. rapa* (CrGC# 1–31) lines to rapid cycling *B. oleracea* (CrGC# 3–1). After vigorous screening in the field, broccoli lines with good seed yield were released, as well as in cauliflower, cabbage and Chinese cabbage. Both the cytosteriles, the Ogura and Anand, have the typical *ms* genes inherited in a similar manner to that seen in onions and carrots. This means that the sterility requires the appropriate 'S' sterile cytoplasm plus *msms* genes. All *B. oleracea*, *B. napus* and *B. rapa* have the *msms* genes, but no fertility-restoring genes, and act as maintainers, while *Raphanus* does have fertility-restoring genes. Obtaining the male-sterile version of an inbred line only requires back-crossing to obtain the desired inbred female parent. Incompatibility can, however, upset this programme, if the recurrent parent has not previously been selected for self-fertility, which is not always so easy to find since for over 30 years most breeders have been selecting for self-incompatibility. More recently, Sigareva and Earle (1997b) have found they can substitute the nucleus of inbred lines into the appropriate sterile cytoplasm by fusion and avoid the need for the slow back-crossing procedure.

Self-incompatibility

The use of self-incompatibility was the accepted and only successful method of production of hybrid vegetable brassicas until the very recent development of cytoplasmic male steriles. Most crucifer flowers must be cross-pollinated, although a few plants are self-compatible, but limited numbers of seeds will set from self-pollination. Pollen is viable and will achieve fertilization in most cross-pollinations, with the exception of a small percentage for which the incompatibility specificity of the plant functioning as the female is the same as the incompatibility specificity of the pollen. Maximum seed set does not occur in self-pollinations if there are identical incompatibility traits in the stigma (female) and pollen (male). When such male and female *S* allele specificities are identical, the self-incompatibility acts to prevent pollen from germinating on and growing into the stigma and style.

The incompatibility specificities of brassicas are controlled by one locus, called the *S* gene. About 50 alleles, S_1, S_2, S_3, ... S_{50}, each giving one specificity, have been identified. Thus homozygous S allele genotype S_1S_1, S_2S_2, etc. have the incompatibility specificities of S_1, S_2 ... S_n for both the stigma and pollen. The incompatibility specificites of the S allele heterozygote brassica plant are complicated because control of the pollen specificity is by the sporophyte, the whole diploid plant, rather than by the individual (haploid) pollen grain. Usually the breeder assigned his own S allele identities, but the Warwick University Department of Horticulture Research International (Warwick-HRI), Wellesbourne, UK) maintains a collection of all known S alleles, which constitutes the internationally accepted nomenclature.

ASSAYING SELF-INCOMPATIBILITY

The level of self-incompatibility can be assessed by the number of seed produced after a self- or cross-pollination, but this takes at least 60 days. Fluorescent microscopy techniques readily display those pollen tubes that have penetrated the style, and this provides a direct measure of incompatibility which can be completed within 12–15 h. Aniline blue stain accumulates in the pollen tubes and fluoresces when irradiated with ultraviolet light. With appropriate light filters under a fluorescent microscope, the tubes are visible, whereas the background of stylar tissue is largely unseen. Penetration by a few tubes indicates incompatibility, while penetration by many tubes indicates compatibility. The rapid results greatly accelerate the evaluation of compatibility levels and the planning of breeding programmes.

Prior to assessing self-incompatibility, breeding for pest and pathogen resistance and for other horticultural characteristics is necessary. Then the evaluation of compatibility should follow. The selected plants can be tested by bud self-pollination and self-pollination of open buds. The resultant seed set or pollen tube penetration from the open flowers will measure the intensity of self-incompatibility. Where an interesting family is established, 11 plants should be selected and cross-pollinated in all combinations to establish the types of dominance and relative degrees of dominance within the family. The details of testing and evaluation of self-incompatibility were reviewed by Dickson and Wallace (1986).

Bud pollination

Bud pollination to overcome self-incompatibility is accomplished by opening the bud and transferring pollen from an open flower of the same plant. Fertilization will not occur where this is done when the bud is very small. At this stage, the style is not receptive, but about 3–4 days before the flower has opened the style and stigma are fully receptive and the self-incompatibility factor has not yet developed, therefore self-fertilization is possible and self-incompatibility can be bypassed.

The bud is opened with a pointed object, such as a mounted needle, and pollen transferred from an open flower. In some large breeding programmes, many, even hundreds, of pollinators can be used to produce selfed seed on self-incompatible plants.

There are two other methods now widely used to overcome the self-incompatibility. The first involves spraying 3–4% sodium chloride solution on to the open flower. It is then left for 20–30 min, the excess salt solution is removed by blowing it off the flower or blotting with a damp cloth or paper towel, and then self-pollination can occur. The salt removes the inhibitor from the stigma, permitting self-pollination. A second and more efficient method if many plants are involved is to self-pollinate the open flowers, and then place the plants in a closed unit or room into which 5% carbon dioxide can be admitted. Then the self-incompatibility factor is overcome and this allows the production of seed following self-pollination. Single plants can be self-pollinated and the flowers enclosed in a plastic bag; it is inflated by exhaling into it, using a straw, and then finally closed. This may be repeated again later in the day. The effect is similar to enclosing the plant in a room and adding carbon dioxide.

When the only method of production of hybrids was by the use of self-incompatibility, then self-fertile plants were usually discarded. Now with the interest in male sterility, the self-fertile plants are the desired ones and the self-incompatible plants are discarded. It is equally important, however, to establish that the plants are homozygous for self-compatibility and not just setting seed because they are heterozygous for *SI* genes.

Production of F_1 hybrids

Inbred brassicas, as with most other out-crossing crops, tend to lose vigour and, although a single cross will restore vigour in the F_1, the seed production on the inbred parents may be uneconomically low. For this reason, whether hybrid production involves self-incompatibility or male sterility, three-way crosses are often used. If self-incompatibility is used, and if both parents are self-incompatible $S_1S_1 \times S_2S_2$, then seed from both parents can be harvested. Alternatively, an open pollinated line can be used as the pollen parent and seed only saved from the self-incompatible line. Where the original S_1S_2 hybrid is self-incompatible, then it can be used as a male or female line together with a self-incompatible C line, and again all the seed may be harvested. That is, however, more complicated and requires deeper understanding of the levels of self-incompatibility and the relationship of each inbred line to the others involved.

With male sterility, the A or male-sterile line Smsms is maintained by back-crossing to a self-fertile B line Nmsms (S = sterile and N = normal or fertile cytoplasm). After several generations of back-crossing, the A line will resemble the B line, but there will be some inbreeding depression. For production of hybrid seed, the A line is crossed to a further unrelated line C. Only the seed on the A line can be saved. Three-way crossing could be made by crossing the A × C hybrid to a further line, D, and this may result in higher seed production levels since the female parent will now be a hybrid.

Seed production success depends on issues such as whether the two parents flower at the same time. This can easily become a problem where the parents may require different degrees of vernalization and exposure times to produce flower stalks and flowers.

Self-pollination

Self-pollination of any of the brassicas can be obtained by brushing or shaking the open flowers if the plant is self-compatible. If the plant is self-incompatible, the buds can be opened 1–4 days before they would open naturally and be bud-pollinated. If the buds are too small, then bud pollination is usually more difficult and unsuccessful, and if left too late then the incompatibility factor may have developed and few seeds will be produced. A young bud 3–4 days prior to natural opening is preferred as the self-incompatibility factor will not have developed, but the bud is still large enough for the stigma to be receptive.

Usually six or eight buds can be opened and pollinated at one time on a raceme, with a pointed object such as a toothpick or forceps, and pollen from an older open flower can be transferred to the stigma and seed obtained. Pollen can be transferred with a brush or finger or using the open flower itself to brush the anthers across the stigmatic surface to transfer the pollen.

Williams (1980) suggested the use of 'bee sticks', constructed by using the hairy abdomen of bees mounted on a toothpick as pollen carriers.

Cross-pollination

When the pollination has a breeding objective other than as a study for self-incompatibility, then the bud should be emasculated using buds expected to open in 1 or 2 days, eliminating the possibility of selfing. The desired pollen is then transferred to the stigma in the same manner as for bud pollination. If a male-sterile or self-incompatible plant is used as a female parent, then an open flower or one with a protruding pistil can be used. When the cross-incompatibility between two plants is being assayed, the open flowers need not be emasculated. The major advantage of hybrids with vegetable brassicas is the uniformity of plant size, shape and maturity, as well as the protection that hybrids give to a seed company that produces the seed. Vigorous open-pollinated cultivars can be just as productive as hybrid ones, although they may lack desirable ancillary benefits such as uniformity of maturity.

SEED INCREASES

Hand pollination is best performed in the greenhouse or in a large screened cage to eliminate insects. If cross- or self-pollination is desired in the field, cheesecloth bags can be used to enclose the blossoms of one or two plants. It is preferable to enclose several plants in a screen cage 2 m high. The plants should be held away from the cage walls to avoid pollination through them by visiting bees and other insects. Bees are the best pollinators, but flies can also be used. If large-scale increases are to be made outdoors, then the minimum isolation between lots should be 2000 m or more. Greater isolation is needed if one lot is downwind from another or the location of the hive results in bees crossing one lot of plants to reach the second.

All brassicas can be dug after growing in the field, potted and moved to the greenhouse after a period of vernalization that is required, except for broccoli and summer cauliflower and some Chinese cabbage. Digging American-type cauliflowers (Snowball forms) is more difficult, but success can be high if the plants are dug with a root ball and potted with minimum root disturbance. If the temperatures are high, then placing the plants in a refrigerator for a couple of weeks while they are re-rooting helps survival. European and Australian cauliflower survive transplanting much more readily, with >95% success rates. Increasing cauliflower by tissue culturing the curd is also used, but this slows seed production by at least 6 months; however, this should eliminate the loss of selected plants.

Asexual propagation can be used for cabbage by cutting off the head and allowing the lateral buds to develop. The buds can then be excised, rooted on a moist medium, vernalized and allowed to flower in small pots. The buds from Brussels sprouts can be treated in a similar way. For broccoli, cuttings can be taken from the lateral branches, and rooted in a mist bed after being dipped in a hormone rooting powder. Rooting starts in 10–14 days and the plants develop rapidly thereafter.

VERNALIZATION

Many brassicas have a vernalization requirement which induces flowering. In practice, all brassicas need vernalization, but the temperature at which this occurs can be quite high, >25°C, for most broccoli and some summer cauliflower. Under most conditions, summer cauliflower and broccoli will flower without any artificial vernalization. Under hot summer conditions in the eastern USA, some Dutch cauliflower cultivars can be so late that they only mature after the first frosts, which can destroy them. This indicates that these cauliflowers require a cooler temperature for vernalization and, if dug with the curds in a prime market stage, flowering may be accelerated by storage for a month at 5°C.

Biennial crops, such as some cauliflower types, cabbage, kale, kohlrabi, collards and Brussels sprouts, all require 2–3 months of vernalization below 10°C. Two months may complete the vernalization, but 3 months is usually preferable. Insufficient vernalization extends the time between removal from the vernalization condition to flowering by as much as 2 or 3 months. This results in an actual delay in time of flowering and often in incomplete or poor flowering or even devernalization after the seed stalk has been initiated. Consequently, an adequate duration of vernalization is essential. In most cases, 5°C is the most ideal temperature for vernalization. Lower or higher temperatures result in the need for a longer vernalization period. Within *B. oleracea* crosses, the more annual habit is dominant to the more biennial, but there is a continuous range in this variation, indicating polygenic inheritance. Most brassicas can be vernalized after they reach a stage of maturity in which the stem is at least 10 mm in diameter, preferably 15 mm. Prior to that stage, the plant may not respond to reduced temperatures. Seed of *B. rapa* and *B. napus* can also be vernalized.

PREPARATION OF PLANTS FOR BREEDING FOLLOWING SELECTION

Most vegetable brassicas are relatively easy to transplant, and will withstand considerable abuse during selection and evaluation phases prior to being allowed to grow to maturity and flower. All the vegetable brassicas can be transplanted in a bare rooted state without damage, except for cauliflower. Cauliflower does not regenerate roots very readily. If, however, a cauliflower at market stage is dug carefully, preferably from wet or moist soil, then a root ball can be maintained. The plant is then placed in a large pot with moist potting soil or compost medium and very few selections will be lost. About half of the leaves should be removed to reduce desiccation by transpiration. Plant breeders at Cornell University, New York State, USA, have rarely lost >5% of the plants during the past 20 years. If it is hot (>27°C) in the glasshouse, then it is desirable to place the freshly potted plants in a cool place at 10–20°C for 2–3 weeks. If the glasshouse temperature will not exceed 27°C, then the plants should develop and produce seed. If the cauliflower is a late summer or winter type, it may need varying lengths of cold storage before it will develop flowers from the curd.

The whole broccoli plant can be dug and potted, or cuttings taken from them, dipped into an auxin-based hormonal rooting medium and placed in a mist chamber for 10–14 days, by which time they should be rooted and ready for potting. Cabbage, Brussels sprouts and kale can be dug, some of the leaves are removed and the plants potted up, prior to being vernalized, or the cabbage head can be removed and adventitious buds allowed to develop. These and buds of Brussels sprouts can be removed and placed in moist soil where they will root while being vernalized.

BREEDING RESISTANCE TO PATHOGENS

Comprehensive details of methods used to screen brassicas for disease resistance are provided by Williams (1981). The author also lists some of the rapid cycling accessions of *Brassica* and their specific genes for use by breeders. Rapid cycling *Brassica* forms are described in more detail later in this chapter. Plant breeders from across the world have been active in incorporating resistance to the many pathogens of vegetable brassicas. Resistance in most individual cultivars is limited, however, to a few pathogens. Consideration of the degree of success of resistance breeding to pathogens, pests and disorders is given in Chapters 7 and 8.

RESISTANCE AND DOMINANCE

As with some other crop families, such as the Cucurbitaceae, many resistances are dominant, although this is often due to major genes at specific loci. This holds true for white rust (*Cystopus candida*), black leg *Leptosphaeria maculans*), black rot (*Xanthomonas campestris*), *Alternaria* spp, downy mildew (*Peronospora parasitica*), powder mildew (*Erysiphe cruciferarum*), *Fusarium* yellows (*Fusarium oxysporum* f.sp. *conglutinans*), soft rot (*Erwinia carotovora*) and turnip mosaic virus (TuMV). As such, these can be used in hybrid production, where one parent can be resistant to some pathogens and the other resistant to alternative pathogens, but the hybrid can have good tolerance to all pathogens (Chapter 7).

JUVENILE VERSUS MATURE PLANT RESISTANCE

In some cases, breeders have assumed that both seedling and mature plant resistance resulted in resistance throughout the life of the plant and in all environments. It has become accepted that with many pathogens of brassicas, and other crops, there may be both juvenile and mature plant resistance and these traits may be expressed separately. If this fact is not taken into account, the results can be confusing and misleading. For example, resistance to downy mildew, black rot, *Alternaria* and white rust all respond differently with plant age. It is also common for the expression of resistance to be affected by the environment in which a crop is grown (Chapter 7).

SELECTION TECHNIQUES FOR SPECIFIC CHARACTERS IN VEGETABLE BRASSICAS

Cabbage head shape

The desirable cabbage head shape in commerce has changed from pointed, flat or round to almost exclusively round. Pointed head is dominant to round. Many genetic factors, however, influence head shape. Selection is best made by cutting the head vertically through the core; this allows selection for head shape, core length, diameter and solidity, or leaf toughness, leaf configuration within the head and leaf or rib size. The cut core will heal and lateral buds develop from it so that selected plants can be saved for seed production.

Heading versus non-heading

A distinguishing character of cabbage is the development of several wrapper leaves surrounding the terminal bud. As these leaves develop, the head will become solid and develop a head or heart. Heading is recessive to non-heading. The F_1 between a non-heading or raceme type and a heading cabbage will be intermediate.

Depending how wide the cross is, two genes (n_1 and n_2) or more are involved in heading. Likewise, the loose head of a savoy type is recessive to the hard heading of a smooth-leaf cabbage. By the time a second back-cross is reached within *B. oleracea*, the head type can be regained, provided reasonably large populations are used.

Head leaves

In a cross between lines with few and those with many wrapper leaves, few is dominant. There are modifying factors. The number of head leaves is also confounded by whether the cabbage is an early or late maturing type, the early maturing types having fewer leaves. The evaluation and selection are made following vertical splitting of mature heads with a knife.

Size of head

Head size is a quantitative trait and may be related to hybrid vigour, but not necessarily as open pollinated lines can be just as large. Cabbage is responsive to day length, and in northern latitudes (such as Alaska) the cabbages can become very large, weighing as much as 20 kg. Likewise, the further south or the shorter the day length, as in winter, the smaller the head. Agronomic issues such as crop spacing and row spacing confound head size. Dense cropping naturally results in smaller heads.

Plant height

There is a major gene '*T*' for plant height. Most cultivars are recessive in this respect. Again there are, however, genetic modifiers. A long stem is an undesirable characteristic as the plant may fall over from the weight of the head. For machine harvesting, however, the stem should not be too short and it is important that the head stands upright. Generally, in *B. oleracea*, tall is dominant to short, but plant height involves several genes acting additively.

Core width

Wide core appears to be dominant to narrow. A narrow core is desirable, but less important than a shorter core. A wide core is usually softer or more tender than a thin core. Tough cores result in 'woodiness' caused by the deposition of lignified tissues in the fresh and processed products, and are not desirable.

Core length

Core length is controlled by two incompletely dominant genes for short core. A short core of $<25\%$ of the head diameter would be desirable.

Frame size

Older cultivars generally had large frames and basal leaves. In adapting cabbage to mechanized harvesting and in the effort to develop high-yielding cultivars, the trend has been towards smaller frames. The compact fresh market cultivars generally have smaller frames than processing types. An adequate frame is needed, however, for photosynthesis, but expression is also influenced by factors such as season, spacing and fertility. Adequate water and nutrition early in the plant establishment and development will result in a larger frame, which also may cause the plant to develop tipburn (Chapter 8) if water is withheld or is in short supply during maturity.

Head splitting

Three genes control head splitting, and these act additively with partial dominance for early splitting. According to Chiang (1972), narrow sense heritability for splitting was 47%. To evaluate cultivars for splitting, they must be allowed to grow to full maturity. Long-cored cultivars usually split at the top of the head, while short-cored cultivars tend to split at the base.

Storability

Late, slow growing cultivars are most suitable for storing. Dry matter of the better storage types is higher (9–10%) than that of standard and early cultivars (6–7%). The best storing cultivars usually have finely veined leaves. Eating quality is usually inferior because the leaves are tougher and harder. Selection for long-term storage is made most effectively by cutting the head from the stem.

At the end of the storage period, the selected heads can be placed on moist potting mix and roots will grow out from the base of the head. Alternatively, the lateral buds can be allowed to develop and then be removed and rooted and allowed to produce an inflorescence and flowers. Cabbage used for coleslaw should be white internally, while cabbage for the fresh market should preferably be green.

Winter cultivars

The best cultivars for growth in more southern regions in the winter, and which may be subject to considerable cold, should be bolting resistant, and require a long cold induction before flower stems will develop. They are often slightly savoyed. Selection can best be done in areas where they will be grown and subject to cold induction. They should be planted earlier than for the commercial crop so that they are larger when the cold develops and are subject to longer periods of cold. Plants that do not bolt rapidly under such conditions must be bolting resistant and can be saved for seed production.

BROCCOLI BREEDING

Heat tolerance

Susceptibility to high temperature damage is one of the major problems in broccoli. This makes the crop agronomically more suitable for cultivation in cool, moist climates. Heat at harvest time is not critical, except for the fact that high temperatures will reduce the period over which the crop is marketable. About 3–4 weeks prior to the head being marketable is the critical period when the growing point is differentiating to become reproductive (Chapter 4). Bjorkman and Pearson (1997) studied the development stage at which the temperature response is most critical. The cross-section of the developing reproductive structure is shown in Fig. 2.4. It is most sensitive at the straightened and bowed stages, which is a relatively short period of 4–5 days. Leaf bracts can enlarge and grow through the head.

When the plant is sensitive to heat, the buds will become enlarged and grow through the head, due to elongation of the sepals, and the degree of enlargement is quantitative and influenced by the amount of heat (Fig. 2.5). Additionally, the secondary inner whorls of buds may turn yellow, especially the inner buds of the whorl, causing a condition called 'yellow eye' or 'starring'. Yellow eye tends to be dominant to normal bud formation. Also, leaflet extension may be encouraged in the head. Non-leafiness tends to be dominant to leafy. All three defects are enhanced or triggered by heat, and can render the crop unmarketable. The best screening method is to grow the plants under normal ideal conditions. They are then moved to a hot environment (35–25°C day/night) for 10 days when the growing point is at the critical 'straightened and bowed' stage before returning the plants to a normal growing regime at 20–25°C. Efforts are being made to map and find

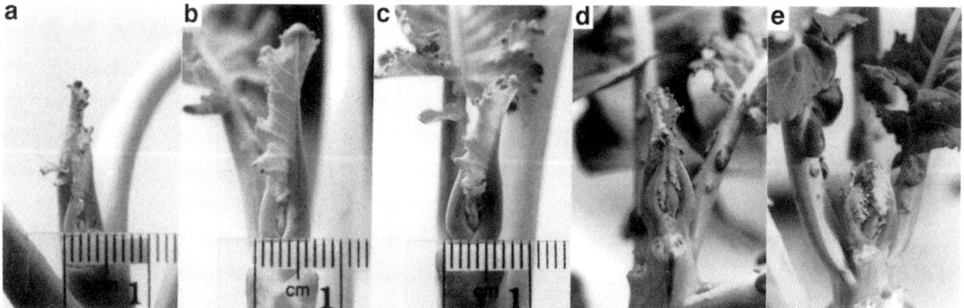

Fig. 2.4. External appearance of the shoot tip of broccoli (calabrese, *Brassica oleracea* var. *italica*) during the transition from vegetative to reproductive growth. The plants can be screened for heat tolerance in the straightened (b) or bowed stages (c). When the plants reach the crown (d) and head (e) stages, it is too late for effective screening and the vegetative stage (a) is too early (M.H. Dickson).

Fig. 2.5. Illustration of 'yellow eye' or 'starring' in broccoli (calabrese, *Brassica oleracea* var. *italica*) (M.H. Dickson).

quantitative trait loci (QTLs) for heat tolerance or susceptibility to permit breeding for heat tolerance.

The plants can be grown in the field in a location where high temperatures are expected when the plants are maturing, and subsequently field selections are made for resistance to heat injury. Some seed companies have bred vigorously for heat tolerance, and selections show considerable tolerance compared with most broccoli hybrids. Yang *et al.* (1998) developed hybrid 'Ching-Long 45' with heat tolerance from the cross of broccoli and an inbred involving Chinese kale (*B. oleracea* var. *alboglabra*).

Branching

Branching is also influenced by temperature, and a hot summer increases the degree of branching. Branching is dominant to non-branching. In the most extreme non-branching type of broccoli, the plant is similar to a cauliflower in that no lateral buds develop in the leaf axils and, if the head is removed, no lateral buds will develop. Under these conditions, if the head is removed, no seed can be produced on the saved plant. The degree of branching varies, however, depending on the cultivar and the temperature. High temperature increases the degree of branching. Moderate branching is probably desirable, but excessive branching is not.

Storage ability

The need for improved harvest efficiency in broccoli has resulted in screening of cultivars for concentration of yield and uniformity of harvest as well as postharvest, maintaining quality. McCall *et al.* (1996) stored broccoli at 2°C for 14 days and thereafter at 12°C for a further 19 days. The culivars kept well at 2°C, but deteriorated rapidly at 12°C. Modifications of this system have been used to select individual heads in segregating populations for quality and then for storage ability, and considerable variation was apparent (Chapter 8). The cultivar 'Marathon' has better shelf life than many other cultivars. Much more selection and breeding for shelf life is needed, and preliminary observations show that there is the opportunity to develop more superior cultivars with potentially extended shelf lives.

Other characteristics

A solid stem is desirable in broccoli, and cutting the terminal head off the plant reveals the internal stem structure, and will permit selection. The plant should not be too tall, but a very short plant is not desirable, while most very tall plants do not have solid or large heads. Tallness is dominant to intermediate and short stem, but can be stabilized at any desired height. Cabbage produces the tallest plants at flowering stage compared with the other common vegetable brassicas. In broccoli and cauliflower, stem elongation is more limited, with little or no elongation of broccoli beyond the natural flower position in the head at market stage. The broccoli head should be dome shaped, and heavy at maturity. The trend is for new cultivars to have small buds or beads. Large bud is dominant to small bud. So far, small buds and heat tolerance have been hard to combine.

Increasingly the world land area available for cropping is being curtailed by soils becoming saline. The maritime origin of *B. oleracea* (see Chapter 1) should make cabbage a valuable crop for use on soils where salinity is becoming a problem. The effectiveness of cabbage in this respect might be increased if salt tolerance was increased still further. This might be achieved by transferring the bacterial gene *betA* which codes for the biosynthesis of glycinebetaine into cabbage, as suggested by Bhattacharya *et al.* (2004) using cv. 'Golden Acre' in India. Glycinebetaine is a well established and effective biostimulant compound that promotes plant growth and health when used as extracts from seaweed.

CAULIFOWER BREEDING

Cauliflower can be divided into several subgroups (see Chapter 1) dependent on their origin from Australia, Europe (especially Italy), India and the USA.

There are wide further subgroupings of cauliflower from different regions within European countries, such as Italy, The Netherlands, Germany, and Cornwall in south-west England. In general, the early cauliflowers such as the Snowball types are self-compatible, while progressively the later cultivars are more self-incompatible. The Indian cauliflowers that are mostly early are quite self-incompatible. The Australian cauliflowers that are of intermediate to late maturity are quite self-compatible. The Australian cauliflowers are usually very large.

Careful handling of selected cauliflower plants when being dug for seed production is essential. Digging mature plants with root balls disturbed as little as possible, along with removal of about half the leaves, usually resulted in >95% survival in the Cornell University breeding programmes. The curd will eventually elongate and produce flower stalks and plenty of flowers. If, however, the seed crop produced is not too heavy, and if the plant remains healthy, a plant will continue to develop flowers from undeveloped sections of the curd for several months.

Some breeders have had trouble with saving field selections, and grafting curd portions on to young stock plants has been fairly successful and does not delay seed production appreciably. Alternatively, tissue culture has been a means of vegetative propagation. Small pieces of the curd are removed from the parent plant, surface sterilized and small pieces 1–2 mm in diameter are placed on paper wicks in nutrient solution. Roots and shoots develop and can be transferred to pots and grown to produce new plants (Crisp and Gray, 1975a, b, c).

Heat tolerance

Cauliflower is heat susceptible, but not to the same extreme degree as broccoli. Some of the Indian cauliflowers such as 'Pusa Katki' have the greatest heat tolerance, but these are early and have relatively poor quality curds. If the temperature is high, then the curd may develop bracts (green bracts corresponding to auxilary leaves are usually present in the curd, Fig. 2.6) which make it unmarketable. Bracting is recessive, and heritability was reported at 73%. The curd may also produce true green leaves in the head.

The curd may also turn purple if exposed to the sun; this may be due to development of small purple buds, which are similar to the small white velvety buds which develop in some curds and is called 'ricey'.

Growing the plants under hotter than ideal conditions will allow for selection for heat tolerance. Solidity of the curd is also desirable, and continued selection for this trait is required as soft curded plants will bolt more readily and have a selective advantage. Solid tends to be recessive, although it is a quantitative factor, and there may be more than one genetic route to solidity, so that crosses of two solid curds does not necessarily mean the hybrid will be solid.

Fig. 2.6. Illustration of leaf bracting in cauliflower (*Brassica oleracea* var. *botrytis*) (M.H. Dickson).

Curd colour

The most desired colour in cauliflower is a white or cream curd. In many cauliflowers, if the curd is exposed to the sun, it will turn brownish yellow; this is especially so for the Snowball type. Cauliflowers are protected from the sun, either by wrapping the leaves over the head and tying them, or by developing plants with long leaves, which protect the curd from exposure. The upright leaf character is partially dominant, with a heritability of 67% and three loci. This is particularly true for some of the later autumn and winter type cauliflowers, but has been bred into some early types. The other alternative is to breed persistent whiteness into the plant. A cauliflower, Plant Introduction (PI) 183214, from Egypt had this character (Dickson and Lee, 1980) and remained completely white even when exposed to full sun. Several recessive genes are involved and, unless large numbers of F_2 plants are grown (>250), there is little chance of obtaining persistent white curds.

The curd of persistently white plants will remain white even after extensive elongation (Fig. 2.7). Unfortunately, the original PI 183214 was very soft curded, and obtaining persistent white curds that are solid has been difficult.

There are several curd colours, especially in Italian cultivars 'Autumn Giant' and 'Flora Blanca'. Italy is regarded as a centre of origin for the cauliflower (Chapter 1). Colours such as bright green (due to two genes),

Fig. 2.7. Illustration of persistent white and normal cauliflower (*Brassica oleracea* var. *botrytis*) curds extending towards flower production (M.H. Dickson).

orange, purple, and yellow or golden are also available. The purple colour can be due to precociously developed flower buds that have developed on the curd and can assume the purple pigmentation of the buds. The green curds are also more inherently frost tolerant than white curds (Gray, 1989). Orange curd is dominant to white (Crisp *et al.*, 1975). The inbred orange cauliflower, however, is very small and weak, and orange cauliflower can only be grown commercially as a hybrid. The degree of colour is quantitative, however, and is influenced by the whiteness of the normal white curd parent.

The white curd parent should be similar to the Snowball types, namely with a tendency to turn brown when exposed to the sun. White curd parents result in an orange curd hybrid that is pale. The orange colour is due to β-carotene pigments, and these can range from >400 μg/100 g of tissue in a deep orange curd to about 3 μg/100 g in a very white curd (Dickson *et al.*, 1988).

SWEDE BREEDING

Brassica napus is a cool season vegetable crop primarily grown in the northern countries of Europe and Canada, where it is also called rutabaga. The swollen root is pale orange fleshed with the upper half being off white or slightly yellow and the lower half brown. Desirable characters are uniform interior colour, smooth round exterior shape, smaller rather than large leaf attachment and good storability. They have been sources of resistance to some physiological races of *Plasmodiophora brassicae* (clubroot), and these have been used in interspecfic crosses. Also, the interspecific progeny from crosses to *B. oleracea* have been partially male sterile. In the latter, however, flowers are often fertile and the level of fertility is inconsistent from year to year, so that this has not been a satisfactory source of male sterility.

Turnip mosaic virus can be a major problem (Chapter 7), and there are wide differences in cultivar reaction. There are also different reactions among resistant lines, which are inherited independently. These are symptomless for necrosis or mosaic development. Various isolates differed in virulence, but caused similar host reactions.

Cabbage maggot or root fly (*Delia brassica*) (Chapter 7) is a major pest problem in swede, and breeding efforts to control it have had varying success. There does not appear to be immunity to this pest in either *B. napus* or *B. oleracea*, but some selections are much less susceptible than others.

BRUSSELS SPROUT BREEDING

Sprouts vary from early (90 days) to late (120 or more) days from planting to maturity. Flavour varies widely from quite bitter to sweet, with the latter obviously being preferred. Frost, however, results in improvement in flavour just as it does in root crops such as parsnip. Some new hybrids have much improved flavour without being chilled. Almost all sprouts now grown are hybrids, produced using self-incompatibility. The major disease problem in the USA is *Fusarium* yellows (*F. oxysporum* f.sp. *conglutinans*), and some of the newer hybrids are resistant (Chapter 7). Susceptible hybrids can develop serious rotting, especially of the lower buds.

Buds can vary from small to large at maturity, but mid-sized buds are preferred. Likewise, the plants can vary in height. Taller plants have higher yield potential, but unless the crop is hand harvested, it is difficult to obtain uniform sprout maturity along the whole length of the stem.

Sprouts that are planted too close together are difficult to harvest, while wide spacing will reduce yield. Easy detachment of the sprouts themselves is a very desirable trait, as is easy removal of the leaves. Both can vary from easy to hard to pick.

Dark green sprouts are preferred to pale sprouts. Sprouts are susceptible to all the pathogens causing problems with other *B. oleracea* crops, but, because they are grown in cool climates, some of the impact of diseases is often reduced. In particular, downy mildew (*P. parasitica*) and aphids can be problems (Chapter 7). Ellis *et al.* (2000) tried to select for aphid resistance. Some glossy leaved sprouts are resistant to aphids and lepidopterous pests, but the glossy leaf is not acceptable to the consumer.

CHINESE CABBAGE BREEDING

Chinese cabbage (*B. rapa*) types (Table 2.1) are as diverse as *B. oleracea*. Chinese cabbage is susceptible to almost all the pathogens, pests and environmental stresses affecting *B. oleracea*. Telekar and Griggs (1981) discuss issues of specific importance in the breeding of Chinese cabbage. The evolution of Chinese cabbage forms is shown in Fig. 1.6. Most types are readily cross-compatible. As with *B. oleracea*, hybrids have been made using self-incompatibility, but recently there has been an increasing interest in using male sterility, of both genic and cytoplasmic types.

NEW AVENUES OF *BRASSICA* DEVELOPMENT

The wide genetic diversity and flexibility of *Brassica* taxa discussed in Chapter 1, and referred to in this chapter, has provided many opportunities for both natural and man-driven evolution. In domestication, these characteristics have resulted in a vast range of morphologically different *Brassica* crop types often individually tailored to particular local and regional husbandry systems. In Europe, *B. oleracea*, and in the Orient, *B. rapa* especially have been intensively selected over many centuries in cultivation, producing plant forms that are markedly different in their horticultural and culinary traits. Botanically, however, they are closely similar and frequently difficult to characterize on strictly taxonomic criteria. Frequently they may be divided solely by horticultural characteristics (Toxopeus, 1979).

In the 20th century, members of the family Brassicaceae took on new roles as ideal model forms for use in basic scientific research and more latterly as educational tools. The very genetic flexibility and diversity that allowed generations of horticulturists to hone new crop types as divergent as Brussels sprouts and cauliflower offer new sources of benefits to mankind as a route to fundamental genetic and evolutionary information. Indeed, it could be argued that the value of this family to mankind is set to increase substantially because of the roles of its members in the molecular and genetic biological revolutions which will be a central part of scientific endeavour in the 21st century. It might not stretch credulity too far to compare the emerging opportunities for the Brassicaceae as the plant science equivalent to stem cell research in human and animal biology.

BIOLOGICAL MODELS

When research attempts to understand the fundamental processes of nature irrespective of whether these are biological, chemical, mathematical or physical, the initial vehicle used to translate theory into experimentation must be as simple as possible. Simplicity of operation allows the principles that are being investigated to be identified and analysed in ways that are uncluttered by other extraneous and possibly unrelated processes.

In biology, the complexities associated with crop plants can, for example, be added at later stages when research moves downstream into applied areas and results in useful applications. Initially, the fundamental biologist seeks plant or animal systems that are easily amenable to being cultured in confined and controlled environments with a rapid, simple, ephemeral life cycle and small mass. These organisms become 'model forms' around which enormous amounts of scientific knowledge are assembled and theories developed. Zoologists have used the fruit fly, *Drosophila melanogaster*, since the early 1900s as a convenient vehicle for genetic studies. Similar detailed studies have been made with the yeast fungus, *Saccharomyces cerevisiae*, and the nematode *Caenorhabditis elegans*.

Whyte (1960) identified what he termed 'Botanical Drosophilas'. These did however, include several complex crop plants such as cereals, soybean, peas, *Kalenchoe blossfeldiana*, *Perilla* spp., and the simpler weeds *A. thaliana* and cocklebur (*Xanthium pennsylvanicum*). Where crop plants are used in the early stages of the translation of theory into experimentation, it is because a particular plant has some specific trait that is amenable to laboratory studies. Thus much use is made of spinach (*Spinacea oleracea*) in photosynthetic research because it has large and relatively accessible chloroplasts, and photoperiodic studies have centred on the *Chrysanthemum* because its responses are clear cut and amenable to manipulation.

Much of our knowledge of plant morphogenesis has come from studies of carrot (*Daucus carota* var. *carota*) cells in culture because they are suited to this form of husbandry where single root cells exhibit clear totipotency. *Antirrhinum* spp. and *Petunia* spp. have formed admirable vehicles for the analysis of primordial differentiation and growth in flowers.

The family Brassicaceae contains two outstanding models, the thale cress, *A. thaliana*, and a series of genotypes derived from crosses of culinary brassicas with *B. alboglabra*, which introduced traits allowing rapid cycling from seed to flower and back to seed. These latter have become known as the 'Wisconsin Fast Plants™' because they were discovered at the Madison Campus of Wisconsin University, USA.

ARABIDOPSIS THALIANA – BOTANY AND ECOLOGY

The genus *Arabidopsis* contains a small number of species of slender annual or perennial herbs. Thale cress, *A. thaliana*, is common throughout northern Europe, living on dry soils in walls, banks, hedgerows and waste places. It is believed to be native to the Old World, although its exact geographical origin in unclear. It has been collected or reported as present in many different regions and climates ranging from high elevations in the tropics to the cold climate of northern Scandinavia and including locations in Europe, Asia, Australia and North America. A related species, *A. suecica*, is restricted to Scandinavia, and kindred genera in the Tribe Sisymbrieae include *Sisymbrium* species themselves (the rockets) and *Isatis* (woad) (Stace, 2001).

Botanically, *A. thaliana* is an ephemeral with simple erect stems varying from 50 to 500 mm tall; usually the stems are rough and hairy bearing simple hairs. The basal leaves form into a rosette, each having an elliptical to spathulate shape, and each leaf is stalked, with a toothed edge, especially at the distal end. The flowers are small, usually <3 mm diameter, formed of white petals in the characteristic cruciform pattern. Each inflorescence is composed of 4–10 flowers; the fruits are dry siliquae. In miniature, therefore, *A. thaliana* resembles the floral structure of many other often much larger brassicas described in this book.

Once mature, the dry seed may be stored for several years with little loss of germination capacity, but once moistened they germinate rapidly. In optimal conditions, each plantlet produces a rosette of leaves, forms a flower stalk and blossoms within 4 weeks of sowing. The rate of development and time to flowering are controlled by the prevailing environmental conditions. A shortage of nutrients, for example, leads to a decrease in regeneration time but decreases the seed set. *Arabidopsis thaliana* flowers more rapidly in long days or in continuous light, while short days retard flowering significantly, with differing genetic constitutions in various morphotypes (Table 2.2).

The commonly used ecotypes such as 'Columbia' and 'Landsberg' have a regeneration time of approximately 6 weeks. Since most types of *A. thaliana* continue forming flowers and seeds over several months, it is possible to harvest in excess of 10,000 seeds from each plant. The plant grows satisfactorily in either soil or compost and can be cultured on nutrient agar or liquid sterile media. Consequently, it is admirably suited to growth in artificially lit controlled environment chambers and 'growth cabinets' or 'growth rooms' well divorced from traditional field husbandry.

USE OF *A. THALIANA* IN GENETIC AND MOLECULAR STUDIES

This small cruciferous plant has taken on the mantle of being a ubiquitous research tool in studies of many aspects of plant and, increasingly, animal biology. Proofs of concept studies almost invariably utilize *A. thaliana* and, in consequence, there is an enormous literature describing every facet of its germination, growth and reproduction in genetic and molecular terms. It was originally identified as an experimental tool by Laibach (1943) in Germany following detailed investigations of its genetics that started in 1907. Its haploid chromosome number was identified as $n = 5$; a collection of ecotypes

Table 2.2. Examples of the morphological mutants of *Arabidopsis thaliana*.

Character	Effect
Angustifolia and *Asymmetric* leaves	Narrow or asymmetric leaves when homozygous
Glabra and *distorted* trichomes	Remove or change the shape of the leaf hairs
Erecta and *compacta*	Alter the disposition and size of stems
Brevipedicellus	Reduces the pedicels when homozygous
Apetala-1	Reduces petals
Eceriferum	Changes the morphology of the epidermal wax

was established and mutant forms resulting from exposure to chemical and physical mutagens were catalogued.

There is a set of recessive mutants that cause homeiotic variations that can be used in studies of developmental biology (Pruitt *et al.*, 1987), and a group of colour variants make useful visible markers. Lethal recessive mutants can be employed for developmental, biochemical and physiological studies. Resistance to herbicides and a range of plant pathogens offers practical agricultural and horticultural applications. *Arabidopsis thaliana* is thus well characterized genetically; a wealth of useful and interesting mutations have been induced, analysed and mapped. These characteristics alone do not distinguish *A. thaliana* from other experimentally useful and genetically well studied plants such as the tomato (*Lycopersicon esculentum*) and maize (*Zea mays*).

It is the small size, short generation time, high seed set and ease of mutagenesis in *Arabidopsis thaliana* that make it easier and faster to induce, select and characterize new mutations which attracts scientists of many biological disciplines. Added to this are very attractive cellular and nuclear characteristics. *Arabidopsis thaliana* is attractive for both quantitative and qualitative genetics because of its small genome size and flexible genomic organization (Somerville, 1989; Griffiths and Scholl, 1991; Meyerowitz and Somerville, 1994; Meyerowitz, 1997). It has the honour of being the first plant where the entire genome has been sequenced.

The advantages offered by *A. thaliana* include the following.

- Small size – hence ease and simplicity of cultivation in controlled environments where space is at a premium for availability and cost. More than 100 plants can be grown in the space occupied by this open book.
- Short life cycle – regeneration from seed to seed takes between 4 and 6 weeks, which offers short repeatable and reproducible experiments.
- Perfect flowers – these are self-fertile so that pollen transfer is simplified with no requirement for complex procedures to ensure cross-pollination, and the flowers tend not to be open pollinated. Self-fertilization exposes recessive mutations in homozygous form in the M_2 (mutation two) generations following the application of mutagens.

- Ability to produce large amounts of seed – up to 10,000 seeds can be harvested from one plant. This provides large populations very quickly from a single cross. In turn, this minimizes space requirements and makes the statistical analysis of populations easier with the large numbers of individuals available.
- A very compact genome – 70,000–100,000 kb per haploid genome with only 10% of highly repetitive sequences. The genome is amongst the smallest of the higher plants with the haploid size of about 100 Mb of DNA. This makes *A. thaliana* an ideal organism for genetic and molecular studies such as mutant analysis, molecular cloning of genes and the detailed construction of a physical map of the genome.
- *Arabidopsis thaliana* can be genetically engineered with relative simplicity using either *A. tumefaciens*-mediated transformation or direct particle bombardment of the tissues.

Arabidopsis thaliana has the smallest known genome among the higher plants. Both DNA reassociation kinetics and quantitative genome blotting experiments indicate a haploid genome size of approximately 70,000 kbp. This genome size is consistent with data from studies of the nuclear volume. *Arabidopsis thaliana* is more resistant to ionizing radiation than many other angiosperms. This property correlates with its genome size. The genome size is only five times larger than that of the yeast fungus (*S. cerevisiae*) and only 15-fold larger than the common intestinal bacterium, *Escherichia coli*. The contrast of the *A. thaliana* genome with those of other higher plants frequently used in molecular and genetic studies is striking (Table 2.3). The 'genetic baggage' of many higher plants is thought to be an evolutionary insurance against environmental changes. Fewer than 25% of a complex organism's genes may be required for growth and reproduction. Possession of multiple copies increases the chances that it will posses a variant (allele) that will make the difference between surviving or succumbing to an environmental stress such as drought.

The significance of this small DNA content for molecular genetics is that a genomic library of *A. thaliana* chromosomal fragments is easier to make and simpler and more economical to screen. Only 16,000 random λ clones of 20 kb average insert size must be screened to offer a 99% chance of obtaining any *A. thaliana* fragment. In contrast, in tobacco (*Nicotiana tabacum*), 370,000 clones from a similar library would have to be screened to have a similar chance of success; the comparable number for pea (*Pisum sativum*) is 1,000,000 and for wheat (*Triticum aestivum*) is 1,400,000 λ clones.

It is thus rapid and inexpensive to screen *A. thaliana* genomic libraries repeatedly, and this is important for experiments involving chromosome walking. A second advantage of the small genome size is the enhancement of signal relative to noise in genetic gel blotting experiments. This enhancement allows the detection of weak signals from heterologous probe hybridizations

Table 2.3. Comparison of genome sizes.

Species		Genome size (base pairs)
Brassicaceae		
Thale cress	*Arabidopsis thaliana*	1.0×10^{8}
Oilseed rape (canola)	*Brassica napus*	1.2×10^{9}
Graminae		
Rice	*Oryza sativa*	4.2×10^{8}
Barley	*Hordeum vulgare*	4.8×10^{9}
Wheat	*Triticum aestivum*	1.6×10^{10}
Maize (corn)	*Zea mays*	2.5×10^{9}
Leguminosae		
Vining and dried culinary pea	*Pisum sativum*	4.1×10^{9}
Soybean	*Glycine max*	1.1×10^{9}
Solanaceae		
Potato	*Solanum tuberosum*	1.8×10^{9}
Tomato	*Lycopersicon esculentum*	1.0×10^{9}
Hominid		
Man	*Homo sapiens*	3.2×10^{9}

After Adams (2000).

and, in consequence, permits the detection of hybridization between very divergent probes and an angiosperm genome.

Arabidopsis thaliana has a low content of repeated sequences. Those elements that are repeated are set at far distances from each other. This characteristic is different from that of other angiosperms studied so far; for example, in tobacco (*N. tabacum*), the mean length of uninterrupted single-copy DNA is 1.4 kb and in pea (*P. sativum*) it is 0.3 kb. Thus, chromosome walking experiments are made conveniently with *A. thaliana*, whereas with other angiosperms the interspersed repeats make it difficult to select single-copy probes for each step in chromosome walking.

The ability to transform *A. thaliana* genetically allows for detailed analysis of gene function and expression. In addition, it permits the functional complementation of a mutant phenotype with cloned genes which is the critical final stage in isolating genes by mutational analysis. Many *A. thaliana* genes have been cloned and characterized; this provides genes for many different types of experiment and it is possible to develop a picture of genomic organization and its evolution.

ADVANTAGES OF *ARABIDOPSIS THALIANA*

The main objective is to clone genes that have been identified by mutational analysis. The combined efforts of many laboratories have resulted in a completed outline map of the *A. thaliana* genome. Detailed maps of morphological and biochemical markers such as RFLPs and RAPDs are available for much of the genome. Approximately 90% of the total *A. thaliana* genome lies within 0.8 Mbp of an RFLP marker and 50% is within 0.27 Mbp of an RAPD marker.

The genome of *A. thaliana* lacks much of the repetitive DNA that riddles the genomes of other angiosperms. It does, however, contain a complete set of genes controlling developmental patterns, metabolism, reactions to differing environmental cues and resistance to pests and pathogens.

Arabidopsis thaliana was the first angiospem for which the entire genome was sequenced (The Arabidopsis Genome Initiative, 2000). This achievement now presents unlimited opportunities to mine new knowledge for functional and metabolic processes controlled by the 25,000 genes present in *A. thaliana* (The Arabidopsis Genome Initiative, 2000; http://www.nature.com/genomics).

Information from the analysis of this genome is being entered into a publicly accessible database along with details of the proteins that are coded for further along the chain of nuclear and cellular processes (The Arabidopsis 2010 Project; http://nasc.nott.ac.uk/garnet/2010.html). Similarities between plant and animal biology are highlighted by the information obtained from the Arabidopsis projects (Leitch and Bennett, 2003). The boundaries between these sectors of biology largely disappear, and there are distinct possibilities that knowledge obtained from *Arabidopsis* can be applied to the study of animal and human genomes and to understanding their nuclear and cellular processes in health and disease (Sanderfoot and Raikel, 2001).

Major efforts are underway to complete physical maps of the *A. thaliana* genome consisting of the overlapping sets of cosmids and yeast artificial chromosomes (YACs). Two powerful tools have resulted from the analysis of mutants in *A. thaliana* and these complement map-based cloning methods. The first, T-DNA tagging, relies on using 'seed transformation' or the transformation of zygotic embryos with *Agrobacterium* which allows T-DNA to be used as a mutagen. This method has been used successfully in the isolation of several genes. It becomes of greater value as larger populations of transformed plants with low levels of somaclonal variation become available. The second approach is a method of genomic subtraction which is used to clone genes that correspond to deletion mutants. It has, for example, been used to clone the *A. thaliana ga-1* locus.

The research knowledge obtained using *A. thaliana* can be of the 'pure-blue skies' nature on the one hand and very applied near market on the other. In an example of the former, Lolle *et al.* (2005) suggest that genetic information may be stored outside the chromosomal DNA as ancestral RNA sequence caches. Such a process would question a generally accepted and fundamental tenet of Mendelian genetics. At the other end of science, Montgomery *et al.* (2004) applied mechanical brushing techniques to *A. thaliana* plants in order to induce stress and consequent dwarfing. The successful development of mechanical dwarfing techniques offers opportunities to improve the robustness of seedling propagation in businesses raising transplants (Chapter 3), allowing greater probability of success when the plants reach their final field stations.

RAPID CYCLING *BRASSICA* TYPES

Rapidly reproducing forms of *Brassica* named 'Fast Plants™' have been bred for use in association with inexpensive growing systems developed from recycled plastic containers. This combination offers an effective and efficient set of tools for biological studies at all levels of complexity from the research bench to school students. Depending on the genotype used, Fast Plants™ will germinate in 1 day, grow and flower in 2 weeks and produce viable seed in little more than 28 days.

Changes in the patterns of growth and development can be studied over 24 h periods, with physiological changes monitored hourly. The rapid life cycle, small size and ease of growth of these plants make them ideal for investigations into genetics, reproduction, physiology, ecology and growth (Greenler and Williams, 1990). Fast Plants™ resulted from a project that screened >2000 *Brassica* accessions obtained from the USA Department of Agriculture (USDA) National Germplasm System (Williams and Hill, 1986). A few plants of each species screened flowered in significantly shorter times compared with the average. These faster flowering genotypes were developed further, providing populations tailored for use in experiments made under controlled conditions. Combining genes from several early flowering forms provided material with even greater levels of accelerated flowering.

Rapid reproduction traits were combined with other characteristics such as diminutive size that allowed large numbers of plants to be grown in standardized laboratory and classroom conditions. In practice, the fast-flowering types of *Brassica* and, more recently, *Raphanus* are cultured at 24°C with continuous illumination of 250 $\mu m/s/m^2$ at high population densities.

Test populations were bred by interpollinating several early flowering types within each *Brassica* and *Raphanus* species. The criteria used in the selection process included characteristics such as: minimum time from sowing to flowering; rapidity of seed maturation; absence of seed dormancy; small plant size; and high levels of female fertility. Populations of approximately 300 plants were used in each cycle of reproduction, and 10% of the population that flowered most quickly was selected and mass pollinated to produce the next generation. Once the reduction in the average number of days to flowering was stabilized and when >50% of any individual population flowered within a 2–3 day period, the selection process for that population was stopped. Each population fulfilling these criteria was then increased by mass pollination and designated as a 'rapid-cycling base population' (RCBP). The flowering and growth characteristics of each base population are given in Table 2.4. Curves describing the transition from germination through to flowering and seed formation for six *Brassica* species are shown in Fig. 2.8. This transition is shown pictorially in Fig. 2.9 for *B. rapa*.

The RCBPs are relatively homogeneous for plant habit and time to flowering; they retain considerable variation for other traits such as their response to pathogens causing disease. They respond strongly to changes in

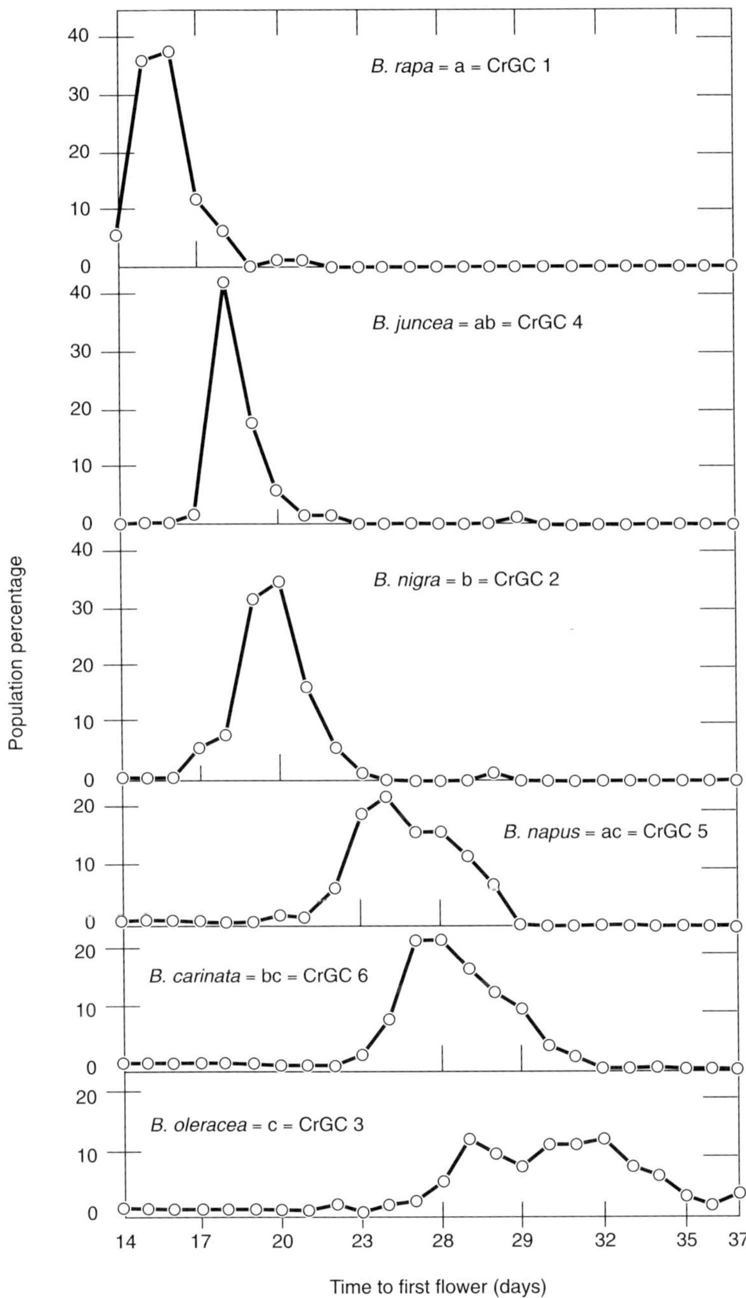

Fig. 2.8. Illustration of the speed of growth and seeding cycles in rapid-cycling *Brassica* populations (RCBPs) of six *Brassica* species when grown under standard conditions (P.H. Williams).

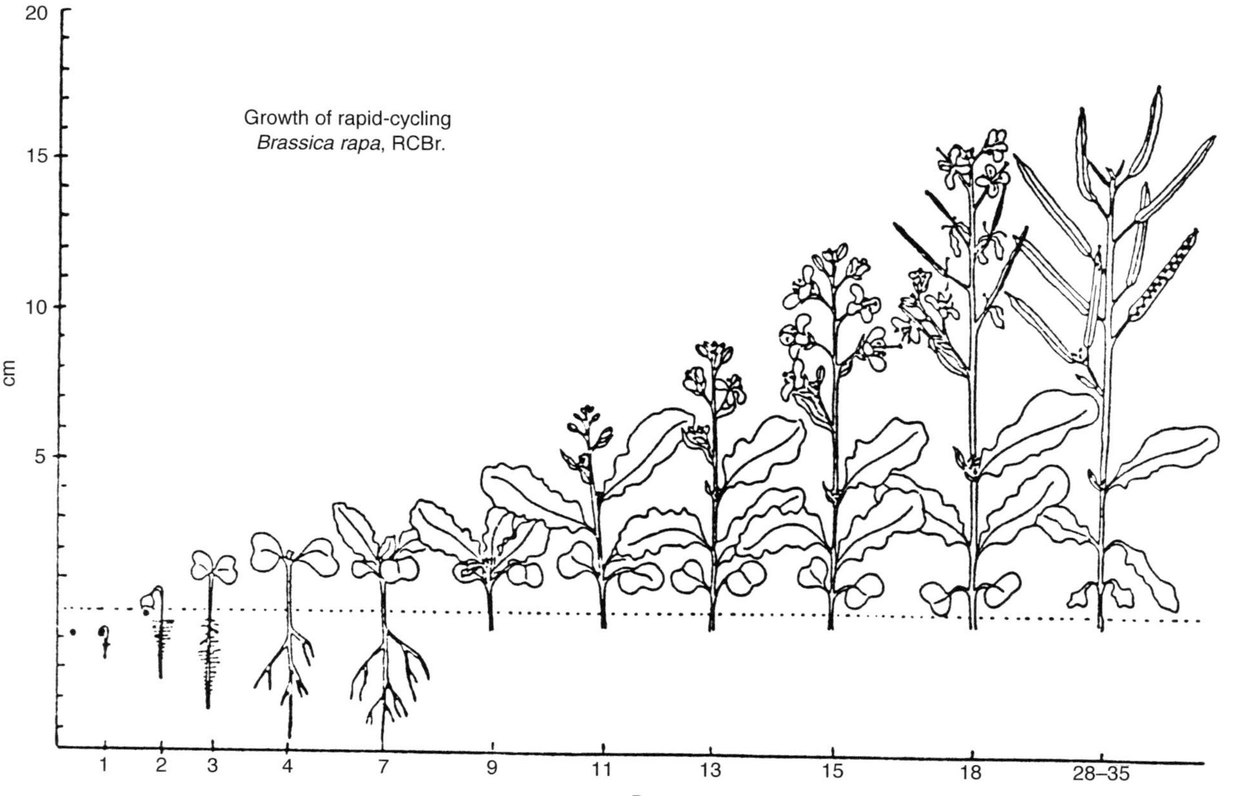

Fig. 2.9. Stylized representation of germination, growth, flowering and seed setting in a rapid-cycling *Brassica* population (RCBP) *B. rapa* (P.H. Williams).

Table 2.4. Phenotypic characters of rapid-cycling *Brassica* and *Raphanus* base populations.

CrGC stock number	Species	Genome	Days to flowering (SD)	Length (cm) to first flower (SD)	Seeds per plant (SD)	Days per cycle	Cycles per year
1	*B. rapa*	Aaa	16 (1)	11.9 (3.1)	78 (54)	36	10
2	*B. nigra*	Bbb	20 (2)	27.1 (4.9)	69 (49)	40	9
3	*B. oleracea*	Ccc	30 (3)	22.6 (5.3)	18 (21)	60	6
4	*B. juncea*	Abaabb	19 (1)	29.6 (4.0)	107 (46)	39	9
5	*B. napus*	Acaacc	25 (2)	35.3 (7.1)	76 (53)	55	6
6	*B. carinata*	BCbbcc	26 (2)	41.7 (6.6)	67 (46)	56	6
7	*R. sativus*	Rrr	19			48	7

Cultural conditions: 24°C and continuous light; CrGC = Crucifer Genetics Cooperative; data are expressed as mean values.
The cytoplasmic genome is indicated by upper case and the nuclear genome by lower case; a = 10 chromosomes; b = eight chromosomes; c and r = nine chromosomes.
By permission of Professor P.H. Williams, University of Madison-Wisconsin, USA.

the cultural environment; where greater space for foliage and root growth is provided, the plants become much larger, forming abundant quantities of seed.

These genotypes may be used for the safe repositories of genes for future use in molecular and conventional genetic breeding studies. Self-compatible stocks of the three diploid species are available, together with tetraploid forms of *B. oleracea*, *B. rapa* and *R. sativus* obtained by colchicine treatment of diploids, the latter being used to form triploids and subsequent trisomic and aneuploid forms.

A range of mutant forms with traits such as gibberellin responders, dark green dwarves, elongated internodes, chlorophyll deficiencies and anthocyanin suppressors offer characteristics for studies of inheritance and as molecular markers. As an aid for plant breeders, a range of tester incompatibilities has been generated. Cell and protoplast lines of RCBPs of *B. oleracea* and *B. napus* regenerate simply and easily, and hence can be used with haploid embryo cultures of the natural polyploid species (*B. carinata*, *B. juncea* and *B. napus*) and offer great potential for use in breeding and genetic transformation studies.

In addition to serving as repositories for nuclear genes, the RCBPs have shown especially useful properties for introducing the nuclei of various species into cytoplasms expressing distinct phenotypes, such as male sterility and triazine herbicide resistance. Because the cytoplasmic traits are transmitted only through the female line, the nuclei of a species can be introduced into the cytoplasm of any other species by making crosses between selected parents (see Fig. 1.1). Normally after making interspecific or intergeneric crosses, the amount of seed set becomes very low and floral morphology and nectary functions become abnormal. In the RCBPs, normal levels of seed production and floral characteristics can be restored. An example of the potential of rapid-cycling *Brassica* stocks may be seen in breeding lines carrying multiple resistance to fungal, bacterial and viral pathogens. Resistance to *P. brassicae* (the causal agent of clubroot disease) from *B. rapa*, to powdery mildew (*E. cruciferarum*) and to viral pathogens from Chinese cabbage (*B. rapa*) and cytoplasmic male sterility from *Raphanus* spp. (radish) were combined. This provided a source for multiple pathogen resistance and cytoplasmic male sterility for superior Chinese cabbage lines that was constructed in under 2 years (Chapter 7).

EDUCATIONAL TOOLS

The RCBPs offer a source of stimulating and interesting living material for use in classroom education for all levels of age and attainment. Most biology courses make use of model animal materials such as *Drosophila* spp. (fruit fly). There is little plant material that can be used for courses in botany, horticulture, agriculture, forestry and general science education that matures sufficiently quickly so as to allow experiments to be completed within the 6–8 weeks required by many school and higher educational curricula. Use of RCBPs in such classes allows students to explore aspects of plant growth, development and reproduction, physiology, genetics, evolution and ecology. The RCBPs are well suited to this purpose since their hallmark is remarkably rapid development.

The plants can be induced to flower in 14–18 days, they are small and compact and may be grown at high population densities (e.g. 2500 plants/m^2) under continuous illumination in the classroom. The ease of cultivation and array of interesting morphologies makes these plants attractive models for both teachers and students. Fast Plants™ are becoming integrated into national and international educational programmes particularly in the USA and UK, offering 'hands-on' experience of the principles of plant biology.

BRASSICAS AS FUNCTIONAL FOODS

There is increasing recognition of the scientific basis for the adage 'We are what we eat'. In other words, our food is not simply a source of energy but provides many other essential components for human health and welfare. The *Brassica* vegetables are becoming increasingly seen as contributing substantially to the long-term health of consumers. In addition to providing high levels of fibre, vitamins and minerals, the brassicas contain glucosinolates (see Chapter 8) that are hydrolysed into isothiocyanates by the action of myrosinase enzymes when plant cells are damaged by chewing. Isothiocyanates are well recognized as potent inducers of mammalian detoxication (phase 2) and antioxidant enzyme activity that protects against tumorigenesis when tested in rodents suffering from mammary, stomach and prostrate tumour models. Sulphoraphane (4-methylsulphinylbutyl isothiocyanate) derived from broccoli (calabrese) (*B. oleracea* var. *italica*) is especially effective against tumorigenesis, in reducing hypertensive stress and curing *Helicobacter pylori* infections (Fahey *et al.*, 1997, 2002). Dietary intake of isothiocyanates can be satisfied by eating a range of *Brassica* green vegetables and from broccoli seed and sprouts. Plant breeding programmes aiming to increase isothiocyanate contents in hybrid *Brassica* lines are described by Mithen *et al.* (2003) in the UK and by Abercrombie *et al.* (2005). Increasing the isothiocyanate content of *Brassica* sprouts and seed offers a cheaper alternative because open pollinated cultivars may be used (Farnham *et al.*, 2005). A rapid-cycling base population of *B. oleracea* was grown hydroponically with closely controlled temperature, photoperiod and radiation environments by Charron and Sams (2004). High temperature (32°C as compared with 22°C) increased the formation of glucoraphinin, a precursor of sulphoraphane, by fivefold. Production of total glucosinolates tended to be promoted by both low and elevated temperatures, which supports the view that the original function of these compounds in brassicas is in support of mechanisms that counter environmental stresses.

Significant quantities of glucosinolates are found in ornamental cabbage and kale forms of *B. oleracea* (Kushad *et al.*, 2004), which may also present opportunities for hybridization. Other health-promoting compounds such as lutein (3*R*, 3′*R*, 6′*R* β, ϵ-carotene-3, 3′ diol) and β-carotene are found in kale, especially those derived from Portuguese land races as reported by Kopsell *et al.* (2004), with high levels of variation that could be exploited by plant breeders.

THE CRUCIFER GENETICS COOPERATIVE, INTERNATIONAL COLLABORATION AND *BRASSICA* SYMPOSIA

The development of rapid-cycling *Brassica* genotypes prompted the formation in 1982 of the Crucifer Genetics Cooperative (CrGC) in the USA, as a vehicle for distributing seed and information amongst research workers and others around the world. Since 1982, CrGC has held regular meetings to discuss research findings. In 1994, the 9th CrGC workshop was held in Lisbon, Portugal in collaboration with the 1st International *Brassica* Symposium sponsored by the International Society for Horticultural Science (ISHS), Leuven, Belgium. This arrangement of combined international *Brassica* symposia and CrGC workshops has successfully led to meetings in Rennes,

France (1997), Wellesbourne, UK (2000) and Daejeon, South Korea (2004). Between these combined meetings, the CrGC continues with biennial meetings in the USA and elsewhere. While considerations of oilseed rape (*B. napus*) are largely beyond the scope of this book, it is worth noting that there is also a series of international Rapeseed Congresses which run parallel to the vegetable *Brassica* meetings. These meetings and organizations are mutually supportive and, in collaboration with the *Crucifer Newsletter* (published by INRA, Rennes, France), offer means by which scientists who are interested in *Brassica* and related genera are able to exchange knowledge very rapidly. There are also *Brassica* websites at: www.brassica.info, www.bbsrc.ac.uk, www.brassicagenome. org and www.brassica-resource.org.

3

Seed and Seedling Management

SEED DEVELOPMENT

Brassica seed development is divided into three phases: cell division and early expansion in the first 14 days after anthesis, followed by reserve accumulation in the next 14–49 days and finally a period of dehydration over the final 7 days (Norton and Harris, 1975; Dasgupta and Mandal, 1993; Gurusamy and Thiagarajan, 1998). The processes associated with seed development and maturation are illustrated in the sequence of figures taken from the detailed studies of Gurusamy and Thiagarjan (1998), who used cauliflower as a model system (see Figs 3.1–3.8).

The geographical origin of parent genotypes producing the seed influences the detailed timing of each of these phases. Cauliflower, for example, originated in temperate, cool climates, hence its seed requires longer periods for development from flowering to maturity.

Seed pods (siliquae) change colour from green to pinkish yellow and brown at about day 49. The seed pods lengthen and increase in width (Fig. 3.1) as they mature. There are fresh weight increases up to about 4 weeks after anthesis and steadily increasing dry weight as the pods desiccate (Fig. 3.2), these changes are mirrored by changes to the seeds (Fig. 3.3). As the seeds mature, their internal cellular membranes strengthen and become more capable of retaining solutes within the cell. Membrane strength or integrity is measured by placing a sample of seed in distilled water and determining changes in electrical conductivity as solutes leak into solution.

The quantities of free amino acids and sugars in cells increase from the start of seed formation up to 21 days after anthesis. Thereafter, free solutes decline with accelerating speed as the seeds mature. The decrease in free, soluble compounds in the seed results from their transformation to bound, insoluble, forms over the period of development. The activity of de-hydrogenase enzymes can equate with seed quality and subsequent seedling vigour, and is inversely related to seed moisture content (Khattra *et al.*, 1993).

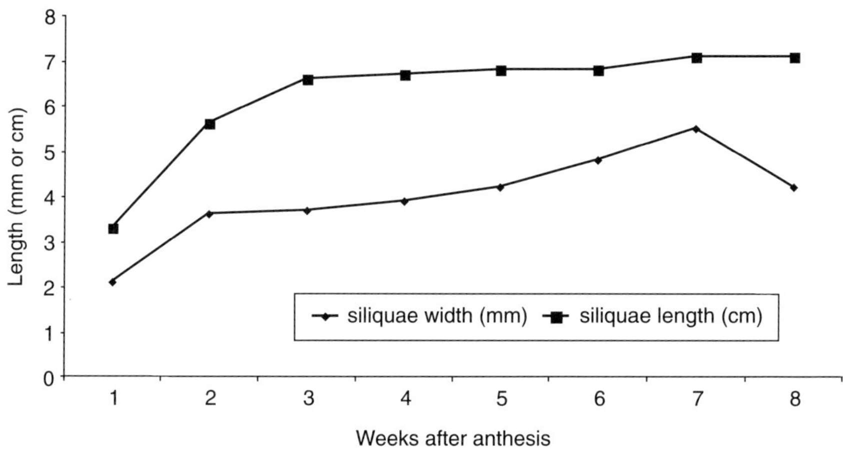

Fig. 3.1. Illustration of changes in length and width of cauliflower (*Brassica oleracea* var. *botrytis*) siliquae after anthesis (Gurusamy and Thiagarajan).

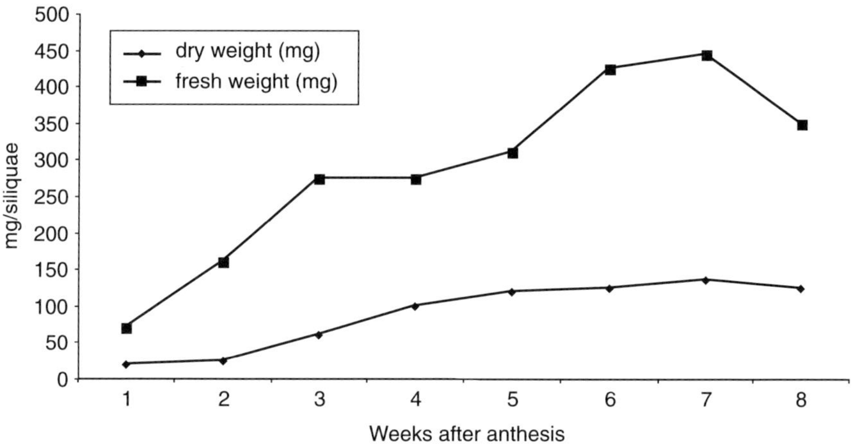

Fig. 3.2. Illustration of changes in fresh weight and dry weight of cauliflower (*Brassica oleracea* var. *botrytis*) siliquae during development (Gurusamy and Thiagarajan).

Enzyme activity increases slowly with advancing seed maturity, reaching a maximum at 56 days after anthesis (Fig. 3.4).

Oil content increases steadily up to harvest, whereas there is an initial increase in sugars which then drops sharply 4 weeks after anthesis (Fig. 3.5). Protein and free amino acid content increase linearly in the first 28 days after anthesis and slowly thereafter (Fig. 3.6).

Staining seed samples with 2,3,5-triphenyl tetrazolium chloride, which turns red when in contact with actively metabolizing tissues, provides a

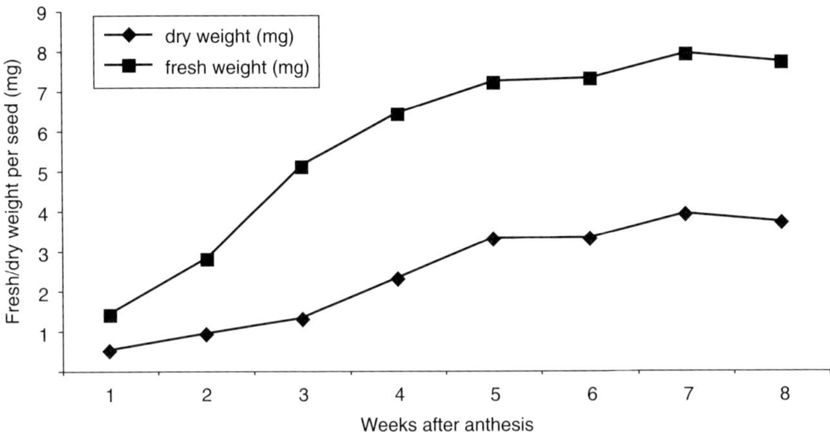

Fig. 3.3. Illustration of changes in fresh weight and dry weight of cauliflower (*Brassica oleracea* var. *botrytis*) seeds during development (Gurusamy and Thiagarajan).

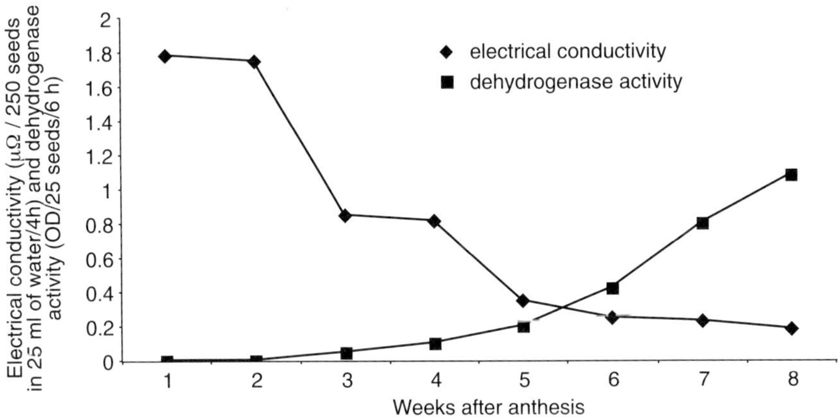

Fig. 3.4. Illustration of changes in electrical conductivity and dehydrogenase activity in developing cauliflower (*Brassica oleracea* var. *botrytis*) seeds (Gurusamy and Thiagarajan).

further means of assessing seed quality and subsequent potential seedling vigour (Burris *et al.*, 1969).

PHYSIOLOGICAL MATURITY

The length and width of maturing *Brassica* pods increase towards maturity, coinciding with seed swelling as they accumulate mass and storage compounds and as the moisture content falls.

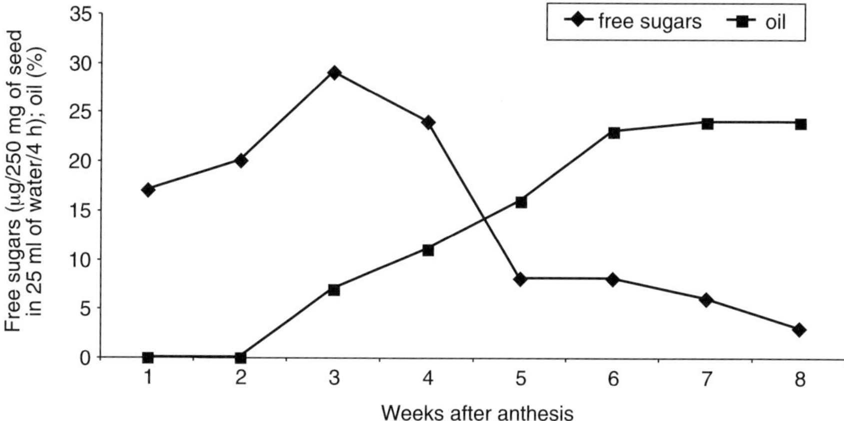

Fig. 3.5. Illustration of changes in free sugars and oil content of cauliflower (*Brassica oleracea* var. *botrytis*) seeds during development and maturation (Gurusamy and Thiagarajan).

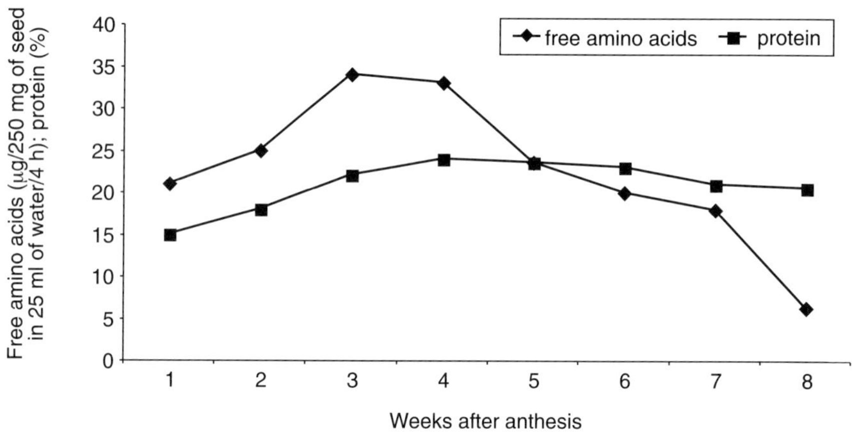

Fig. 3.6. Illustration of changes in free amino acids and protein in cauliflower (*Brassica oleracea* var. *botrytis*) seeds during development and maturation (Gurusamy and Thiagarajan).

Physiological maturity is reached when the seed achieves maximum dry weight, and this marks the end of the swelling phase of the seed (Shaw and Loomis, 1950). In cauliflower, physiological maturity coincides with the reduction of seed moisture content to about 40% at 4 weeks after anthesis (Fig. 3.7). This may represent a normal part of maturation (McIlrath *et al.*,

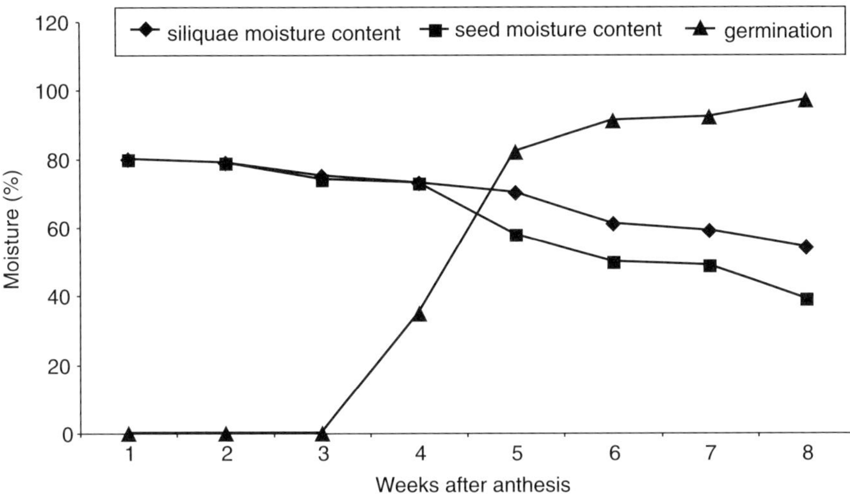

Fig. 3.7. Illustration of changes in moisture content and germinability of cauliflower (*Brassica oleracea* var. *botrytis*) seeds during development (Gurusamy and Thiagarajan).

1963) and again is influenced by the environment in which the seed-producing parent plant is growing (Sreeramulu *et al.*, 1992). The capacity for germination is initiated by week 3 after anthesis and increases steadily thereafter (Fig. 3.7).

The precise timing of these phases depends on the particular *Brassica* species, its cultivars and the environments in which the seed crops are grown. Studies of broccoli cv. Waltham-29, an open pollinated form, showed that it matures more rapidly than cauliflower. Maximum seed weight was reached 42 days after pollination. Abscission of the seed from the placenta (funiculus) started 42–49 days after pollination and coincided with a sharp decline in fresh weight. Chlorophyll content of both seed testa and siliquae declines from 42 days after pollination, and they acquire a tan and reddish-brown coloration, respectively, by 56 days. This is a common phenomenon with dry fruits, such as pods, where desiccation coincides with maximum dry weight accumulation. In mustard, for instance, there is a complete loss of chlorophyll from the seed coat (testa) by 42 days after pollination, while in broccoli the majority of pods (siliquae) have shed (dehisced) their seed by 84 days after pollination (Jett and Welbaum, 1996). Typical changes in the appearance of seed pods during development are shown in Table 3.1, using red cabbage (*B. oleracea* var. *capitata*) as an example.

Seed maturity is not an absolute prerequisite for the onset of germination. The *Brassica* embryo can be capable of germinating during development (Bewley and Black, 1994; Baskin and Baskin, 1998). Cauliflower germination

Table 3.1. Moisture content, siliquae and seed characteristics of red cabbage (*Brassica oleracea* var. *capitata*) during seed development.

Days after full bloom	Moisture content (%)	Siliquae characters	Seed characters
27	85	Waxy, dark green	Light green, liquid endosperm
33	80	Waxy, green moist inside siliquae	Light green
44	60	Light green-yellow, slightly moist	Dark green, shiny, still moist inside, fibre beginning to form along the inside length of the siliquae
50	55	Light green-yellow, more fibrous	Dark dull green, not moist inside
54	50	Yellow-green, fibrous	Brown and black
56	40	Yellow, fibre covering entire inside of siliquae	Black, beginning to wrinkle
61	10	Yellow-white, brittle, dehiscent	All seeds black

After Still and Bradford (1998).

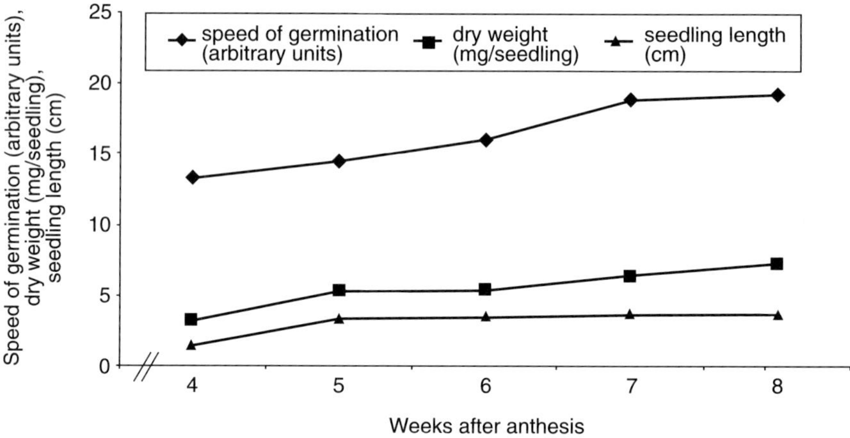

Fig. 3.8. Illustration of changes in the speed of germination, seedling dry weight and seedling length of cauliflower (*Brassica oleracea* var. *botrytis*) during development (Gurusamy and Thiagarajan).

may behave in a precocious manner. The speed of germination increases as seeds approach maturity (Fig. 3.8).

The maximum germination rate for cabbage (*B. oleracea* var. *capitata*) was reported at 48 days after pollination (Still and Bradford, 1994). Mature seeds require sufficient nutrients and other stored reserves to support germination and post-germination activity until the seedling is capable of independent growth. Generally, the speed at which seedlings grow and increase in dry weight correlates with the maturity of the parent seed (see also Ellis *et al.*, 1987). Harrington (1972) suggested that physiological maturity of seeds is defined as the developmental stage at which they achieve maximum viability and vigour. At this point, nutrients cease entering the seed from the parent plant and ageing begins.

Physiological maturity is the most appropriate time to harvest seed (Ellis *et al.*, 1987). In cauliflower, the period between 49 and 56 days after anthesis is the point of physiological and harvest maturity. After this point, there are reductions to seed and siliquae mass and dimensions, with deterioration commencing due to oxidation, volatization, desiccation and loss of nutrients. In comparison, mustard (*Sinapis alba* L.) reached physiological maturity 60 days after pollination (Fischer *et al.*, 1988). Immature mustard seed harvested without drying germinated as early as 14 days after pollination, although desiccation tolerance was not attained until 35 days after pollination (Ahuja *et al.*, 1981).

GENETIC CONTROL OF SEED DEVELOPMENT

The pattern of physiological events in seed development follows identifiable sequences of gene expression. Proteins, carbohydrates and lipids build up in seeds during embryogenesis, resulting in increased dry mass. The accumulation phase is terminated by ovule abscission and marked by maximum dry mass. Over 20,000 distinct genes are expressed at any one time during embryogenesis (Goldberg *et al.*, 1989), and there are a number of recognizable patterns. Expression continues after ovule abscission controlling the synthesis of storage compounds, preparation for desiccation, prevention of premature germination and the establishment of dormancy. It is suggested that a post-abscission programme of gene expression is completed before subsequent maximum seed vigour can be attained.

SEED SIZE AND MATURITY

In *Brassica* spp., determination of seed maturity is complicated by the indeterminate growth and extended flowering periods of the parent plants (Jett and Welbaum, 1996). Overall inflorescence development continues for a considerable period of time within individual racemes and between different racemes. The processes of flower opening, pollination and fertilization each requires an extended period of time, causing the fruits and seeds to mature in rotation. Harvesting too early results in poor quality, immature seed, while delayed harvesting leads to possibly 50% seed losses due to pod shattering.

Seeds are released when the ripe fruits dehisce or 'shatter' (Dickson and Wallace, 1986).

Where the entire raceme is harvested mechanically, seeds at different stages of maturity and differing in viability and vigour are inevitably mixed together. Seed maturity is associated with the presence of growth regulators, particularly abscissic acid (ABA) (Still and Bradford, 1998). This compound suppresses germination early in development, stimulates the accumulation of storage reserves, imposes dormancy and determines the sensitivity of germination to the seed water potential (ψ).

Each broccoli seed pod for example, contains 10–30 seeds and all are at slightly different stages of maturity from the others. In some *Brassica* species, up to one-third of the carbon in developing embryos was fixed by photosynthesis in the siliquae walls. It is not, however, known whether assimilates are supplied uniformly to developing ovules (Opeña *et al.*, 1988). Little is known of the effect of seed position within the *Brassica* pod and its subsequent impact on vigour. In crops such as soybean (*Glycine max* (L.) Merr.) and runner bean (*Phaseolus coccineus* L.), seeds produced at the stylar end of the pod, i.e. those formed first, are heavier than those formed at the peduncular end (formed last and closest to the flower stem). Generally, it appears that where fruits contain less than three seeds, then their position has little or no influence on subsequent seedling vigour. Where there are more than five seeds per fruit, then the position in the pod begins to influence subsequent seed size and the vigour of growing seedlings substantially.

Seeds are often sorted commercially and sold on the basis of size; larger seeds are perceived to be more vigorous and have greater financial and agronomic value compared with smaller ones (Liou, 1987). The effects of seed size on emergence, growth and yield of many vegetables is well documented (McDonald, 1975). In broccoli, for example, seed size has been correlated positively with seedling growth rate. Seedling dry weight, stand establishment and final crop head yield are improved by using larger, potentially more vigorous seeds (Heather and Sieczka, 1991). General rules concerning seed and seedling vigour have been difficult to establish. For instance, no differences were detected in the germination rate of large cabbage seed compared with small ones (Jett and Welbaum, 1996). There is a tendency for larger seed to produce larger seedlings, however, at least in the early stages of growth.

Maturity is not an absolute prerequisite for germinability; the embryo may be capable of germinating during development (Bewley and Black, 1994). In their study of cauliflower, these workers found that germination occurred precociously. The capacity to germinate and its speed increased, however, as the seeds matured. Importantly, mature seeds contain sufficient nutrients and other stored reserves to support germination and post-germination growth.

SEED QUALITY

Seed that attains maximum viability and vigour is physiologically mature, and thereafter deterioration begins (Powell and Matthews, 1984). Seed should be harvested ideally when it has attained physiological maturity and has not started to deteriorate. Reaching greatest dry mass is generally correlated with optimum maturity and quality as quantified by maximum germination in controlled tests. This state may not correlate with the maximum seedling vigour under field conditions. For growers of vegetable *Brassica*, the difference between the capacity to germinate and seedling vigour has crucial significance. Seed is very expensive and forms a considerable element in the variable costs of growing each crop. The costs of propagating excess plants where the seed merchant makes an erroneously low estimate of viability place a substantial financial burden on the grower.

Simply expressing the quality of seed by germination and purity values fails to reflect the potential field performance and the cost–benefits of the products from one seed house compared with another. In consequence, attempts have been made to establish laboratory tests that quantify the future potential vigour of seedlings in the field and translations of this into 'useable plants' that are placed in the field.

SEED VIGOUR

Seed vigour may be defined as 'the sum total of the properties of seed which determine its potential activity and performance during germination and seedling emergence' (adapted from Perry, 1978). Seeds are subjected to varying degrees of water and temperature stress during their development, maturation, harvest and storage during seed production, and then again during rehydration and germination following sowing. Those with high vigour are capable of rapid germination and the establishment of healthy seedlings, leading to efficient crop production.

Plants respond to environmental stresses by forming a series of specific proteins. These include heat shock proteins (HSPs) and late embryogenesis abundant proteins (LEA proteins) (Bettey and Finch-Savage, 1998). These proteins also form in the absence of external stress as part of the normal developmental processes. They are found at specific stages in the growth cycle, especially during dehydration such as in pollen and seed formation. The HSPs prevent protein aggregation and, by acting as molecular chaperones, ensure that protein folding is correctly completed. Late embryogenesis abundant proteins have conserved elements existing as amphiphilic α-helices that may be associated with desiccation protection.

Both types of protein increase during the later stages of seed development; this is when seed vigour also increases. Hence there can be a positive direct correlation between the occurrence of these two characteristics. It has yet to be established whether stress protein content contributes directly to seed vigour. This question was investigated with batches of *B. oleracea* L. var. *capitata* cv. Bartolo (F_1 hybrid white cabbage) using seed lots grown in different years, representing the product of interactions with a range of environmental conditions and possessing high levels of viability. Hence any differences between them would relate to vigour characters and the ability to withstand stressful conditions.

It is thought that HSP17.6 contributes to the performance of cabbage seed under stressful conditions and, therefore, is an important component of vigour. The seed was subjected to 'rapid ageing' processes (42°C, 10 days, 10% moisture content). The HSP17.6 was unaffected, and seed with the highest concentrations were better able to withstand subsequent stresses.

Since vigour results from a combination of characters, seed batches with high vigour relative to others as determined by one set of tests may have lower vigour when this is tested using a different set of environmental conditions. Thus measurement of a single characteristic does not provide a reliable estimate of seed vigour. A combination of tests is essential for the reliable prediction of seedling vigour.

EXAMPLES OF GERMINATION AND VIGOUR TESTS

Standard germination tests

These are performed according to internationally accepted rules established by the International Seed Testing Association (ISTA) using prescribed temperature regimes. Example regimes could be either continuous 20°C or alternating 20/30°C during dark and illuminated periods. Lighting regimes are also standardized. Radicle emergence is recorded daily for the first 5 days of incubation, and counts of normal and abnormal seedlings are made after 10 days by suitably qualified seed analysts.

Controlled deterioration test

The moisture content of seed is adjusted to 24% and they are sealed in laminated polyethylene–aluminium foil pouches and held at 1°C overnight allowing moisture to equilibrate throughout the seed sample. The sealed pouches are then placed in a water bath at 45°C for 24 h (Matthews and Powell, 1987). The standard germination test is then made at 20°C and normal seedlings counted. The measurement of chlorophyll content of seed has been suggested as a means of determining seed maturity and quality

(Jalink *et al.*, 1999). Chlorophyll molecules emit fluorescence; measuring the strength of these emissions provides plant physiologists with a very powerful tool when studying responses to environmental stress (Maxwell and Johnson, 2000). Patterns of the distribution of vigour within populations of deteriorated white cabbage (*B. oleracea* var. *capitata*) were analysed by Dell'aquilla *et al.* (2002). This permitted viable and deteriorated batches of seed to be identified electronically and separated automatically. Seed quality is controlled by several genes; studies of seed of *Arabidopsis thaliana* (see Chapter 2) which have been subjected to controlled deterioration are beginning to reveal the mechanisms by which this trait is expressed and the means by which it can be improved (Tesnier *et al.*, 2002).

Determination of K_i

Seed is brought to 18% moisture content, sealed in laminated polyethylene–aluminium foil pouches and held at 1°C for 3 days allowing moisture to equilibrate throughout the sample. The pouches are placed in a germination cabinet at 40°C and sampled daily, for 10 days, to perform a standard germination test at 20°C. A germination test is also carried out directly after moisture equilibration (Ellis and Roberts, 1981).

Emergence tests

Trays are filled with commercial compost and each sown with 100 seeds at 40 mm depth and covered by a standard weight of compost. Subsequent seedlings are grown either under ambient conditions, at standardized temperatures, e.g. 20°C, or at an elevated temperature, e.g. 30°C. Emerged seedlings are removed and counted on a daily basis. Temperature profiles are monitored with thermistors inserted directly into the compost.

SEED PURITY

Control of seed purity is vital for all crops, but is of particular importance with vegetable brassicas where the uniformity of the plant stand is essential for the future profitability of the crop. The detection and elimination of siblings is of particular importance where F_1 hybrids are predominantly used as cultivars as with most *Brassica* crops. The sibling problem arises because *B. oleracea* possesses a single locus, multiallelic, sporophytic incompatibility system. Plant breeders usually maintain their lines by bud pollination (see Chapter 2). All incompatibility alleles are not equally effective, however, and varying amounts of self-fertilization may take place within inbred lines that are homozygous for S alleles. Additionally, the incompatibility reaction may be weakened by environmental factors, and the ratio of selfing to crossing can be affected by the behaviour of pollinating insects and the availability of foreign pollen (Wills *et al.*, 1979).

Sibling plants are at best far less productive for the grower than the hybrid cultivar, and represent a failure by the seed-producing company. Elimination is achieved by testing hybrid populations or by producing complex multiparent hybrids where any siblings may also be productive. This may not always be feasible and increases the cost and complexity of a breeding programme. Quality control has generally been achieved by sampling hybrid populations as either seed, cotyledons or leaves, followed by visual inspection or by biochemical testing such as the use of isoenzyme markers. Visual inspection is not satisfactory since the morphological differences may require a long time period over which they emerge. Isoenzyme studies are suggested as quick, simple and accurate, but again may not always be applicable since the high level of inbreeding within the parent lines of a hybrid cultivar means that they may express similar isoenzyme bands. Use of DNA probes could circumvent this problem. Image analysis also offers a further solution.

Image analysis has the advantages of being non-invasive, open to automation and is increasingly interactive and user-friendly. Studies have shown that use of image analysis with cotyledons or early adult leaves gives a more accurate evaluation of the number of siblings when compared with isoenzyme analysis. Development of image analysis offers the seed industry opportunities for much improved quality control. Sibling detection and vigour estimations may be made based on mean seed size and population variance. Image analysis may also be applied to monitor seed swelling and expansion, thereby establishing automated critical quality control systems in the early stages of seed development. Image analysis of seed area is identified by Dell'aquilla (2003) as a useful marker for studies of vigour and the effects of deterioration and salt stress. Commercial development requires refinement to the image analysis techniques, subsequent statistical analysis, greater precision in the selection of biological parameters to be analysed (seed part or organ/tissue of the developing seedling) and enhanced user-friendly computerized interfaces.

SEED ENHANCEMENT

'Seed enhancement' is an industrial term originating in the seed industry that has recently acquired scientific use, and covers beneficial techniques applied to seeds between harvesting and sowing. The objective is to improve germination and seedling growth, and raise the efficiency of seed delivery and associated materials at the time of sowing. Enhancement covers three areas: pre-sowing hydration (priming), seed coating and seed conditioning.

Seed Priming

Seeds require water, oxygen and a suitable temperature for germination. Water uptake follows a three-phase pattern with an initial rapid uptake or imbibition (phase 1), followed by a lag period (phase 2) and then a second increase in water uptake associated with seedling growth (phase 3). Seeds are tolerant of desiccation during phases 1 and 2 but frequently intolerant of it in phase 3 (Taylor *et al.*, 1998).

Water uptake may be either uncontrolled or controlled. In the former, water is freely available and not restricted by the environment. The seeds may be soaked or placed on moistened blotters. Soaking can involve the total immersion of seed in water with or without artificial aeration. Provided the seeds are viable and not dormant, with a sufficient oxygen supply available and a suitable temperature, germination will proceed. The process is arrested at a specific time, inhibiting the onset of phase 3.

Homogeneity of field establishment from either direct-drilled crops or emergence from seed sown into modules increases the ultimate uniformity of crop growth and concentrates the harvest into a shorter period of time, increasing efficiency and decreasing costs. Where brassicas are grown below or above their optimal temperatures for germination and establishment, direct-seeded crops tend to be drilled at high rates, leading to the need for subsequent thinning to achieve efficient final stand densities. Alternatively, seed priming techniques may be employed.

Priming is a controlled hydration process followed by redrying that allows the metabolic activities of germination to commence but not reach the stage of radicle emergence. Priming using polyethylene glycol (PEG) increased the germination rate of smaller seeds compared with larger ones. When seeds are rehydrated after priming, the germination rate is increased and, in some cases, the temperature range for continued germination may be expanded. Two forms of priming treatments are available, either osmotic or matric. In the former, seeds are incubated in solutions of either low molecular weight inorganic salts such as potassium nitrate, sodium chloride or potassium phosphate or high molecular weight, non-penetrating solutes such as PEG at a water potential that is low enough to inhibit germination.

This technique has been applied successfully to several vegetable *Brassica* crops. In particular, it will increase seedling vigour in cold, moist soils. Matric priming utilizes moistened solid carriers such as the clay mineral, vermiculite or calcium silicate to hydrate seeds. In several vegetable crops, matric-primed seeds germinate more rapidly compared with those that have been osmotically treated.

To achieve osmotic priming, seeds are placed on two layers of filter paper saturated with the priming agent and sealed in containers, such as Petri dishes, for up to 7 days in darkness. For matric priming, water and calcium silicate are mixed thoroughly and placed in a sealed container for 24 h before adding the seeds. The containers are rotated every 12 h to ensure the uniform mixing of the carrier and seed for 7 days. After priming by either method, the seeds are washed for 2 min in tap water to remove either the osmotica or the solid carriers, rinsed in distilled water and blotted dry. The seed is then dried with forced air at 37°C for 2 min and finally reduced to a moisture content of 5–6% on a dry weight basis by being placed over silica gel in a desiccator. The seeds may then be coated with thiram (tetramethylthiuram disulphide) or other approved fungicides and insecticides, and sealed into tin-foil packages or held in glass or plastic bottles at 4°C.

Aerated hydration for periods up to 24 h at 20°C was used by Mehra *et al.* (2003) as an adjunct to priming with PEG for *B. juncea* and *B. rapa* seed. There appeared to be increased resistance to water and salt stress in *B. juncea* and an extension of the range of temperatures for successful germination for *B. rapa* when aerated hydration was used.

Improving the prediction of seed viability and vigour in relation to stage of development in the silique has major commercial significance. Seed taken from several parts of broccoli (*B. oleracea* var. *italica*) pods at different times after pollination was dried and primed by Jett and Welbaum (1996). Rapid drying increased the viability of seed harvested between 35 and 42 days after pollination. Large seeds failed to germinate more quickly than small ones but produced seedlings of greater dry weight. Priming increased the percentage germination of seeds harvested 42 days after pollination. The germination rate of seed harvested before 42 days after pollination was also improved by priming. This overcame differences between mature and immature seed, and might offer a means of reducing the variation within batches of mechanically harvested seed batches.

Seed Coating

Seeds vary greatly in size, shape and colour. Many seeds are small and irregular, making their separation and precision placement in the field difficult. Additionally, seeds require protection from pests and pathogens. Seed coating is used to allow mechanical sowing in precise patterns. This achieves uniformity of plant spacing and provides carriers for plant protection agents. Seeds may be both pelleted and film coated. Pelleting is defined as the deposition of a layer of inert material that transforms the original shape and size of the seed. The shape is changed to spherical with increased weight and improved opportunities for precision placement. Film coating retains the original shape and size of the seed with minimal increase in weight. Since *Brassica* seeds are naturally spherical, film coating is more frequently used. Coatings may contain polymers, pesticides, biological agents such as bacteria, coloured markers or dyes and other additives.

Seed Conditioning

Harvested seed is seldom pure and contains undesirable materials including poor quality seed that needs to be removed. Conditioning has two objectives: (i) removal of contaminants such as other crop or weed seeds and inert materials which results in pure samples of the desired cultivar; and (ii) elimination of poor quality seed that may be immature, damaged or of an undesirable size. Conditioning is usually carried out with a series of automated grading stages controlled by microprocessors. Further improvement is achieved by exploiting specific physical characteristics such as seed colour. Colour-based sorting has improved greatly over the past couple of decades using optical systems, microprocessors and detection systems.

Colour sorters that sense light from 360° have been developed. Light detectors which can also quantify reflectance at ultraviolet (UV), near infrared (NIR) and fluorescence wavelengths have expanded the capacity of colour sorters. Ultrasound technology is now being added to these systems, providing sorters that can discriminate between seeds by differences of size, shape, texture, colour, lustre, mass, hardness and mechanical damage. Ultimately, such systems will identify the presence of diseased seeds before symptoms are expressed visually. It has been demonstrated that while the peak value of the ultrasound waves decreased, the slope and bandwidth values increased where seeds were invaded by a pathogen.

Seed enhancement offers systems for the detection and removal of low quality seed from larger lots. Pre-sowing hydration elevates seed moisture content and reactivates cellular functions. Cell membrane integrity may be assessed indirectly by measurement of solute leakage. Seed coatings offer systems that hold compounds in close proximity to single seeds, and colour sorting separates high and low quality seed. In sequence, pre-sowing hydration, seed coating and conditioning are applied to enhance seed lot quality in *Brassica*.

A projected quality enhancement system can be developed for *Brassica* seeds based around the presence of sinapine (Lee *et al.*, 1997). This fluorescent alkaloid forms naturally during *Brassica* seed development, it has a yellow colour at high alkaline pH (>10) and leaks from non-viable but not from viable seeds during germination. Measurement of sinapine leakage from *Brassica* seed has proved more accurate at predicting germination success than electrolyte conductivity tests. Single seed tests for sinapine leakage can form the basis for quality control in *Brassica* seed lots. First, the seeds are hydrated by soaking in water for 4 h or primed in PEG solutions for 24 h. Sinapine leakage was greatest from the non-viable seeds. Freshly hydrated seeds are then coated with a filler containing finely ground cellulose. This acts as an adsorbent, trapping the sinapine leachate in the coating. Coated seeds are dried and sorted by UV light into non-fluorescent and fluorescent individuals. Seed conditioning improved germination for cabbage (*B. oleracea* var. *capitata*), broccoli (*B. oleracea* var. *italica*) and cauliflower (*B. oleracea* var. *botrytis*). Determination of sinapine leakage is advocated by Devi *et al.* (2003) as a means of determining germination percentage and subsequent field emergence in preference to using measurements of electrical conductivity for mustard (*B. juncea*) seed.

The efficiency of upgrading seed lots by exploiting sinapine leakage is compromised if the seed coat restricts leakage from non-viable seed. Non-viable seeds were undetected, causing false negatives where the solute leakage failed to diffuse through the seed coat into the cellulose. The cutin content of the seed coat may affect permeability, retarding sinapine diffusion. Soaking the seed in a dilute solution of sodium hypochlorite (NaClO) enhanced leakage.

Seed germination at low temperatures can be predicted using thermal time models. In broccoli (*B. oleracea* L. var. *italica*), for example, priming lowered the mean thermal time to germination but had little effect on the minimum temperature required for actual germination. Primed seeds germinated more quickly because of their lower thermal time requirement. Priming advanced germination more rapidly per unit of thermal time compared with non-primed seeds, but did not reduce the minimum temperature for germination.

Similar effects are found with tomato (*Lycopersicon esculentum* Mill.) and onion (*Allium cepa* L.) which have no dormancy requirement and hence the minimum temperature for germination is genetically controlled, showing little variation within a particular species. Where priming lowers the minimum temperature for germination, this is a response to the substitution of priming for after-ripening, as in musk melon (*Cucumis melo* L.), overcoming some part of the dormancy requirement and expanding the temperature range for germination.

Root growth is more sensitive to variations above and below the optimal growing temperature compared with radicle emergence. Hence poor stands of direct-drilled crops may be due to a lack of root growth rather than an impairment of radicle growth. Matrically primed seed tends to germinate more quickly than osmotically primed seed. It has been noted that the calcium content of matrically primed seed increases development (Jett *et al.*, 1996).

GERMINATION

Studies of cellular and molecular events during seed germination provide information on the processes that are activated in the cell nucleus during the transition from a quiescent to an active state. Upon imbibition, the initial events are DNA, RNA and protein synthesis (Bewley and Black, 1994). The major component of the early DNA synthesis comprises a DNA repair process taking place in the first hours of germination (Osborne, 1983). Both RNA and protein synthesis have been demonstrated in leek seed (*Allium porrum*) at higher rates in germinating primed seed compared with untreated controls, while the rate of synthesis correlated positively with seed vigour (Bray *et al.*, 1989). Work with *B. oleracea* var. *capitata* cv. Bartolo (white cabbage) showed that DNA replication is not a prerequisite for radicle protrusion and the initial extension growth. The embryo in cabbage is sufficiently differentiated in dry, quiescent seed in order to produce a plantlet without further DNA replication or cell division following imbibition. Further seedling development, including root growth and root hair formation, however, appears to be dependent upon DNA replication.

The processes of DNA replication and accumulation of β-tubulin, the latter being necessary for cytoskeleton development, are two parallel but independent events during seed germination (Górnik *et al.*, 1997). Respiration increases rapidly during imbibition, corresponding to developing mitochondrial activities. Oxygen uptake has been correlated with rates of germination, seedling growth, emergence in the field and ultimately with crop yield. In the white cabbage cv. Bartolo (*B. oleracea* var. *capitata*), oxygen consumption increases at imbibition and at germination (Bettey and Finch-Savage, 1996). These changes reflected increasing oxidation of carbohydrate reserves via respiratory pathways. Relative differences in the activity of key enzymes in these pathways correlated with the germination rates (T_{50} = time to 50% germination) where there were large differences in seed vigour. Enzyme activities equated with differences in the flux through glycolysis where seed lots differed substantially in vigour, however, they were not determinants of seed vigour.

SEED DORMANCY

Some *Brassica* species exhibit postharvest dormancy; although this is thought to be of very limited duration it can vary in time from 0 to 140 days depending on the genotype concerned. Storage at 10–70% relative humidity promoted the release from dormancy during storage (Watanabe, 1953; Tokumasu *et al.*, 1975). Conversely, Opeña *et al.* (1988) removed dormancy by a brief period of dry storage.

In broccoli (*B. oleracea* var. *italica*) (Jett and Welbaum, 1996), germinability was first apparent in 10% of fresh (undried) seed by 28 days after pollination. Germination percentages increased in days 42–56 after pollination when essentially all fresh seeds were germinable. Seed drying increased germination at 42 days after pollination. In some species, it has been proposed that dehydration provided a switch, changing gene expression from a developmental programme to a germinative one. In broccoli, this did not appear to happen since dehydration was not a prerequisite for germination.

SEEDLING PROPAGATION

The plant breeding and seed industry has refined *Brassica* seed into a defined and reliable resource. This makes it a major item of cost in the crop balance sheet. As a result, its initial propagation into seedlings has also become refined and diverged as the separate and specialized industry of plant propagation. Frequently this is quite distinct from the production industry that produces the finished item that is sold to retail consumers. The separation of propagation and production enables growers to concentrate on the business of growing crops, leaving the earliest growth stage from seed sowing to transplant production to others. It also means that crops occupy field space for the minimum of time, with opportunities for multiple cropping during the season. This reduces the overhead cost of land for each crop (Wien, 1997).

The grower will specify what cultivars are to be propagated and their sequencing, and may also purchase the seed directly from the seed house. Thereafter, the responsibility for the first 6–8 weeks of the crop's life is handled in very intensive and environmentally controlled glasshouse conditions. Single seeds are dispensed mechanically into the individual units of multicelled trays, normally with approximately 300 cells per tray, with each containing 15–25 ml of propagation medium. Peat-based media have been mainly used that are supplemented with macro- and micronutrients and possibly agrochemicals. The germination phase may be separate from the growing phase. Optimum temperatures for germination of 20–25°C are provided in controlled environment cabinets where the propagation medium is maintained in a moist condition. Normally, germination will be completed in 8 days when the plumules begin to appear. The plants are then moved into glasshouses. The propagator aims to produce very uniform, sturdy transplants normally about 10 cm high and with 3 or 4 leaves according to a schedule agreed with the grower. This delivery schedule may be disrupted by periods of inclement weather when the grower may be unable to accept supplies of transplants. It is possible to retain the transplants for short periods in the glasshouse when they may be treated with applications of potassium nitrate to avoid shortages of nutrient in the compost. In some instances, transplants may be placed in unlit stores for short periods. Storage for 6 days at 10°C of cabbage (*B. oleracea* var. *capitata* cv. Matunami) is described by Sato *et al.* (2004). Stored plants were held under water stress (–1.0 to –1.2 MPa), and this encouraged the closure of stomata and reduced the loss of carbohydrates, minimizing seedling etiolation. Stressed plants recovered more quickly when returned to illuminated conditions (see also Fitter and Hay, 1987).

Developmental Physiology

Physiology describes and measures the genetically driven processes of growth and reproduction, and relates these to environmental factors such as temperature, radiation, photoperiod and nutrient availability. It strives to understand the interaction between genotype and environment. Crop physiology has immediate practical significance by permitting the prediction of growth and maturity rates as affected by changing patterns of weather and other events. Managing *Brassica* crops is notoriously difficult because of the impact of periods of high or low temperatures accelerating or retarding maturity. Increasingly, detailed knowledge of the phases of plant growth and their alteration by temperature is permitting horticulturists to plan and predict planting and harvest dates for *Brassica* crops. This aspect of physiology has dominated research work with brassicas for the past generation, and forms the main focus of this chapter. Fruits of this research mean that harvesting schedules have been made more reliable for growers, and supermarket buyers can formulate their promotion of particular commodities to the retail customer with increased confidence. Commercially, such knowledge should mean that profits along the food supply chain are maintained and possibly increased.

For many crops, the use of accumulated temperature values above a specified base (thermal time) provides a unifying time scale by which progress to maturity can be monitored. This method is applicable only when the rate of maturation is a linear function of temperature (Monteith, 1981). In cereals and grasses, for example, the rate of leaf appearance and the reciprocal of the time to anthesis (flowering) are linearly related to temperature (Gallagher, 1979). Leaf appearance in some dicotyledonous crops, such as sugarbeet (Milford *et al.*, 1985), is also related to temperature. Regrettably, defining the relationship of *Brassica* genotypes to their environment and predicting maturity has proved less straightforward.

BRUSSELS SPROUT (*B. OLERACEA* VAR. *GEMMIFERA*)

The imperative for improved understanding of growth and maturity in *Brassica* crops began with Brussels sprouts in the early 1960s as their production for quick-freeze processing took an increasing share of the market especially in the early and mid-season periods. The processing companies demanded regular and predicable supplies entering their factories so that the production lines operated at maximum efficiency as had previously been achieved for vining peas and green beans. Plant breeders contributed towards increased uniformity of growth through the production of F_1 hybrid cultivars. This advance supported the achievements of physiologists in understanding the manner in which the physiological efficiency of the crop could be increased.

Brussels sprouts (*B. oleracea* var. *gemmifera*), like other *Brassica* spp., are photoperiodically day neutral, and flowering is induced at low temperatures. In heated conditions, the plants remain in a vegetative stage, as do cuttings taken from them. Plants sown or transplanted into the field in the spring first pass through a juvenile stage when the plant cannot be induced to flower.

An adult stage is reached in summer when the axillary buds (sprouts) which are the ware crop develop. If the plants are exposed to low temperatures in the autumn and winter, normal flowering begins in the following (second) spring (see Fig. 4.1). The degree of flowering and rapidity of bolting are determined by plant age and the length of cold exposure (Stokes and Verkerk, 1951). Older plants form larger numbers of axillary buds and hence produce more flowering shoots. Plants given longer periods of cold break bud more quickly and produce more flowers, compared with plants offered only the minimum period of cold necessary for flowering.

During the juvenile phase, an increasingly large proportion of dry matter is transferred into the stem. This swells as storage products accumulate. The ratios of leaf and root to stem dry weights reach maximum values in the juvenile stage and thereafter remain approximately constant as the plants age (see Fig. 4.2).

The formation of axillary buds (sprouts) and changing morphology at the apical growing point accompany the maximum accumulation of storage materials in the stem.

Growth stages

The growth stages of Brussels sprout are as follows

Juvenile

1. The stem apex is flat and very small; perhaps four rudimentary leaves and three primordia present.

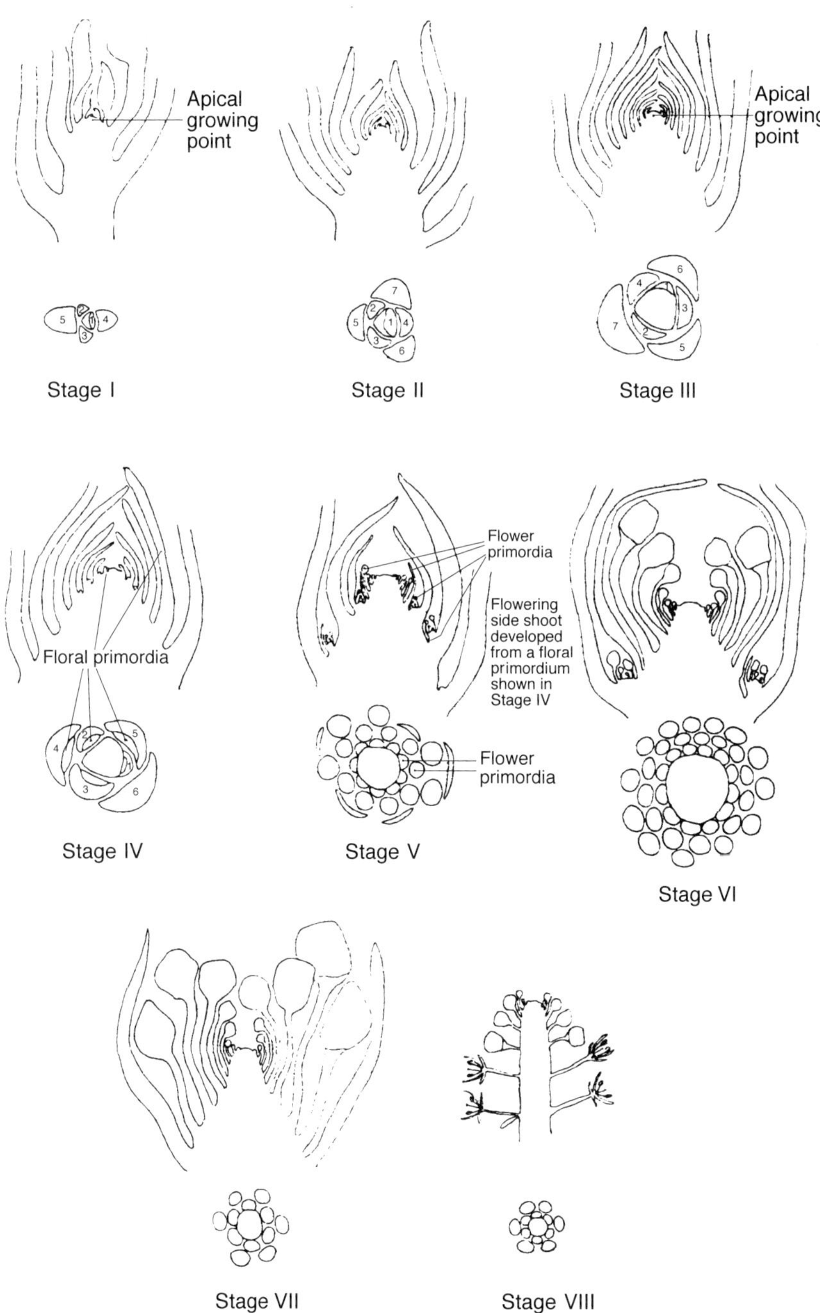

Fig. 4.1. Illustration of the developmental stages of the apex of Brussels sprouts (*Brassica oleracea* var. *gemmifera*) (Stokes and Verkerk).

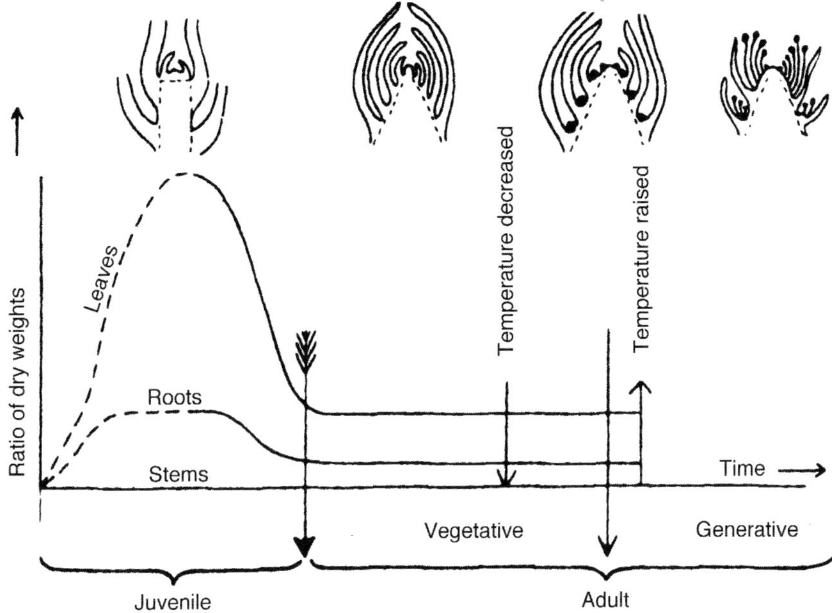

Fig. 4.2. Illustration of the changes in the structure of the apical growing points and the distribution of growth during the juvenile, vegetative adult and generative stages in Brussels sprout (*Brassica oleracea* var. *gemmifera*) (Stokes and Verkerk). Dry weights of leaves and roots are represented as multiples of stem dry weight.

2. At puberty, the stem apex becomes pointed, the growing point enlarges and the apical bud becomes swollen by the increasing numbers of leaves and primordia. Once the growing point becomes dome shaped, Brussels sprout plants will flower following an exposure to a period of cold.

Adult

3. Rapid enlargement of the growing point and top bud; the growing point becomes a globular structure on the apex of the stem and the leaf primordia enlarge.

4A. This stage is reached once the plants have been subjected to cold, the first floral primordia become visible in the axils of the leaves. Primordia are present on the actively growing buds; usually only those in the apical buds continue to full development.

Buds lower down the plant will develop much later and, if given sufficient exposure to cold, they can develop into flowering shoots. At this point, the triangular arrangement of the apex is lost and the shape becomes rounded as progressively larger numbers of initials segregate round the growing point.

Although this is the first indication of potentially generative structures, the change is reversible since if the plant is returned to warmer temperatures the buds will form only leafy shoots.

Plants resulting from the reversal of environment from cold to warm retain the phyllotaxis of a generative plant; they bolt but form no flowers and the apices of the leafy shoots remain in stage 3 (above).

4B. As more buds segregate from the apex, the development of flowering buds takes precedence over that of leaf primordia, until the latter show solely as rudimentary lobes and eventually completely disappear.

5. At this stage, only flower primordia segregate from the apex.

6. The apical meristem reaches maximum size containing large numbers of flowers at varying stages of development and 30 or so leaves. The bud is visible to the naked eye when the leaves are removed.

7. This is the flower bud stage where the buds open out displaying individual flowers prior to bolting. The apical meristem decreases in size.

8. This is the bolting stage where the apical meristem is very small and only a few initials are cut off together; the growing point is permanently retained but no flowers are formed. Some primordia become arrested in growth so that the plant eventually dies with a terminal growing point still present.

The change to the flowering state is accompanied by alterations to the general morphology of the plant. This follows a reduction in petiole length and the production of long strap-shaped leaves towards the top of the plant. These leaves have an increasingly wider angle with the stem (after Stokes and Verkerk, 1951).

In Brussels sprout, the number of leaves determines the number of potential bud sites for the formation of the crop of harvestable sprouts. During crop growth, there are two distinct phases of leaf formation. In early crop growth, the rate of leaf formation is faster than in later periods. Leaf initial formation increases with later planting, but the lower rate of leaf formation is reached earlier with later planting.

In the first phase of crop growth, planting density has no influence on the number of leaves formed. Subsequently, the number of leaves formed at high cropping densities lags behind that formed at wider spacing. Late and higher density planting restricts leaf formation and ultimate ware sprout yield, but these factors appear to operate independently of each other.

Leaf area index (LAI, leaf area per unit area of land, Watson, 1947) reaches maximum values (5–6) at between 80 and 100 days after transplanting and then decreases to values of 1–2.5 in the second half of the crop growing period and through to harvesting. Development of LAI is faster with crops that are planted later but is heavily influenced by planting density. Specific values of LAI are attained more quickly in earlier planted crops.

Stem length is greatest in the early crop growth phases with later plantings. The rate of increase of stem length decreases most quickly with later plantings compared with early and second planting dates. Planting density has little effect on the ultimate crop height. Dry matter production at harvest was highest with early plantings, although crops planted very late in the season (i.e. late June/early July) showed the fastest initial accumulation of dry matter. Sprout bud initiation is greatly influenced by planting date. The buds start appearing between 21 and 32 days earlier with later plantings, i.e. further into the calendar year, and are unaffected by plant density. Bud dry weight was always lower with later planting dates. Buds finally formed in approximately 80% of the leaf axils. Increasing plant density reduces the final fresh weight per bud.

Brussels sprout crops intercept as much as 90% of available radiation at LAI 3.5–3.8; this stage is reached 50–70 days after transplanting. LAI values of 3.5 are attained most quickly by later transplanted crops. This had little practical value, however, since radiation use efficiency (RUE) declines within these crops linked with the seasonal reduction in incoming radiation. Crops transplanted early in the season make the most efficient use of the environmental resources available to them. The grower's most effective strategy is to develop high LAI sufficiently quickly to exploit maximum available incoming radiation during the early summer months. Where planting dates are similar, bud initiation commences more quickly in early maturing cultivars compared with those maturing mid- and late season (Fisher and Milbourn, 1974).

Bud initiation tends to start earlier with low crop planting densities, commencing when the total amount of assimilates produced can no longer be utilized completely by the growing apex. Such assimilate 'saturation' may trigger a hormonally regulated start to bud growth. Bud initiation continues up to harvesting, leading to a spread of bud ages up the plant. This resultant spread of bud maturity affects postharvest quality and shelf life. Traditionally, the sprout buds were selectively harvested by hand labour 'picking-over' the crop several times per season. The advent of machine harvesting does not allow grading in the field. Produce which is mechanically harvested in the field and destined for the fresh market needs to be graded for size and maturity mechanically and by the human eye in the packing shed.

High yields are achieved most efficiently by planting as early as field conditions will permit. This is constrained by the possible incidence of low temperatures which encourage vernalization and flowering in preference to sprout bud formation (Everaarts *et al.*, 1998).

CAULIFLOWER (*B. OLERACEA* VAR. *BOTRYTIS*)

Summer (early and late) and autumn maturing genotypes

The cauliflower is a short caulescent plant with shoot tips composed of young leaves and leaf primordia situated around an apical dome and separated by expanding internodes. In comparison with most *Brassicas*, the shoot tip components of the cauliflower are large, and easy to detach, measure and analyse following environmental changes. The grower's cauliflower consists of a large immature inflorescence (the curd), formed at the stem tip after a period of vegetative growth.

The growth phases of both cauliflower and broccoli (calabrese) can be divided into the following (see Fig. 4.3):

0 = vegetative state;
1 = dome-forming stage;
2 = early curd-forming stage (initiation of inflorescence primordia);
3 = intermediate curd-forming stage (increase of inflorescence primordia or flower buds);
4 = late curd-forming stage (new initiation of inflorescence primordia around the first-order inflorescence primordium);
5 = sepal-forming stage (initiation of flower organs in the floret);
6 = stamen- and pistil-forming stage;
7 = petal-forming stage;
8 = petal-elongation stage.

Normally, the inflorescence primordia of cauliflower do not develop beyond the state of primary protuberance (stages 3 and 4) and increase only in number during the curd-forming and thickening stages. Only after the curds have matured and the peduncles of the inflorescences have elongated is flower bud development resumed, leading to the formation of floral organs in the floret such as sepals, stamens and pistils.

The mature curd of cauliflower is composed of a single flower stalk and numerous first order inflorescences which branch several times and whose tips are shortened considerably. There are numerous inflorescence primordia in a state of primary protuberance on the surface of a cauliflower curd.

The field development of cauliflower plants from transplanting to harvesting may be divided into three physiologically distinct phases which respond differently to ambient temperature.

Juvenile phase

Juvenility is recognized where plants grow and form leaves at a temperature- and size-dependent rate but cannot be induced to reproduce. The end of juvenility is identified in the cauliflower by the initiation of curds at the main stem apices. This morphological change may relate to the plant having produced a critical number of leaves, and this varies with different genotypes (Hand and Atherton, 1987; Booij and Struik, 1990).

The presence of leaf numbers ranging from 8 to 19 marks the end of juvenility in different cultivars. The initiation of new leaves continues until the stem apex changes from vegetative to generative growth at curd initiation. The rate at which leaves are initiated is not uniform and increases in the latter stages of juvenility.

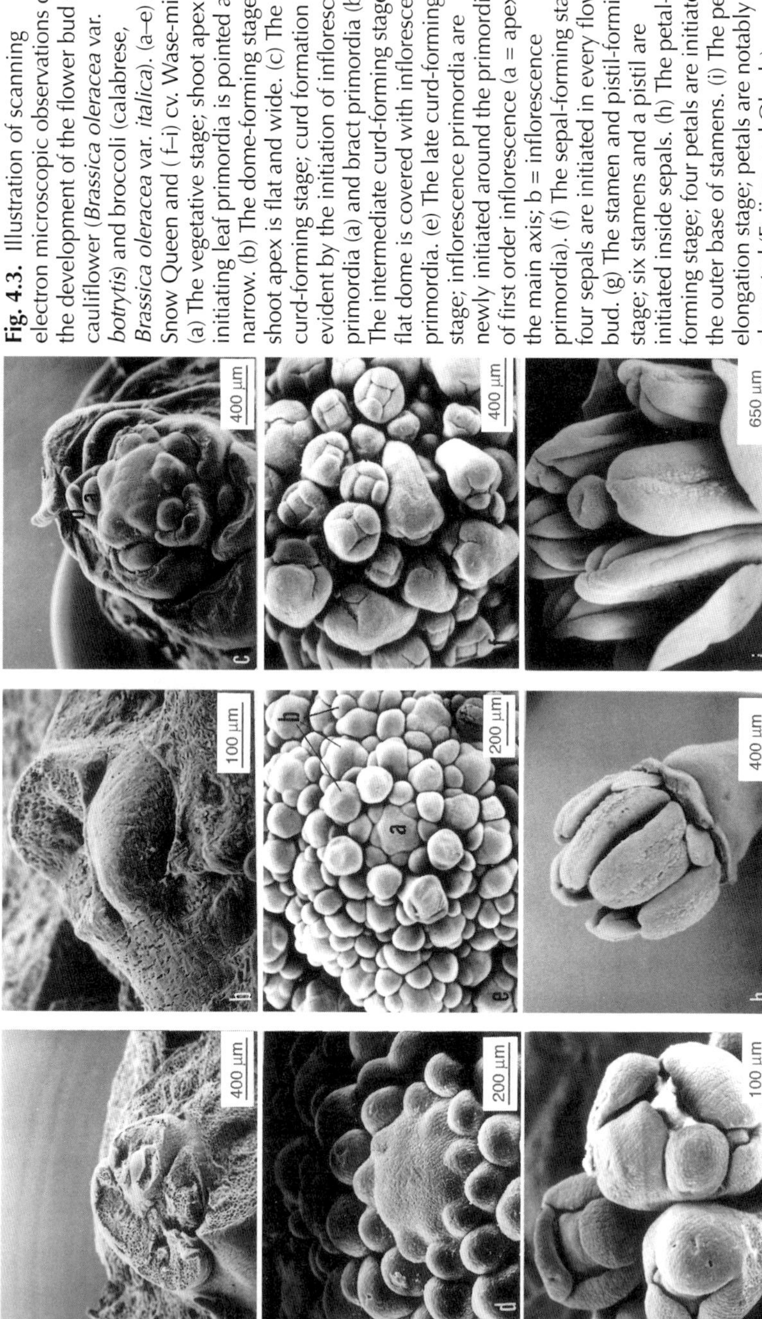

Fig. 4.3. Illustration of scanning electron microscopic observations of the development of the flower bud of cauliflower (*Brassica oleracea* var. *botrytis*) and broccoli (calabrese, *Brassica oleracea* var. *italica*). (a–e) cv. Snow Queen and (f–i) cv. Wase-midori. (a) The vegetative stage; shoot apex initiating leaf primordia is pointed and narrow. (b) The dome-forming stage; shoot apex is flat and wide. (c) The early curd-forming stage; curd formation is evident by the initiation of inflorescence primordia (a) and bract primordia (b). (d) The intermediate curd-forming stage; flat dome is covered with inflorescence primordia. (e) The late curd-forming stage; inflorescence primordia are newly initiated around the primordium of first order inflorescence (a = apex of the main axis; b = inflorescence primordia). (f) The sepal-forming stage; four sepals are initiated in every flower bud. (g) The stamen and pistil-forming stage; six stamens and a pistil are initiated inside sepals. (h) The petal-forming stage; four petals are initiated at the outer base of stamens. (i) The petal elongation stage; petals are notably elongated (Fujima and Okada).

Leaf initiation rate is related to temperature (Wiebe, 1972a) and the rapidity with which growth restarts after transplanting. This shift, however, is not linked to the end of juvenility according to Booij and Struik (1990). Booij (1987) identifies the juvenile phase as being similar in both direct-seeded and transplanted crops. The cauliflower reaches a 'physiological age' before curds are initiated following a period of lower temperatures. Some research suggests that the rate at which physiological age is attained may be manipulated by seed treatment, since Fujime and Hirose (1979) enhanced curd induction by applying cold treatment to germinating seed.

Curd induction or initiation phase

After juvenility, there follows a phase of curd induction for which relatively low temperatures are required. Factors controlling initiation have been investigated by Wiebe (1972a, b, c), Wurr *et al.* (1981a), Hand and Atherton (1987) and Booij and Struik (1990). The term 'vernalization' is sometimes used to describe the response of cauliflower initiation to temperature. Vernalization implies that cool temperatures hasten curd initiation. Wurr *et al.* (1990a), in studies over three seasons, showed this may not always be the case (see Fig. 4.3).

Cool temperatures in spring are well recognized to delay curd initiation more than warmer temperatures in summer. In practice, this effect is well recognized in older traditional cultural systems since warmth applied to pot-raised glasshouse-grown early spring cauliflowers accelerated 'buttoning', i.e. curd initiation was accelerated but resulted in small, immature, low quality curds. The growers' practical objective was to move these plants into the field as early as soil conditions would permit and prevent premature curd initiation.

The stimulus of curd initiation in early summer maturing genotypes is thought to be quantitative. The early summer group form marketable curds in the period June–August following transplanting in March. Transplants must be retained in the vegetative juvenile state without curd initiation during propagation. If curds form before transplanting then they will fail to reach marketable size and quality in the field. The juvenile state is maintained by retaining the transplants in modules at 0–10°C and limiting the availability of inorganic nitrogen. Curd maturity could extend over a period of 2–6 weeks (Atherton *et al.*, 1987). Salter (1969) correlated the period of curd maturity to the time over which such plants initiated curds.

Closely synchronizing curd initiation would have immense implications for the improvement of harvesting and land use efficiency. Atherton *et al.* (1987) showed that in a comparison of the cvs Perfection and White Fox, curds initiated most quickly on smaller plants held at an earlier stage of development after they had been exposed to temperatures of 5 or 10°C for 4 weeks. The most marked response to treatment was in the earlier maturing cv. Perfection. In this cultivar, the optimum temperature for 'vernalization' was

5.5°C, expressed as the reciprocal of leaf number at curd initiation (Roberts and Summerfield, 1987).

Temperatures outside the range 0–22°C are considered as inhibitory to curd initiation. In cv. Perfection, chilling was required for 10–14 days and, although carbohydrates increase or accumulate at the tip in this period, the simple process of accumulation of dry matter does not explain the 'vernalization' effect. The immediate effect of chilling is to stop leaf initiation at the shoot apex; growth and dry matter accumulation continued in existing leaves of the apex for about 4 days during chilling. Rapid accumulation of dry matter began in the apical dome after 10 days of chilling, coinciding with the period when a minimum duration of cool temperatures promotes initiation. It would appear that chilling is associated with the suppression of leaf initiation and growth in the apex, by diverting dry matter accumulation to the apical dome and permitting its development following floret initiation. Atherton *et al.* (1987) showed that the dry matter content of the shoot tip responds to light intensity. This effect declined where plants were transferred from natural illumination in the glasshouse to relative shade (60 W/m^2). Hence, it is probably the strength of the apex acting as a resource sink rather than the requirement for an absolute accumulation of specific quantities of dry matter that determines the onset of curd initiation. Optimum temperatures of 10–12°C have been reported from controlled environment experiments (Wiebe, 1972b) to be required for curd induction.

More recently, Wurr *et al.* (1993) reported 9.0–9.5°C as the optimum temperature for curd initiation in cv. White Fox. Their findings are based on data from field experiments and the use of an asymmetrical temperature response function to construct estimations. Where plants produce greater numbers of leaves during the curd induction phase, this is accompanied by a delayed rate of induction (Fujime, 1983).

Temperature sensitivity relates to the 'earliness' of genotypes; later maturing cultivars are more sensitive to higher temperatures and have greater final numbers of leaves. Daily variations in temperature, however, appear to be of little significance. Comparison of diurnal variation and constant temperatures with comparable mean values produced similar results in attempts to predict curd initiation. In practice, counts of leaf number will provide the grower with information concerning the time of curd induction and temperature conditions during that phase. The relationship between mean temperature and the final number of leaves is dependent on the specific genotype.

Curd growth phase

This phase follows on from curd initiation. The diameter of the curd increases with temperature, up to a maximum (Wiebe, 1975; Wurr *et al.*, 1990b). The contrasting effects of temperature on development in different growth phases are likely causes of the major problems encountered by producers in maintaining the continuity of maturation of cauliflower curds in the field.

Ultimately, the duration of curd growth is fairly constant and independent of eventual weight at harvest. Lower temperatures increase the duration of curd growth. The relationship between radiation and curd growth is connected with the time of transplanting or direct seeding, since with plantings made later in the season, daylength (in Europe) is declining and radiation quantity is reduced. Consequently, the duration of curd growth and its interaction with incident radiation could be an expression of declining daylength which determines maturity as related to elongation of the inflorescence.

Use of a degree-day system (a product of temperature and time above a specified critical value) failed to improve the precision of estimating cauliflower maturity. Salter (1969) established a good relationship between temperature and time by measuring weight changes during curd growth. Temperature and time products, however, failed to predict harvest dates with sufficient precision for practical use. Developing sound, reliable relationships between the duration of curd growth and the curd initiation date would allow harvesting time to be predicted 6–8 weeks in advance. Achieving this requires the establishment of the date of curd initiation, and this is determined by the termination of juvenility.

Some authors have suggested that a model of the early development of cauliflower could be based on meteorological observations and then used to predict curd initiation. After this point, there is a simpler relationship between crop development and temperature and, therefore, the date of harvest can be predicted with greater accuracy (Wurr *et al.*, 1990b).

Modelling the early period of development and combining it with a harvest prediction model would give growers an early warning of disturbances in their production schedules (Grevsen and Olesen, 1994). These Danish authors describe the duration of the juvenile phase by a temperature sum starting at transplanting and using the base of 0°C. The temperature sum requirements (thermal time in Kelvin days (Kd)) for six genotypes ranged from 26 to 83 Kd compared with 250 Kd for two genotypes in similar Dutch experiments.

Leaf initiation is well accounted for at temperatures above 0°C. Counting the number of leaves that have been initiated provides an indication of the end of juvenility. A strong description of leaf initiation (R^2 values of 0.9) was obtained when a gradual acceleration in the leaf initiation rate with increasing leaf number and rising temperature values were used. The end of the juvenile phase occurred when the plants had 12 leaves initiated in these Danish experiments compared with 17–19 leaves in the Dutch experiments. The duration of the curd induction phase in the model is described by using linear responses to temperature that are symmetrical below and above an optimum temperature. A common optimum temperature for curd initiation was estimated to be 12.8°C for two cultivars in the Dutch experiment. The base temperature was estimated to be 0°C and the maximum temperature, therefore, is taken to be 25.6°C. The best fits of data in the combined model of juvenile and curd induction phases show R^2 values from 0.4 to 0.6.

This model requires validation that takes account of season of production (the genotypes used were late summer and early autumn maturing types). There will be differences for each model dependent on the location of production areas, and most probably microgeographical variations within such regions. Hence, ultimately, validation of such models must be done at the farm and field level to provide the standard of confidence that is required for crop scheduling in order to meet market demands. Commercially, the market itself imposes constraints on when the producer will harvest a cauliflower crop. There are demands for greater levels of supply at particular times within the week. This market-driven constraint requires the integration of short-term crop storage with the scheduling of crop maturity and harvesting.

Salter and Fradgley (1969), using open-pollinated cultivars of autumn-maturing cauliflower, showed that transplanted crops matured more slowly and the length of the maturity period would be extended compared with direct-drilled cultivars. Competition for light during plant raising and increased age of the transplant also lengthened the maturity period once the transplant had reached the field. The use of graded seed for direct-drilled crops achieved a small but none the less valuable improvement in mature curd uniformity and shortened the length of the maturity period.

Forms of curd abnormality resulting from the effects of temperature are shown in Fig. 4.4.

Influence of crop nutrition

Manipulation of the supply of inorganic nitrogen to cauliflower plants may substitute for chilling and restrict growth in terms of leaf area and shoot size. In warm non-promotive temperatures, nitrogen starvation delayed curd initiation in terms of both time and the increase in the number of leaves formed before curds began to develop. Nitrogen restriction did not affect the 'vernalization' rate in cool inductive temperatures. Nitrogen limitation reduces the dry matter contents of the apical dome and young leaves (see Chapter 5).

Influence of carbohydrate content

Carbohydrates have long been associated with 'vernalization' processes. They increase in the shoot tip as the process advances. Further 'vernalization' may only proceed in excised buds and embryos when an external supply of sugars is available. Atherton *et al.* (1987) postulated that feeding sucrose to the shoot tip of intact cauliflower might partially substitute for the low temperature stimulus for curd initiation. They suggested that the events

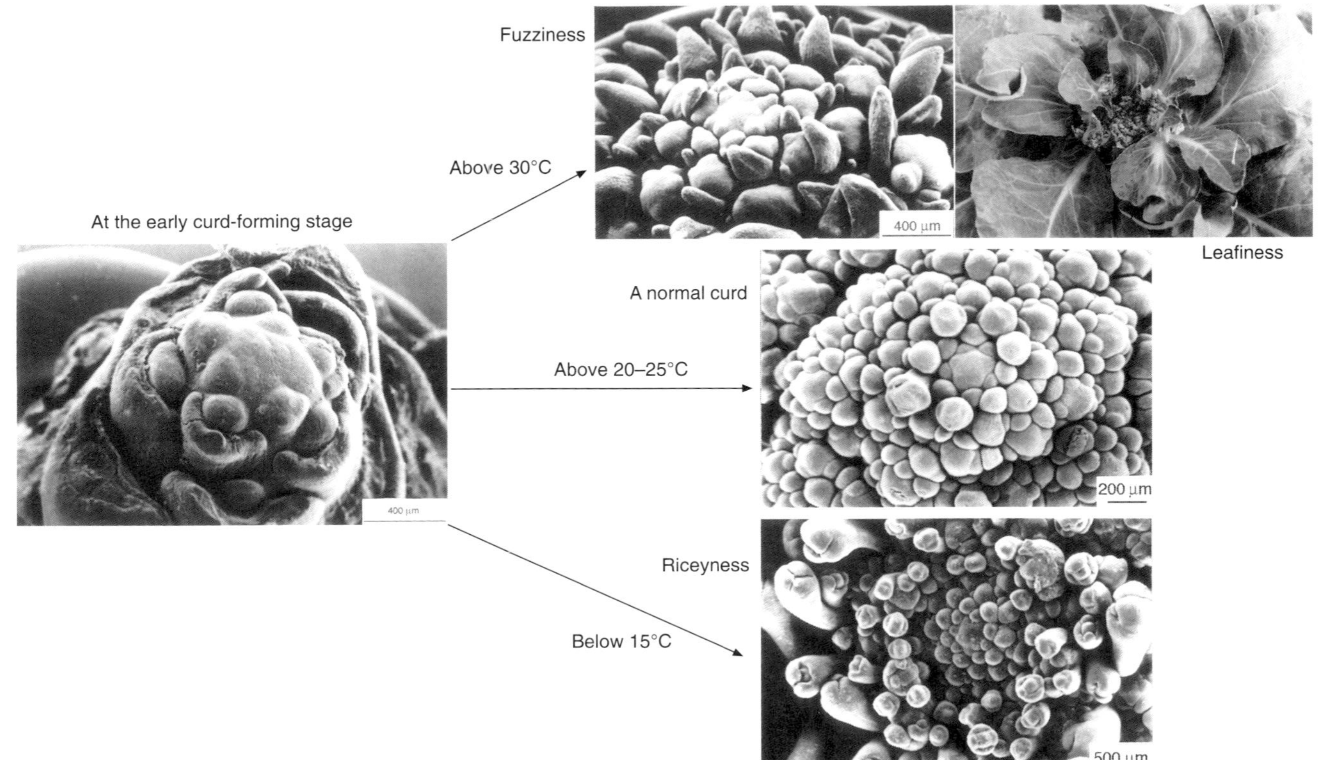

Fig. 4.4. Illustration of abnormal curds where cauliflower (*Brassica oleracea* var. *botrytis*) cv. Snow Queen was grown at various temperatures after curd formation (Fujima and Okuda).

leading to curd initiation in cauliflower include preferential redistribution of carbohydrates to the apical dome at the expense of young leaves, leaf primordia and adjacent stem tissues. This phenomenon is also reported for *Sinapis alba* (Bodson and Remacle, 1987).

Development of curds (cauliflower) or spears (broccoli/calabrese) is a developmental event as distinct from growth processes which involve the accumulation of dry matter and increasing leaf area. For both crops, models which aid in the prediction of maturity dates exist based on the use of accumulated thermal time values and logarithmic values of curd diameter (Salter, 1969; Hand, 1988; Wurr *et al.*, 1990a, b, 1991). Where curd diameter is known, then the thermal time required for maturity may be estimated using mean temperatures obtained from meteorological records for a particular locality (see Fig. 4.5).

Influence of genotype

More recently, Pearson *et al.* (1994) have attempted to formulate models which predict the time required for curd initiation and the duration or rate of curd growth of summer and autumn maturing cauliflower (*B. oleracea* var. *botrytis*). Optimum temperature for curd growth varied with genotype, as reported by Wurr *et al.* (1990b) who derived a regression model based on constant thermal time between curd initiation and the attainment of a specific curd diameter.

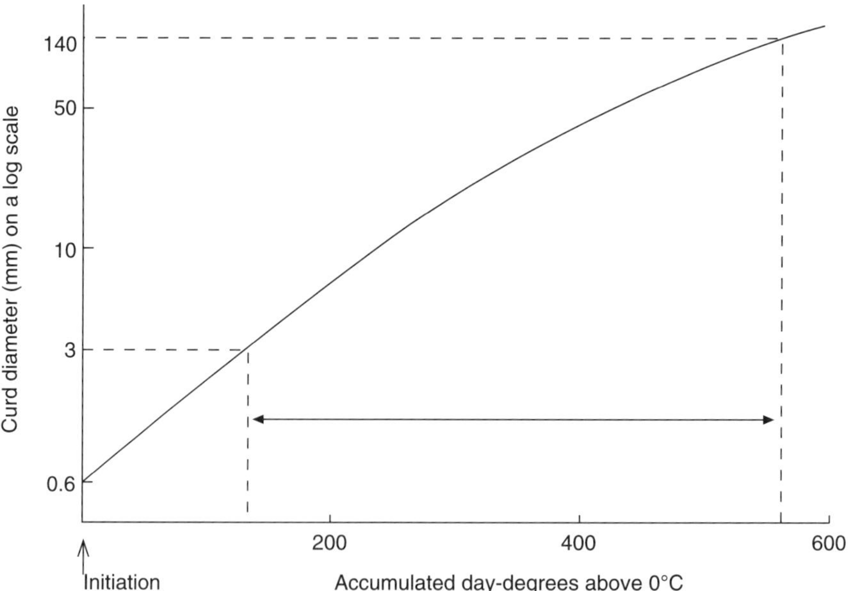

Fig. 4.5. Illustration of the prediction of curd size in cauliflower (*Brassica oleracea* var. *botrytis*) (Wurr, Fellows, Sutherland and Elphinstone).

Variations from this model were apparent with different genotypes. The later-maturing types required more thermal time to reach maturity. Currently, the time taken to reach maturity is controlled by a combination of genotypic and environmental factors, and major variations are due to differences in the time taken to satisfy the requirements for curd induction. A sequence of cultivars is used to improve crop continuity, but these do not always mature in the expected order. Identifying the genes involved in the regulation of maturity and the manner in which they exert their action would help plant breeders produce cultivars with more controlled induction and, therefore, more predictable orders of maturity (D.C.E. Wurr, personal communication).

Overwintered cauliflower (Roscoff and Walcheren types)

The overwintered crop is transplanted into the field in mid to late summer, forming a large framework of leaves in the juvenile phase. Only very limited and, in some cases, rudimentary information is available concerning curd initiation by winter cauliflower. Information available for these crops in general compares poorly with summer and autumn types where recent research has made significant advances in defining the parameters controlling plant juvenility and maturation, and the subsequent initiation and development of curds, and spears in broccoli. Early evidence produced by Wellington (1954) identified an empirical correlation between the time of curd initiation and temperature during the vegetative juvenile phase. This led to variations of up to 2 months in the maturity of overwintered cauliflower cultivars depending on seasonal weather conditions. Later quantification showed that plants maintained above 15.6°C remained vegetative, and the time of exposure to cool conditions was cumulative (Haine, 1959). The most likely reason for such variations is differences in the time of curd initiation. More recently, Wurr *et al.* (1981b) showed that the temperature threshold below which curd induction proceeds is genotypically controlled. Wurr and Fellows (1998) used two Roscoff cauliflower selections, 'December/January' and 'March' (ex CODEBRIC Seed Group), and showed that both light and temperature are important regulators of leaf production. Raising the cultural temperature by growing under polythene covers increased the time to curd initiation in 'December/January' by up to 93 days and in 'March' by up to 57 days.

Leaf number at curd initiation varied between 32 and 112 depending upon treatment in 'December/January', and varied in 'March' by 40–122 leaves. The Roscoff types normally have higher leaf numbers at the time of apex phase change compared with summer and autumn types. During the juvenile growth phase, stem apex diameters in both cultivars expanded linearly with temperature up to approximately 0.2 mm. Leaf number at this stage varied between 23 and 28. 'Vernalization' was most rapid in the earlier cv. 'December/January' between 12 and 16°C, with an optimum of about 14°C, whereas the optimum for 'March' was slightly lower than this value and increasing the time at temperatures in excess of 16°C delayed curd initiation.

For overwintered cauliflower types, for instance, low light intensity and the periods of cold treatment affect the extent of juvenility. Thus overwintered cauliflower can form in excess of 100 leaves before curd initiation commences, whereas with summer/autumn cauliflower types juvenility may end once 17–22 leaves have formed (Wurr *et al.*, 1994). Once this is achieved, then growth towards initiation is determined by temperature. The 'vernalization' rate then increases with temperature to a maximum, and thereafter declines (Atherton *et al.*, 1987; Wurr *et al.*, 1988) and, as suggested by Nieuwhof (1969), temperatures in excess of 23°C prevented curd initiation in annual cauliflower.

The implications of climate change on cauliflower

Analyses of potential effects of global climate change suggest that an elevation of +1°C at temperatures below 15.5°C will reduce the time taken to cauliflower curd initiation. Where temperatures are above 15.5°C, then time to curd initiation will be likely to increase since curd initiation is reduced at elevated temperatures, hence crop duration from plantings later in the season is likely to be extended. Elevation of carbon dioxide content increases curd weight (biomass) especially dry matter content, but has much less effect on curd diameter which can be slightly reduced in size (Wheeler *et al.*, 1995). Specifically for the UK, Hulme *et al.* (2002) suggest that warming will be greater in summer and autumn compared with winter and spring and that there may be greater warming during night time in winter and daytime in summer. Such changes are likely to have a particular influence on curd induction in winter cauliflower, which has a relatively low optimum temperature (Reeves *et al.*, 2001). The impact of climatic change will also vary depending on which part of the developmental cycle of cauliflower is considered. The whole growth period of cauliflower combines distinct phases of juvenility, induction and curd growth, as indicated earlier in this chapter, and each responds differently to temperature. During juvenility, there is a linear response; curd induction is described by a gamma (overturning) relationship and that of curd induction is quadratic. These relationships are individually and collectively affected by cultivar genotype; hence the effects of global climatic change on cauliflower maturity will be a practical example of the interaction of genotype × environment in action (Wurr *et al.*, 2004). The flexibility of maturity periods offered by summer/autumn cauliflower genotypes does not exist in the winter-maturing forms. Consequently, these will be affected to the greatest extent by climatic change. The degree of effect varies depending on the severity of climatic change that is postulated. A marked increase in temperature towards the latter part of the 21st century would call for substantial changes in the genotypes of cauliflower grown in those regions where at present winter cauliflower is a predominant crop. There could also be movements of production of current winter genotypes to more northerly locations.

CALABRESE (GREEN BROCCOLI) (*BRASSICA OLERACEA* VAR. *ITALICA*)

Agronomically, broccoli (calabrese) growth may be divided into the two phases: first, from drilling or transplanting to spear initiation and, secondly, from initiation to maturity. Predicting the duration of both growth periods has vital relevance to crop scheduling, for land occupancy and hence rotational growing, and for adherence to marketing schedules. Leaf number defines the stage of crop growth. This is firstly related to genotype and secondarily increases with later planting dates, partly controlled by rising temperatures that accelerate growth. There is some evidence that the time from transplanting to spear initiation varies more than the time from initiation to maturity with earlier season crops, but the reverse is the case for later season cultivars (Wurr *et al.*, 1991).

Opinions vary as to whether broccoli (calabrese) has a 'vernalization' requirement for spear initiation. Possibly this is a facultative cold requirement which means that cooler conditions, while not essential for spear initiation, do accelerate earlier head formation when they are applied. Temperature affects the number of nodes produced prior to spear initiation. Variation in the time from transplanting to maturity between crops emphasizes the practical difficulties of maintaining continuity schedules. Head growth is also affected by solar radiation, reflecting the high planting densities at which broccoli (calabrese) is grown (four- to fivefold higher than may be used with cauliflower). Consequently, interplant competition commences much earlier in the life of a crop.

Morphologically, broccoli (calabrese) follows similar developmental patterns to cauliflower, i.e. juvenile, head induction (initiation) and head growth phases. Head initiation requires the development of 14 leaves, seven of which are visible to the naked eye. Periods of warm weather delay initiation and lead to greater numbers of leaves being formed. There is a strong genotype effect here, and some cultivars such as 'Shogun' are unsuited to warm temperatures, which result in head malformation. The minimum leaf number required before head initiation may also indicate the length of the juvenile phase in different broccoli (calabrese) cultivars and is consistent with late maturing types such as 'Shogun' forming nine leaves in the juvenile phase compared with early maturing types such as 'Emperor' forming seven.

The total number of leaves initiated at the stem apex is directly related to the number of visible leaves (this also applies to cauliflower). This provides a means of determining the growth stages of crops in the field by counting visible leaves. The quadratic relationship between the natural logarithm of head diameter (measured in mm) and a temperature sum from head initiation provided an accurate method for predicting head maturity. A two-dimensional search for the optimal combination of base and ceiling values for the temperature sum expression gave base at 0°C and ceiling at 17°C as limits to the minimum residual sum of squares. The fit of the quadratic expression is more sensitive to changes in the base temperature limits over 0°C.

Head diameter proved to be a better parameter for calculating time to harvest compared with the use of plant or head weight. Some workers such as Pearson and Hadley (1988) used a linear relationship between head diameter and temperature sum as opposed to a curved one. The good fit they achieved with a linear relationship may result from the fact that they were experimenting with smaller broccoli (calabrese) heads. Others (e.g. Wurr *et al.*, 1992) have used logistic models, but the difference in goodness to fit between logistic and quadratic relationships was very small. Indeed, Wurr *et al.* (1992) changed to using quadratic models when investigating interactions between plant density and harvest maturity.

Plant density has a high impact on head size in broccoli (calabrese). Grevsen (1998) reports that densities of 5–10 plants/m^2 are normal Danish cultivation practice in order to achieve harvestable 150–250 g heads with diameters of approximately 90–120 mm. Some other variables which have been examined as components for these models include genotype, solar radiation and planting month; only the latter enhanced the accuracy of the model sufficiently for use in practice.

Wurr *et al.* (1991) suggested utilizing 'effective day-degrees' (Scaife *et al.*, 1987) as a means of predicting time to crop maturity. Effective day-degrees (EDD) ('effective temperature sum') is a combination of temperature and radiation (R) in MJ/m^2/day, DD is the temperature sum (day-degree) with a certain base and ceiling temperature, and a is a factor denoting the importance of radiation relative to temperature:

$$1/\text{EDD} = 1/\text{DD} + a/\text{R}$$

Some research has indicated a high degree of association between temperature and radiation using this equation, especially where high plant densities (22 plants/m^2) are employed. High plant density results in competition for light early in the growth of the crop. Only relatively small proportions of the overall variance are accounted for by including radiation or planting month values in the predictive equation. Hence, these Danish authors advocated use of a simple temperature sum to determine harvest date for commercial purposes.

The Marshall and Thompson (1987a, b) model was applicable to direct-sown crops. This form of husbandry, however, is now largely discarded, in Europe at least, in favour of using module-grown transplants. With direct sowing, the period to maturity is 90–140 days and their model could predict the harvest date for 90% of crops to within ±7 days and for 100% to within ±12 days. Later models (Wurr *et al.*, 1992) started when heads initiated and at 10 mm, and predicted the harvest date to within ±2–3 days using head diameters and weather data. The problem with the Wurr method is that it predicts small head diameters too late and large head diameters too early (comment taken from Grevsen, 1998).

In broccoli (calabrese) inflorescence primordia form at the early curd stage (stage c Fig. 4.3) and thickening stages. The increase of infloresence primordia and flower bud development progress together. Thus the flower buds of broccoli reach the petal-forming stage prior to spear maturation. Cauliflower, by comparison, possesses a mechanism that stops the progression to flower bud development during curd forming and thickening. The flower head of broccoli (calabrese) is composed of organs similar to that of the cauliflower, but their development progresses differently. There are numerous flower buds each initiating sepals, stamens and a single pistil on the broccoli (calabrese) spear surface. This means that the quality of broccoli spears as defined by the retention of a wholly undeveloped green stage is more difficult to preserve. Yellowing of the spear resulting from the appearance of sepals happens far more quickly than in cauliflower curds (see Chapter 8).

Cauliflower and broccoli (calabrese) curd or spear maturity

Predicting when a broccoli (calabrese) or cauliflower crop will mature is vital to modern marketing practice because it determines when supplies of the product will be available and enables the grower or cooperative to adjust marketing strategy in anticipation of crop maturity.

The maturity of a broccoli (calabrese) head (spear) is primarily determined by developmental states of the florets. The head is harvested shortly before the flowers open. Since flowering as in most species follows the cessation of leaf production on the stem, the reciprocal of the time taken from sowing to maturity is essentially an estimate of the rate of development and would be expected to correlate closely with average temperature.

Maturity is also judged by the size of spears. Hence solar radiation which governs crop growth also affects maturity date. There are no reports of daylength (photoperiod) affecting the development of broccoli (calabrese). In broccoli (calabrese), Fujime and Hirose (1979) and Marshall and Thompson (1987a) showed that the concept of thermal time could be applied to the prediction of maturity using a multiple linear regression model. Maturity is defined as the point where half of the spears had been harvested. Using daily records of maximum and minimum air temperature and total solar radiation, maturity could be predicted to within ±7 days for 90% of crops tested over a 4 year period, and within any one year the precision improved to ±5 days. The crops progressed to maturity when the air temperature exceeded defined base temperatures. This is termed the 'active duration'. When the air temperature was below this base temperature, the crops ceased growing and this was defined as 'inactive duration'. For broccoli (calabrese) in particular, growth ceased at a base temperature of 0°C .

Crop duration steadily declined from about 140 days down to about 90 days as the sowing date advanced from late February (calendar day 55) to early May (calendar day 130). The actual date of harvest changed by only 25 days compared with 75 days difference in sowing date. Crop duration remained at 85 days until late June (day 175). Thereafter, the duration increased to around 120 days for mid-July sowings (day 202). In practice, the forecast harvest date and intervals between harvests for successive cropping requires continual revision throughout the season using the current year's weather records combined with data for long-term averages. A dynamic approach to harvest forecasting can reduce the spread of intervals between harvest to 1.8 days (Marshall and Thompson, 1987b).

CABBAGE (*BRASSICA OLERACEA* VAR. *CAPITATA*)

Predicting the maturity of cabbage heads using methods of heat unit accumulation was suggested by Isenberg *et al.* (1975), but has not found acceptance in practice. Estimates of maturity still remain subjective and empirical, based on visual assessments and finger tests of firmness. None the less, the quantitative measurements of yield become more necessary as crop producers require increasingly greater precision in their use of resources and the capacity to provide predicable supplies to their customers. Yield can be established accurately by use of the relationship between head number, size and density (Kleinhenz, 2003; Radovich *et al.*, 2004).

CHINESE CABBAGE (*BRASSICA RAPA* VARS *PEKINENSIS* AND *CHINENSIS*)

Chinese cabbage responds to inductive temperatures from germination; apparently there is no juvenile stage, with flower initials forming when 'vernalization' is completed. Minimum plant raising temperatures of 18°C are required in order to reduce bolting in the field. The growing point height is determined at or before transplanting. A linear model incorporating temperature and solar radiation describes subsequent extension of the growing point.

The processes of 'vernalization' and growing point extension are controlled separately. It is known that stem elongation can occur without flower formation. The influence of solar radiation varies with genotype, for example being greater in the bolting-susceptible cv. Jade Pagoda and less in cv. Kasumi, although temperature has similar effects on both cultivars. Bolting is encouraged by long days. Head development and flower stalk extension appear to be separate processes. The former is driven predominantly by temperature and the latter by temperature subject to an upper maximum and solar radiation (Wurr *et al.*, 1996).

Crop Agronomy

Agronomy provides the scientific rationale that describes how crops respond in terms of the efficiency of their growth, yield and ultimate financial profitability to the husbandry inputs provided by growers. This discipline is the most direct and strongest expression of the relationship between research and commercial production. It is at the forefront of the drive to raise crop yields worldwide in order to cope with continuing and escalating increases in the human population and loss of productive land. Regrettably, over the past 20 years, an attitude has emerged, especially in developed nations, that there is little 'new science' which agronomists can offer, especially in comparison with the opportunities coming from genetic and molecular studies. This attitude shows signs of moderating as the products of genetic modification come into the market place and basic husbandry knowledge is required to verify their effectiveness and environmental safety, and safely exploit their properties.

Agronomists also play a leading role in responding to increasing public demands for conservation of the environment and the avoidance of soil and aerial pollution while still increasing global food supply, quality and security. These demands can only be met by careful and measured changes to husbandry systems validated in advance by reliable, robust research and development. In turn, agronomists require ever more sophisticated methods with which to achieve a clearer understanding of the interaction between crops, other organisms and the environment. Society's demand for increased yields and quality that maintain food security achieved with diminishing environmental impact from less land has gained the umbrella title of 'sustainability' (see Chapter 6). This term now bridges across the entire spectrum from organic crops produced in the absence of soluble mineral fertilizers and synthetic pesticides, to high productivity integrated resource management approaches to food production. Agronomists are required to interpret the efficiency of returns from resources used in each of these forms of production. Adherence to 'sustainability' means that measuring productivity solely in relation to units of economic resource ceases to be an effective definition of efficiency. Factors relating to the interaction of particular crops with the entire production system, the environment and onwards through the supply chain to the ultimate consumer need to be added into the equation that defines sustainable efficiency. Increasingly, the agronomist is in demand to integrate economic, environmental and social factors affecting crop production.

Brassicas are the predominant group of field vegetables worldwide harvested for immediate fresh consumption or with the minimum of post-harvest preparation. Increasingly, agronomic research is required to improve the effectiveness of their husbandry and respond to changing market and social demands in the fresh produce food chain.

NUTRITION

The original wild progenitors of cultivated brassicas were capable of surviving in inhospitable arid conditions with minimal availability of nutrients (see Chapter 1). In stark contrast, cultivated brassicas are very responsive to increasing supplies of nutrients, particularly nitrogen and water. Traditionally, they have been used as the leading crop in rotations planted following the application of organic or animal manure, emphasizing their responsiveness to nitrogen fertilization in particular.

This innate responsiveness causes problems where excessive quantities of fertilizers are applied in efforts to boost yields, because physiologically the brassicas are not very efficient in their use of such resources. Nutrient imbalances in the soil and plant tissues lead to toxicity and deficiency syndromes that impair growth, resulting in stress disorders and the development of off-flavours in harvested produce. For these reasons alone, it is important that fertilizers are applied with due regard to the nutrient reserves in the soil and the demands that each growing *Brassica* crop imposes. This requires practical knowledge of the structure and texture of the soil type and previous cropping history of each field, its current nutrient status as determined by soil analysis and the likely response to added resources. Increasingly, the nutrient status of a field is being monitored in great detail so as to identify areas where additional fertilizers are needed and those places where there is already ample supply. Monitoring systems using space satellite facilities and technology are set to revolutionize the specificity with which crop nutrient requirements are determined and satisfied.

Improving the efficiency with which fertilizers are used in vegetable crops requires ecological knowledge of the manner in which biomass allocation, morphological plasticity and competitive abilities interact. This may have particular relevance where crops are subjected to shading either from surrounding plants or when they are grown under plastic mulches. The maintenance of balanced growth requires the adjustment of morphological characters and physiological behaviour to maximize whole plant growth and hence the eventual yield, especially where this is composed of either leaves or swollen roots and hypocotyls. Plants respond to changes in the external supply of nutrient resources by adjusting the relative sizes and distributions of organs. Ecologists have suggested that a reduced nutrient supply leads to a higher root:shoot ratio, thereby compensating for loss in root foraging capacities. Lower light availability due to competitive shading results in greater shoot growth.

The amount of a resource captured by plants is at least partially related to the area or volume of the organ system responsible for obtaining the resource. Hence, species with a higher biomass allocation to the root system should perform better at low levels of soil nutrient supply compared with those allocating more to stems and leaves (Tilman, 1988). Some initial studies have been made with the brassicas by Li *et al.* (1999) who used oriental 'green cabbage' *B. rapa* cv. Natsurakuten-leafy variety and turnip *B. rapa* var. *rapifera* cv. disease-tolerant Hikari.

In general, plants originating from low resource environments have an inherently high root:shoot ratio (Grime, 1979), and reduced nutrient supply leads to a phenotype with active fine roots, high root:shoot ratio, many root hairs and a slow maximum rate of net photosynthesis which maintains balanced growth at low nitrogen availability (Robinson, 1991). Raising the nutrient supply increased the amount of resources allocated to aerial organs but not necessarily to leaf blades. In the case of oriental 'green cabbage', growth increased in the leaf petioles not the leaf blades. When there is competition for light, the plants increase resource allocation for stem growth and expand internodal length. Species with different morphologies invest more resources in those organs that provide greatest advantage in capturing light; thus 'green cabbage' extended the petioles. The ability to compete for resources is not solely a function of biomass allocation patterns but also depends on morphological characters such as leaf area ratio (LAR) and specific leaf area (SLA), indicating the importance of morphological plasticity.

SOIL NUTRIENT CONTENT AND SOIL INDEXING

Reliable and repeatable analyses that measure nutrient reserves present in soil and allow reasonable predictions of the effects of fertilizer application are available for most elements with the general exception of nitrogen. Soil mineral nitrogen (N) levels vary continuously due to the effects of mineralization, fertilizer dressings, leaching, denitrification and plant uptake. In consequence, the timing of soil sampling is critical, and measurements are only of use for crop management decisions immediately after sampling.

Laboratory analyses of soil samples are used to determine available macronutrients such as phosphorus (P), potassium (K) and magnesium (Mg) expressed in mg/l. For practical simplicity, these values are converted to indices (Table 5.1) which indicate the relative quantities of nutrient available to the crop ranging from 0 (deficiency) to 9 (excess). Soils intended for *Brassica* production should be maintained at index = 3 for phosphorus and index = 2 for potassium and magnesium. Soils with these indices require only maintenance quantities of additional fertilizer; below these values, larger amounts are necessary to ensure economic returns from the crop and to restore the nutrient reserves of the soil.

NITROGEN

Nitrogen has very pronounced effects promoting the growth, yield and quality of vegetable brassicas within the limits of crop need. Availability in excessively high quantities reduces yield and quality, increases susceptibility to pathogen invasion (Chapter 7) and leads to physiological disorders (Chapter 8) and delays in maturity. Unlike other nutrient elements, nitrogen requirements are not usually based on soil analyses but on specific demands of particular crops and status in the field, making allowance for residues from previous crops and applications of organic manure.

Normally, consideration is only given to the last crop grown when determining the nitrogen index (Table 5.2), but after lucerne (*Medicago sativa*), clover (*Trifolium* spp.), extended leys (grass crops) and permanent pasture, a longer cropping history should be taken into account. Three levels of soil nitrogen index are used. Fields in index 0 have low nitrogen reserves and require more nitrogen compared with those with index = 1. Index 2 soils have the highest soil nitrogen reserves. This approach may be satisfactory for most agricultural crops, but the increasing complexity of horticultural brassica production has stimulated research seeking greater precision and efficiency in the use of nitrogen (Table 5.2).

Table 5.1. Relationship between available phosphorus, potassium and magnesium determined by laboratory analysis and the soil index system.

Index	Phosphorus (mg/l)	Potassium (mg/l)	Magnesium (mg/l)
0	0–9	0–60	0–25
1	10–15	61–120	26–50
2	16–25	121–240	51–100
3	26–45	241–400	101–175
4	46–70	401–600	176–250
5	71–100	601–900	251–350
6	101–140	901–1500	351–600
7	141–200	1501–2400	601–1000
8	201–280	2401–3600	1001–1500
9	>280	>3600	>1500

Source: Anon (1985).

Table 5.2. Nitrogen index system based on prediction using the last crop grown as an indicator.

Nitrogen index = 0	Nitrogen index = 1	Nitrogen index = 2
Cereals	Beans	Any crop in field receiving large frequent dressings of farmyard manure or slurry
Forage crops removed	Forage crops grazed	
Leys[a] (1–2 year) cut	Leys (1–2 year) grazed, high N[b]	Long leys, high N[b]
Leys (1–2 year) grazed, low N[c]	Long leys, low N[c]	Lucerne
Maize	Oilseed rape	Permanent pasture – average
Permanent pasture, poor quality, matted	Peas	Permanent pasture – high N[b]
	Potatoes	
Sugarbeet, tops removed	Sugarbeet, tops ploughed in	
Vegetables receiving < 200 kg/ha N	Vegetables receiving > 200 kg/ha N	

[a]Ley is a European term for extended cropping with a forage crop, especially grass, which may occupy the land for several seasons; land occupied for 1–3 years would be a 'short ley' and land occupied for ≥ 3 years would be a 'long ley'.
[b]High N, > 250 kg/ha N/year or high clover content.
[c]Low N, < 250 kg/ha N/year and low clover content.
Source: Anon (1985).

EXAMPLES OF CROP RESPONSES TO NUTRIENTS

Brussels sprout crops require up to 300 kg of nitrogen/ha for optimal yield. Total aerial biomass production and nitrogen uptake rates increase strongly with nitrogen applications, mainly because of its pronounced effect on leaf area expansion (Booij *et al.*, 1996). A high nitrogen uptake rate rapidly depletes nitrate reserves in the soil (Booij *et al.*, 1993).

Crop growth of Brussels sprouts has two phases (see Chapter 4) (Abuzeid and Wilcocksen, 1989). In the first phase, biomass increase is mainly by leaf and stem growth, and the second phase is concentrated towards axillary and terminal bud growth. The period of bud growth coincides with increasing leaf senescence. When nitrate reserves in the soil are depleted before the onset of bud growth, its uptake rate is then dependent on nitrogen mineralization from soil organic matter; hence the well-established practice of applying large quantities of organic fertilizer (such as farmyard manure) in advance of growing Brussels sprout crops.

The low availability of soil nitrogen and high rate of leaf senescence during bud growth imply that nitrogen remobilization takes place to support bud growth. Efficient use of nitrogen fertilizers and the maintenance of the environment require an understanding of the manner by which nitrogen is removed from the leaves and distributed to the buds. Nitrogen in the non-marketable portions of this crop (roots, stems and leaves) may remain in the field, depending on the harvesting method used.

Some harvesting techniques require that the stems bearing sprouts are removed to a static bud stripper located at some distance from the crop, whereas other systems may employ mobile strippers which remove the buds in the field, leaving residue behind. Where residues remain in the field, they are a potential source of further soil nitrogen losses through mineralization (Whitmore and Groot, 1994) and subsequent leaching by winter rains into adjacent waterways (Greenwood, 1990). To diminish this risk, the nitrogen harvest index should be high. Remobilization of nitrogen within the plant should be encouraged since it results in a higher nitrogen index in harvested portions of the crop compared with the total biomass. From an environmental perspective, the rapid depletion of the soil and the later relocation of nitrogen during bud growth lowers the risk of leaching from the soil as nitrogen is conserved in the crop.

Booij *et al.* (1997) found that with Brussels sprouts at the onset of bud (sprout) growth, 60–80% of the total biomass has been produced and an equivalent amount of nitrogen taken up. Eventual harvest bud weight and bud nitrogen content were both positively correlated with total biomass and nitrogen content at the onset of bud growth. Partitioning of biomass and nitrogen among the different aerial plant organs was hardly affected by the availability of nitrogen and the timing of fertilizer applications. During bud growth, the leaves senesced rapidly. Biomass and nitrogen in particular were remobilized from the leaves to the buds before abscission. Fisher and Milbourn (1974) and Wilcocksen and Abuzeid (1991) concluded that where leaves were removed before they became senescent (commonly a cultural practice of defoliating particularly early maturing cultivars), remobilization of dry matter from leaves did not contribute to bud growth. During bud growth, loss of nitrogen from leaves was up to 50% of the nitrogen increase in buds. Where nitrogen was applied at the onset of bud growth, it was accelerated and increased greening, with a delay in leaf shedding.

When nitrogen was applied as a split application, half at transplanting and the rest at the onset of bud growth, the nitrogen content of buds was increased. The partitioning of biomass and final bud yield, however, were unchanged compared with treatments in which the entire fertilizer application was made at transplanting.

Booij *et al.* (1997) conclude that to reach an acceptable bud yield, the nitrogen application rate should aim at optimal biomass at the onset of bud growth. Biemond *et al.* (1995) showed that for Brussels sprouts, sufficient nitrogen should be available for unrestricted growth at the beginning of the season. Where this is not the case, bud growth is delayed. Apparently, under these conditions, most of the nitrogen is removed from the field as crop (buds) at harvest, with relatively small amounts of nitrogen remaining in the field as residues. This has environmental benefits since less nitrogen is leached following the mineralization of crop residues. A larger total green leaf area, attained with more nitrogen, resulted mainly from larger leaves; the number of leaves increased only slightly. Larger leaves were the result of higher rates of leaf expansion. Neuvel (1990) showed that when Brussels sprouts were either direct drilled in April or transplanted in May with an application of 300 kg/ha of fertilizer nitrogen, the total dry matter production in October was on average 14.5 t/ha and nitrogen uptake was 335 kg/ha (Everaarts and van Beusidem, 1998). Maximum sprout yield of about 4.5 t/ha dry matter was usually attained with the largest amounts of nitrogen applied, i.e. 300 or 375 kg/ha. Scaife (1988) showed that in Brussels sprouts, pre-planting nitrogen fertilizer resulted in nitrate nitrogen increasing for 43 days from the date of application. Thereafter, it declined to near zero in November at the time of single harvesting.

Broccoli (green calabrese) is another nitrogen-demanding *Brassica* and its availability determines crop productivity and quality. Uptake of nitrogen increases with availability and application rate, as shown, for example, by Everaarts and de Willigen (1999). Recommendations vary between 220 and 270 kg N/ha (Greenwood *et al.*, 1980; Vågen, 2003). In relating these rates to crop growth for broccoli (calabrese) (*B. oleracea* var. *italica*), Vågen *et al.* (2004) showed that crop biomass and leaf area index (LAI) accumulated, intercepted photosynthetically active radiation (PAR) and radiation use efficiency (RUE) increased with increasing nitrogen application, but RUE was not significantly altered in the range from 120 to 240 kg N/ha. Hence applying higher levels of nitrogen increased crop biomass by increasing intercepted radiation rather than by increased RUE. The saturation level of RUE was reached when the sum of applied nitrogen and mineral nitrogen of the soil before planting was about 200 kg/ha.

There was a strong effect of nitrogen application in early plantings where climatic factors were suboptimal. Low temperature and possibly over-saturation of light were the most likely reasons for low LAI, RUE and relative growth rate (RGR) following low or nil nitrogen applications. The nitrogen concentration decreased as total biomass increased.

Two alternative mathematical relationships of critical nitrogen concentration (N_C) for broccoli were suggested by Greenwood *et al.* (1986, 1996); of these, the linear equation fitted best. It gave higher estimates, and both equations were evaluated on the correspondence between relative N concentration $N/N_C = 1$ and maximum relative values for biomass, LAI, accumulated intercepted PAR and RUE. A curvilinear function might, however, produce a higher correlation over the whole range of plant nitrogen concentrations. Radiation use efficiency and accumulated intercepted PAR approached saturation level at a lower relative nitrogen concentration than did biomass and LAI, suggesting that nitrogen was more limiting for biomass production than for PAR interception and RUE at nitrogen levels below the highest application rate. Incident radiation was fully absorbed at quite a low LAI, resulting in low RUE values. In more recent studies, Vågen *et al.* (2004) suggested that crop biomass, LAI and accumulated intercepted PAR increased with enhanced nitrogen rates. The strongest effects of additional nitrogen were obtained in these Norwegian conditions at the early plantings when low temperature and possibly oversaturation of light were the most likely reasons for low LAI, RUE and RGR at low or nil nitrogen applications. The linear relationship suggested by Greenwood *et al.* (1996) provided the best explanation of these data.

Summer- and autumn-maturing cabbage used 3.8 kg N/t of edible cabbage yield, and crop residues contained 300 kg N/ha. Almost 50% of the nitrogen absorbed by cabbage ends up in the crop residues. Cabbage was found to make efficient use of available nitrogen, absorbing >100 kg/ha from unfertilized fields. Uptake was slow at the beginning of the growing season but increased thereafter, continuing into September. Broadcasting nitrogen as opposed to band placement appeared to encourage cabbage growth. As the crops matured, however, their demand for nitrogen declined.

Spring-maturing cabbage and overwintered cauliflower benefit greatly in terms of yield and quality from top dressings of nitrogen. Broccoli (calabrese) and cauliflower are frequently grown intensively as several successive crops occupying one area of land, leading to double – or even triple – cropping with one plant form. In this type of husbandry, there may be macronutrients other than nitrogen carried over between crops. Reductions in the quantities of phosphorus applied to succeeding crops may be appropriate. By comparison, cabbage crops particularly absorb large quantities of potassium, and it may be necessary to increase applications for succeeding crops accordingly.

Efficient use of nitrogen involves accurate estimation of crop nitrogen demand, choice of application method and the timing of application. The effects of band placement, rate of fertilization, dry matter accumulation, yield and uptake have been studied by Salo (1999). The concentration of a particular nutrient in brassicas may increase and decrease during the life cycle of the plant. A decrease is due to greater production of carbon-rich compounds relative to accumulation of nutrient ions such as N, P, K, Mg and Ca. Starch and cell walls are the principal carbon-rich compounds. Early in a growth cycle, nutrients increase due to high rates of uptake by the young roots and the high relative growth rate of younger plants.

As plants mature, the LAI and hence the degree of mutual shading rises, resulting in decreased net photosynthesis per unit leaf area, reduced RGR and declining nutrient concentration. The onset of this decline can be altered by the timing of fertilizer application, possibly avoiding shortages of nutrient in the root zone. Raising nitrogen status increases the nitrogen content of the plant. There may also be parallel increases of phosphorus and cations as nitrogen is raised. Key factors in these processes are the particular nutrient in question, plant growth stage, crop type and environmental variables, including soil fertility and temperature, light interception and water availability.

All fertilizer recommendations are given in kilograms per hectare (kg/ha). The number of kg of nutrient in the standard 50 kg bag of fertilizer is obtained by dividing the percentage of the nutrient by 2 (100/2 = 50). For example, one 50 kg bag of 20:10:10 NPK compound fertilizer will contain 10 kg of nitrogen, 5 kg of phosphate (P_2O_5) and 5 kg of potash (K_2O) (see Table 5.3 for conversion factors between nutrient elements and their oxides). Detailed figures are quoted here for each crop type; more recently (Anon, 2000), there has been a tendency to condense the data for several *Brassica* crops together.

Using the indexing system, standard tables relating to fertilizer application are available for *Brassica* crops. Those shown in Tables 5.4 and 5.5 are applicable to the UK and northern temperate Europe. The interpretation of these tables requires knowledge of the particular soil type involved and its previous cropping history; this is especially true with regard to nitrogen. Generally, highly organic soils supply more available nitrogen than mineral soils. Hence the recommended rates should be reduced by about 10% for peat-based soils.

FERTILIZER APPLICATIONS

General fertilizer requirements for a range of *Brassica* vegetables are given in Table 5.6. These bear out the more specific requirements cited in the other tables indicating that these crops benefit from substantial applications of major nutrients. To avoid damage to the root systems by increasing soil conductivity to dangerous concentrations, it is advisable to apply nitrogen in particular as split dressings, with half applied to the seedbed or at transplanting and the residue about 2 weeks later.

Nitrogen and potassium when available in excess can reduce germination and damage seedling root systems, particularly on dry, sandy soils. Thorough incorporation of fertilizers into the soil before drilling or transplanting is essential. Both potassium and phosphorus may be applied some weeks in advance of seeding or transplanting. There is a danger that nitrogen applied too early in the growing season may be lost through leaching.

Table 5.3. Conversion factors between common nutrient elements and their oxides.

Nutrient element	Oxide	Conversion factor
Calcium (Ca)	Calcium oxide (CaO)	1:1.40
Phosphorus (P)	Phosphorus pentoxide (P_2O_5)	1:2.291
Potassium (K)	Potassium oxide (K_2O)	1:1.205
Magnesium (Mg)	Magnesium oxide (MgO)	1:1.67
Sulphur (S)	Sulphur trioxide (SO_3)	1:2.50

Table 5.4. Specific nutrient requirements for Brussels sprout, cabbage (including Chinese cabbage), swede and turnip.

Crop and soil type	N, P, K or Mg index 0	1	2	3	4	> 4	Top dressing
Brussels sprouts: market picking or single harvest							
Silt and brickearth soils							
Nitrogen (N)[a]	200	150	100	—	—	—	
Other soils							
Nitrogen (N)[a]	300	250	200	—	—	—	
All soils							
Phosphate (P_2O_5)	175	125	75	50	25	Nil	
Potash (K_2O)	200	175	125	60	Nil	Nil	
Cabbage							
Summer, autumn and Chinese							
Nitrogen (N)[a]	300	250	200	—	—	—	
Winter and savoy							
Pre-Christmas cutting							
Nitrogen (N)	300	250	200	—	—	—	
Post-Christmas cutting							
Nitrogen (N)[a]	150	125	100	—	—	—	0–75
Winter white for storage							
Nitrogen (N)[a]	250	200	150	—	—	—	
Spring							
Nitrogen (N)[b]	75	50	25	—	—	—	200–400
Early frame raised							
Nitrogen (N)	250	200	125	—	—	—	60–120
Cabbage all types[c]							
Phosphate (P_2O_5)	200	125	75	50	25	Nil	
Potash (K_2O)	300	250	175	75	Nil	Nil	
Brussels sprout and cabbage							
Sands and light loams							
Magnesium (Mg)	90	60	Nil	Nil	Nil	Nil	
Other soils							
Magnesium (Mg)	60	30	Nil	Nil	Nil	Nil	
Swedes							
Nitrogen (N)	100	50	Nil	—	—	—	—
Phosphate (P_2O_5)	150	100	50	50	25	Nil	
Potash (K_2O)	250	200	150	75	Nil	Nil	
Turnips							
Early bunching							
Nitrogen (N)	150	100	50	—	—	—	
Maincrop							
Nitrogen (N)	100	50	Nil	—	—	—	
All crops							
Phosphate (P_2O_5)	150	100	50	50	25	Nil	
Potash (K_2O)	250	200	150	75	Nil	Nil	

All values are given as kg/ha.

[a]For direct-drilled crops or transplants on sands and light loams, nitrogen in excess of 100 kg/ha should be top-dressed to reduce the risk of damage to seedlings or young plants and applied at singling or within 1 month of transplanting. Extra top-dressing may be required especially on shallow or sandy soils when rainfall greatly exceeds transpiration within 2 months of applying basal nitrogen. Top-dressings are unnecessary if nitrogen is injected either before drilling or between the rows up to 1 month after emergence or transplanting.

Table 5.4. *Continued*

[b] A fully grown crop can use up to 400 kg/ha nitrogen. Smaller crops of greens harvested in early spring may need less than half of this amount. Applications should be in dressings of 100–200 kg/ha nitrogen, and related mainly to growth, but the potential marketing period and weather conditions should also be considered.
[c] When spring cabbage follows a crop leaving substantial residues, reduce the phosphate application by half and the potash application by 60 kg/ha.
Source: Anon (1985).

Table 5.5. Specific nutrient requirements for broccoli (calabrese) and cauliflower.

	N, P, K or Mg index						
Crop and soil type	0	1	2	3	4	> 4	Top dressing
Calabrese							
Nitrogen (N)	250	200	160	—	—	—	
Phosphate (P_2O_5)	150	75	60	60	30	Nil	
Potash (K_2O)	150	100	75	50	Nil	Nil	
Sands and light loams							
Magnesium (Mg)	90	60	Nil	Nil	Nil	Nil	
Other soils							
Magnesium (Mg)	60	30	Nil	Nil	Nil	Nil	
Cauliflower							
Early summer, late summer and autumn, e.g. Flora Blanca and Australian types							
Nitrogen (N)[a]	250	200	125	—	—	—	
Winter, Roscoff types							
Nitrogen (N)[b]	75	40	Nil	—	—	—	60–125
Winter, hardy types							
Nitrogen (N)[b]	75	40	Nil	—	—	—	125–200
All types							
Phosphate (P_2O_5)[c]	175	125	75	50	25	Nil	
Potash (K_2O)[c]	300	200	125	60	Nil	Nil	
All types							
Sands and light loams							
Magnesium (Mg)	90	60	Nil	Nil	Nil	Nil	
Other soils							
Magnesium (Mg)	60	30	Nil	Nil	Nil	Nil	

All values are given in kg/ha.
[a] For direct-drilled crops or transplants on sands and light loams, nitrogen in excess of 100 kg/ha should be top dressed to reduce the risk of damage to seedlings and should be applied at singling or within 1 month of transplanting.
[b] For February cuttings in frost-free areas, apply the top dressing in the late autumn. For crops to be harvested later than mid-February, apply top dressings in mid-February.
[c] When winter cauliflower follows a crop leaving substantial residues, reduce the phosphate by one half and the potash applications by 60 kg/ha.
Source: Anon (1985).

Table 5.6. Nutrient requirements for several *Brassica* and *Raphanus* crops.

Crop	Nutrient (kg/ha)			Comments
	N	P_2O_5	K_2O	
Brassica juncea (leaf mustard)	90–100	90	90	Compost 10 t/ha; nitrogen applied as split application; half as basal dressing and half as side dressing 2 weeks later
Brassica oleracea (cauliflower and broccoli)	NPK depends on soil type, soil reserves and expected yields; top dressings are applied to stimulate head formation			Compost 20 t/ha
Brassica oleracea (Chinese kale)	250 kg/ha NPK (15–15–15)			10–20 t/ha organic manure
Brassica oleracea (head cabbage)	Cabbage crop of 25 t/ha absorbs 140 kg N, 40 kg P, 180 kg K			20–50 t/ha compost or organic manure
Brassica rapa (oriental greens)	Very responsive to N			Yield 30–50 t/ha
Brassica rapa (Caisin)	60–110	40–60	80–100	10–15 t/ha compost, nitrogen applied as split dressing; half as basal fertilizer and half 2 weeks later
Brassica rapa (Chinese cabbage)	120–200	40–60	70–150	Soluble fertilizer applied as split dressing; half at planting and the rest 10–14 days later
Brassica rapa (Pak Choi)	55–75 kg N, 40–80 kg P, 80–110 kg K at planting			55–75 kg/ha applied 14 days after planting
Raphanus sativus (radish)				Apply compost and adequate NPK pre-sowing and N at regular intervals thereafter

Source: adapted from Siemonsa and Piluek (1993).

SOIL PH AND CALCIUM CONTENT

Brassica crops are most productive when grown on land with an approximately neutral pH. The ideal is pH = 6.5 for mineral soils and pH = 5.8 for organic soils. This rule should be altered where soil-borne pathogens are present, especially *Plasmodiophora brassicae*, the causal agent of clubroot disease. Land where even very low levels of infection are present should be raised to pH values in excess of 7.0. *Brassica* crops vary in their sensitivity to acidic pH and the point at which crop productivity begins to diminish, as shown in Table 5.7.

Lime requirements of acidic soils are expressed in t/ha of ground limestone or ground chalk. The amount of lime recommended for soils of similar pH may vary with soil texture and soil organic matter content. Usually, the recommendations aim to maintain the top 20 cm depth of mineral soil to a pH of 6.5 and an organic soil to pH 5.8. Where there are variations in soil acidity across the profile, then larger applications of lime may be needed. Applications of lime in excess of 12 t/ha should be made as several separate dressings. Lime should be applied well before sowing or transplanting. Several months are required for changes in soil acidity to take place. There is an increasing tendency for growers of brassicas using highly intensive systems to use calcium oxide (CaO) as a liming agent. This has the advantage of acting very quickly to alter pH and is applied at about one-third the rate of carbonate forms. Also, the liming effect on pH is lost by the end of the season and this permits growers to plant potatoes in the following year with lower risks of infection by *Spongospora subterranea* and *Streptomyces scabies*, the causes of powdery and common scab diseases, respectively. Normally, *Brassica* vegetables should not be grown immediately after liming a very acid soil (pH <5.0). Where a crop is failing because of slight acidity, however, then some improvement may be achieved by top dressing across a standing crop. This is most likely to be successful where calcium is applied in readily accessible forms such as calcium cyanamide or calcium nitrate. Overliming, bringing the pH beyond 7.5, should be avoided, especially on sands, light loam and organic soils since this can result in induced deficiencies of trace elements such as boron and manganese.

The results of soil analyses reflect the quality of the sampling methods used. Samples must be representative of the area and taken to standardized depth, usually 15 cm. Areas that differ significantly in soil type, previous cropping and applications of manure, fertilizer or lime should be considered as separate samples. Small areas that are known to differ significantly from the rest of the field should be excluded from the main samples and tested separately. A minimum of 25 individual subsamples (auger cores) will be adequate for a uniform area.

Table 5.7. Guide to pH values below which crop productivity is reduced.

Crop	Soil pH
Brussels sprout	5.7
Cabbage	5.4
Cauliflower	5.6
Mustard	5.4
Swede	5.4
Turnip	5.4

Subsampling points must be selected systematically and evenly distributed across the area. This is usually achieved by following a 'W' pattern and taking subsamples along the legs of this pattern at regular intervals. Samples should not be collected from the vicinities of gateways, headlands or close to trees and hedges. Fields used for *Brassica* production should be subjected to soil analysis a minimum of once every 3 years and more frequently where the land is cropped several times in one season.

TRACE ELEMENTS

Depending on soil type, soil pH and crop sensitivity, trace element deficiencies can develop and cause significant crop losses. Deficiencies of trace elements have substantial effects on the yield, quality and storability of *Brassica* crops. Cauliflower and swede are susceptible to boron deficiency, especially when grown on light soils with pH values >6.5. Boronated fertilizers should be used as a matter of routine, or applications made to the seedbed or prior to transplanting at 20 kg/ha borax (sodium tetraborate) or 10 kg/ha Solubor™.

Copper deficiency is encountered less frequently, but may develop on soils with a high organic matter content, sands (especially reclaimed heathland or moorland) and humose soils which overlay chalk. This deficiency is corrected by applying copper oxychloride or cuprous oxide at 2 kg/ha plus a wetting agent at either high or low volume sprays, or treating the soil with 60 kg/ha copper sulphate.

Manganese deficiency develops on soils similar to those prone to copper deficiency and is corrected by high or low volume sprays containing 9 kg/ha manganese sulphate plus wetting agent.

Cauliflower crops are especially prone to molybdenum deficiency, causing typical whip-tailing symptoms of the foliage with reduced lamina and a prominent main vein. This condition is associated with acidic soils, hence the soil should be maintained at pH 6.5–7.5. Where treatment is necessary, fields should receive 300 g/ha sodium or ammonium molybdate, or transplants are drenched with 0.25 g/l at the plant propagation stage.

MINIMIZING FERTILIZER RESOURCE USE

Brassica crops seldom utilize all of the nutrients applied, leading to the excess remaining in the soil and potentially available for leaching into groundwater and ultimately causing pollution hazards. Public concern regarding the use of fertilizers is leading to a search for more effective means of application which diminish the quantities used and focuses them into the crop root zone, leading to more efficient utilization without compromising yield or product quality.

Nitrogen is derived from several sources, both natural and artificial. Once nitrogen is in the nitrate form, then it becomes subject to movement in the groundwater and contributes to contamination. The USA standard has a maximum of 10 mg of nitrate nitrogen in drinking water and the European acceptable daily intake of nitrate nitrogen is 3.65 mg/kg body weight (Anon, 1997). All vegetables, and *Brassica* crops in particular, have high financial values, and are intensively managed, requiring substantial inputs of fertilizer, especially nitrogen and irrigation water. The problem of groundwater contamination may be exacerbated where growers set unrealistic yield, as opposed to quality, targets and attempt to realize these by excessive use of fertilizers. Overfertilization contributes to groundwater contamination, through leaching, and is also a wasteful use of resources by the producer. In California, USA, studies of nitrogen balances in vegetable crops as early as 1984 (Pratt, 1984) showed that nitrogen leaching losses ranged from 90 to 260 kg/ha, demonstrating a wide and wasteful range of application rates of fertilizers. Nitrogen applications tend to be made in response to known crop requirements as opposed to deficits established by soil analysis, as used for other nutrient elements. Consequently, this level of disparity between growers' use of nitrogen is not surprising. Environmental considerations are now forcing growers to look for more efficient use of nitrogen to avoid losses by leaching into groundwater and the statutory penalties beginning to be imposed in parts of Western Europe and North America.

Nitrogen forms and sources

Ammoniacal forms of nitrogen have tended to be recommended for *Brassica* crops due to their lower cost and higher soil retention compared with nitrate nitrogen sources. In sandy soils that are frequently used for *Brassica* crops, however, with limited cation exchange capacity, such retention is less likely. There may be opportunities for the use of controlled-release nitrogen sources that have relatively rapid breakdown characteristics. Alternatively, the use of organic fertilizers may become more attractive, especially as municipal organic wastes are increasingly disposed of through green waste composting procedures.

Long-term field trials in Nova Scotia, Canada compared the productivity of several vegetable crops when offered conventional soluble fertilizers and organic composts (Warman, 2005). The productivity of most vegetables was favoured by applications of organic composts. Responses from high nutrient-demanding cole crops, however, depended on seasonal factors, and overall soluble fertilizers provided greater yields. Caution is therefore needed when making recommendations for the organic production of *Brassica* crops. Results such as those obtained in Norway (Riley *et al.*, 2003) that showed white cabbage was favoured by the use of organic surface mulches should be interpreted with care (see also Chapter 6).

Split applications

Traditionally, *Brassica* crops have received part of their nitrogen requirement by means of top dressing applications, particularly with the longer term crops such as overwintered cauliflower (Hochmuth, 1992). Applications of granular fertilizer that are broadcast and then incorporated in the traditional manner initially only enrich a very small volume of soil surrounding the fertilizer granule. With direct-drilled crops in particular, there is therefore a delay before the seedling roots grow into a zone of nutrient-enriched soil. This delay can lead to short-term nutrient deficiencies resulting in poor early growth that may diminish the final yield and increase the time taken to reach maturity.

Traditionally, this problem has been minimized by retaining high residual levels of potassium and phosphate in the soils and applying excess nitrogen rates during cropping that were beyond the requirement for optimal yields. Such strategies aggravate the environmental problems associated with fertilizer use and impose an additional financial penalty on the grower. Split applications result in a less steep decline in soil nitrogen content. They also result in soil nitrogen content being sustained more effectively through to harvest (see Biemond *et al.*, 1995). In white cabbage, split applications of N resulted in increases of total N, nitrate N, P, K, Ca and Mg that then decreased as plant growth increased. Gardner and Roth (1989) applied nitrogen to white cabbage and cauliflower 2–3 times weekly and even here N concentration declined as the plants aged. It decreased from 1.4% at week 7 to 0.5% at week 15 after transplanting. Broccoli (calabrese) crops in Germany are given nitrogen fertilizer in the range of 300–465 kg N/ha. With split applications, however, only 80–118 kg/ha were needed at transplanting provided top dressings were given within 25 days (Feller and Fink, 2005). The nitrogen content of transplants had little effect on growth and yield, and there was no significant interaction between nitrogen content in the transplant and fertilizer timing.

There is a reasonable presumption that the initial increase in nutrient concentration is caused by the high relative growth rate of young roots and shoots and high uptake by young roots. During growth, constituents such as lignin and polysaccharides (cellulose, hemicellulose and pectin) increase. Hara and Sonoda (1981) showed that in young cabbage, concentrations of total soluble carbohydrates and starch increased during early plant growth.

Nutrient conditioning

Stand establishment of direct-seeded brassica crops is adversely affected by high or low temperatures, and high soil and water salt levels, leading to erratic germination, emergence and variable stands. Transplanting allows more efficient use of expensive F_1 hybrid seed, ensures uniform populations at the desired spacing, improves adult plant uniformity and hence results in earlier harvesting. Increasing uniform maturity raises the percentage of the crop that is cut and minimizes the number of harvests. Successful transplant establishment relies on plants surviving the stresses associated with replanting and rapidly resuming growth, often in an hostile environment. Establishment is enhanced by traditional hardening techniques, the use of chemicals that diminish transpiration, application of growth regulators, pruning or defoliation of leaves, mechanical brushing prior to transplanting, and transplant nutrient conditioning.

In the USA, McGrady (1996) showed that cauliflower cv. Snow Crown seedlings could be germinated in a mixture of equal parts of peat and vermicultite with a limited nutrient supply. The macronutrients were then applied as transplant conditioning nutrient treatments. Seedling fresh weight, leaf area and stem diameter increased linearly in response to the main effects of nitrogen and phosphate. Transplanting shock as measured by the number of yellowing leaves increased with the highest levels of applied nitrogen, but they also encouraged maximum subsequent growth. Effects were also genotype dependent since with cv. Olympus, the greatest percentage of heads were cut in a single harvest from the lower nitrogen treatments.

Transplanting shock can be defined as severe necrosis appearing on true leaves and affecting >50% of the area of one leaf. Recovery from transplanting shock is defined as the resumption of growth as indicated by an increase in leaf number and/or stem diameter. Experiments showed that nutrient conditioning of cauliflower transplants was beneficial to plants grown in arid climates such as the Yuma area of Arizona, USA. The technique can also be applied with benefit to crops grown in cool temperate conditions.

Band placement

Band placement of fertilizers for *Brassica* crops significantly increases their efficiency compared with broadcasting (Everaarts, 1993). Placement provides better availability of fertilizer nitrogen with either a higher yield or similar yield for lower resource input compared with broadcasting. The degree of response is dependent on crop type, seasonal conditions and husbandry systems (Everaarts and de Moel, 1995).

In The Netherlands, the recommended nitrogen rate is 300 kg/ha, minus the mineral nitrogen in the soil layer 0–60 cm at planting. A minimum of 50 kg and maximum of 250 (minus N_{min}) kg/ha should be broadcast at planting and 50 kg/ha at 6 weeks after planting. There may be opportunities to reduce this recommendation to 225 kg/ha at least for sandy clay soils. An additional 25 kg/ha might be applied in situations where additional nitrogen could be beneficial such as with early spring crops where soil nitrogen mineralization is slow. In the UK, the nitrogen applications for cauliflower range between 100 and 300 kg/ha (Greenwood *et al.*, 1980) and these should be made in a similar split manner to those in The Netherlands.

In direct-seeded North American broccoli (calabrese) crops (Bracy *et al.*, 1995), the effects of side dressing and pre-planting applications of nitrogen were compared. Comparisons of nitrogen applied at rates between 134 and 258 kg/ha and total nitrogen rates of 179–348 kg/ha as side dressings detected little effect on overall yield. Similarly, no differences between broadcasting and banding fertilizer applications were found in terms of total yield, but substantial economies in nutrient use were achieved.

Starter fertilizer

An alternative is to use small volumes of liquid 'starter' fertilizer applied close to the transplant as it enters the planting station, making it readily available to the roots emerging from the propagation module. Several studies have demonstrated that this technique increases the rate of the early growth phases of crops and is ultimately expressed in additional yield. Such benefits have been achieved even where the soil has a high residual nutrient status or where ample fertilizer has been applied as broadcast granules (Costigan, 1998).

Starter fertilizers usually contain phosphates of ammonia since both ions become strongly adsorbed to soil particles, resulting in little change to the soil pC. Use of ammonium ion encourages roots to excrete hydrogen ions (H^+) in preference to bicarbonate (HCO_3^-) ions, with resultant acidification of the rhizosphere and increased phosphate uptake (Marschner, 1995); application of calcium nitrate would have the reverse effect. The presence of excess ammonium can be deleterious, however, in reducing potassium absorption that will adversely affect the early growth of seedlings. Hence, inclusion of potassium in starter fertilizers is beneficial. There can, however, be seedling damage where the form of potassium also releases chloride ions, thereby raising the pC. Use of liquid potassium phosphate may alleviate the problem (Stone, 1998). Recent evidence suggests that starter fertilizers offer little advantage where the soil residual phosphate and potassium are high, but provide opportunities for maintaining yields where these values are diminished by lowering applications of granular fertilizers. This is in accordance with the aim of reducing fertilizer use to minimize environmental hazards.

Exploiting variations in the efficiency of nutrient use by different genotypes

More recently, one approach that has been examined for minimizing the nitrogen inputs to *Brassica* crops is the use of cultivars which are more efficient in the use of nitrogen. Field trials in The Netherlands and Germany with cauliflower (*B. oleracea* var. *botrytis*) used an optimum nitrogen supply of 250 kg/ha, composed of the inorganic nitrogen content of the soil (N_{min}) and applied fertilizer nitrogen, and a limited nitrogen supply which was solely the N_{min} value. Yield was measured in terms of total dry matter and quality (percentage class 1 curds). The cv. Marine produced both the highest yield and quality and could be regarded as nitrogen efficient, whereas other cultivars were either nitrogen inefficient or behaved inconsistently across sites and seasons. Reducing the supply of nitrogen increased the number of loose curds and is suggested to encourage bolting as well. Some of the nitrogen-inefficient cultivars exhibited buttoning of the curd. Rather *et al.* (1999) concluded that nitrogen-efficient cultivars achieved a higher uptake capacity through greater root activity and/or made more effective use of nitrogen.

MANIPULATING DRILLING OR PLANTING DATES AND CROP GEOMETRY

Increasing plant population densities is a useful technique for raising yield and potential profits in brassicas. For high-density cole crop production to be successful, however, nitrogen applications should increase to accommodate increased nutrient demands. The use of high-density populations has certain disadvantages. Broccoli and cauliflower yields per unit area normally increase with closer planting densities but are associated with smaller head size. While this may increase the numbers of heads and the total yield, maturity is often delayed and the quality is reduced (Salter and James, 1975).

Manipulating resource use by altering drilling or planting dates is exemplified by the work of Siomos (1999) (Tables 5.8, 5.9 and 5.10) who studied Pak Choi (*B. rapa* L. Chinensis group) grown under unheated plastic in Greece, in three periods: December/January; January/March; and March/May. Only when the temperatures and light incidence were high (period 3) did increased plant density (ten plants increased to 16.7 plants/m^2) raise yield; here individual fresh weight fell but the total number of plants increased. Changing the planting date and within-row plant spacing had little effect on dry matter, total soluble solids and fibre content.

Nutrient requirements alter with changes to crop density and spacing. In the southeastern USA, vegetables are usually planted on raised beds (Parish, 2000), with either single or double rows to each bed. Double rows offer higher yields per unit area, but may be difficult to maintain physically because of the erosion of the sides of the bed caused by localized heavy rainfall. Beds also provide advantages of quicker and earlier soil warming and allow the use of mechanically guided cultivation such as steerage hoeing. In some areas, beds promote the avoidance of soil-borne pathogens such as *P. brassicae*, the causal agent of clubroot disease, because the soils are drier and

Table 5.8. Seeding, transplanting and harvesting dates for Pak Choi cv. Troy F.

				Days from		
Growing period	Sowing date	Trans-planting date	Harvesting date	Sowing to trans-planting	Trans-planting to harvesting	Sowing to harvesting
1	25 Oct 1993	3 Dec 1993	27 Jan 1994	39	55	94
2	23 Dec 1993	28 Jan 1994	28 Mar 1994	36	59	95
3	25 Feb 1994	30 Mar 1994	14 May 1994	33	45	78

Results for Greece were similar to those from The Netherlands and Australia for glasshouse/polythene-grown Pak Choi; after May, the crops tend to bolt under protection.
After Siomos (1999).

Table 5.9. Effect of the growing period and plant spacing on yield (kg/m^2) of Pak Choi cv. Troy F_1.

	Growing period			
Plant spacing (cm)	1	2	3	Mean
15 × 40	5.14cd	5.63c	12.66a	7.81A
25 × 40	4.02d	5.71c	9.29b	6.30B
Mean	4.58C	5.67B	10.98A	

Means followed by different letters are significantly different at the 0.05 level of probability (Duncan's multiple range test).
Comparisons between means are differentiated by upper-case letters.
After Siomos (1999).

this inhibits the movement of primary zoospores towards the host root hairs. In areas of moderate rainfall, the bed structure is retained and greatest yields came from multiple row plots.

Beds are especially useful for rapidly maturing brassicas such as the leafy greens which have become popular for both processing and fresh markets, such as mustard (*B. juncea*), turnip (*B. rapa*) and collard (*B. oleracea* Acephala group). Growing six rows on 2 m wide beds proves very effective, producing higher yields compared with fewer rows on narrower beds.

In cauliflower and broccoli crops, results show that with increasing rectangularity of spacing, i.e. between-row spacing divided by within-row spacing, crop yields are decreased (Chung, 1982). This indicates that it is more advantageous to grow crops in a square formation rather than in a rectangular pattern (Salter *et al.*, 1984; Sutherland *et al.*, 1989). Modification of plant population densities is used to control cauliflower curd weight. A number of studies demonstrate that curd weight decreases with increasing plant density (Dufault and Waters, 1985; Singh and Naik, 1991).

Commercially, spacing varies substantially depending on location, genotype and husbandry systems. For example, in Europe, summer cultivars require

Table 5.10. Effect of growing period and plant spacing on mean daily growth increments (g/m²/day) of Pak Choi cv. Troy F_1.

	Growing period			
Plant spacing (cm)	1	2	3	Mean
15 × 40	93.4c	95.4c	281a	156.6A
25 × 40	73.0c	96.8c	206.5b	125.5B
Mean	89.2B	96.1B	243.8A	

Means followed by different letters are significantly different at the 0.05 level of probability (Duncan's multiple range test).
Comparisons between means are differentiated by upper-case letters.
After Siomos (1999).

much smaller spacing compared with overwintered types. In Western Australia, wider spacing is the norm, thus between-row spacing of 0.75–0.80 m and, for most cultivars, a within-row spacing of 0.40–0.50 m and two rows of cauliflower per planting bed are used. Resultant curd size varies in the range from 0.5 to 2.0 kg. Recent field experiments in Western Australia (Stirling and Lancaster, 2005) demonstrated that plants grown in a four-row configuration produced significantly ($P = 0.007$) higher total yields than control plants grown in a two-row configuration (Fig. 5.1). Within the four-row configuration, a significant ($P = 0.019$) linear trend was observed, with yield falling by 0.3 t/ha for every 0.01 m increase in within-row plant spacing.

Uniformity of curd maturation improved when the number of plant rows per bed was increased from two to four (Table 5.11). The majority of curds from plants grown in four rows were removed in the first two harvests, with only a small proportion of curds remaining at the final harvest. An increase in the uniformity of mature curds was identified in plants grown in four rows, spaced at 0.40, 0.45 and 0.50 m.

Curd weight decreased significantly ($P < 0.001$) when the planting configuration was altered from two to four rows (Table 5.12). Within the four-row configuration, there was a significant ($P = 0.011$) linear effect of plant spacing on curd weight, which decreased with increasing plant density. Average curd weight decreased by 4.1 g for every 0.01 m decrease in plant spacing. There was a significant ($P = 0.003$) decrease in the average diameter of all curds harvested per treatment when plants were grown in four rows compared with two (Table 5.12). Uniformity of curd maturity improved when the row number per bed was increased from two to four. This is an important consideration for cauliflower producers as it has a major influence on variable costs. Crops that mature in unison require fewer harvests, thereby substantially reducing labour and machinery costs.

In Minnesota, USA, as cauliflower populations were increased from 24,000 to 72,000 plants/ha with nitrogen rates held constant at either 112

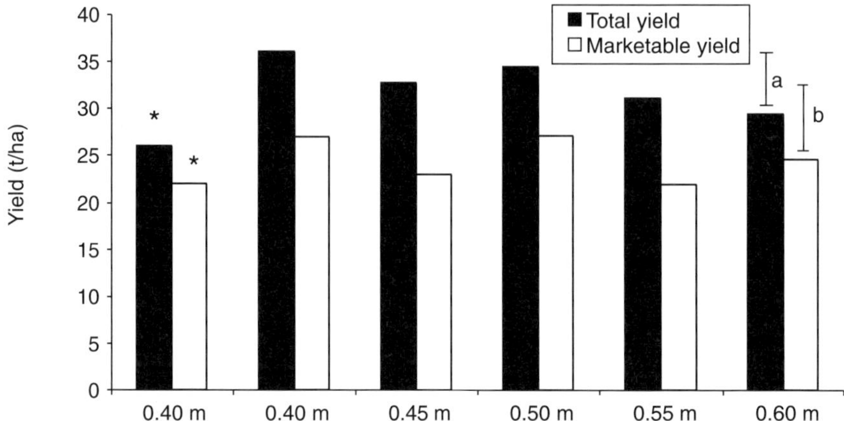

Fig. 5.1. Illustration of total and marketable yield of cauliflower (*Brassica oleracea* var. *botrytis*) cv. Summer Love produced by plants spaced at 0.40, 0.45, 0.50, 0.55 and 0.60 m. *Two row treatment data. Bars indicate the least significant difference between all treatments (5%) = 5.2 (a) and 7.0 (b) (Stirling and Lancaster, 2005).

Table 5.11. Percentages of cauliflower (*Brassica oleracea* var. *botrytis*) curds cut at several harvest dates relative to plant spacing.

Number of rows per 1.6 m bed	Spacing between plants within row (m)	Harvest 1 (%)	Harvest 2 (%)	Harvest 3 (%)	Total (%)
2	0.40	10.79	74.59	14.62	100.00
4	0.40	31.97	65.24	2.78	100.00
4	0.45	34.74	61.72	3.54	100.00
4	0.50	24.53	73.06	2.41	100.00
4	0.55	22.81	69.88	7.31	100.00
4	0.60	25.32	67.63	7.04	100.00
LSD between all treatments (5%)		10.87	15.16	14.46	
LSD between 4-row treatments only		12.16	11.88	11.21	

LSD = Least significant difference.
From Stirling and Lancaster (2005).

or 224 kg N/ha, marketable curd weight decreased in a linear manner at any population. Increasing the nitrogen rate to 112 kg/ha or higher, reduced cull production at 24,000 plants/ha, but not at populations of 36,000 or higher. Cauliflower yields were optimal at 24,000 plants/ha and 112 kg N/ha based on considerations such as reduced cull production, satisfactory curd weights and transplant economy (Dufault and Waters, 1985).

Studies in Minnesota, USA, showed that as broccoli cv. Southern Comet (*B. oleracea* var. *italica*) populations were increased from 24,000 to 72,000

Table 5.12. Average cauliflower (*Brassica oleracea* var. *botrytis*) marketable curd weight (g) and diameter (cm), related to plant spacing.

Number of rows per 1.6 m bed	Spacing between plants within row (m)	Average curd weight (g)	Average curd diameter (cm)
2	0.40	887.71	15.64
4	0.40	723.63	14.92
4	0.45	733.44	14.78
4	0.50	784.76	15.17
4	0.55	750.66	15.00
4	0.60	816.35	15.21
LSD between all treatments (5%)		69.74	0.40
LSD between 4-row treatments only		67.4	0.42

LSD = Least significant difference.
From Stirling and Lancaster (2005).

plants/ha at nitrogen rates of 112, 168 or 224 kg/ha, head weight decreased linearly. Increasing the nitrogen rate from 56 to 224 kg/ha at any population increased broccoli head weight and marketable yields, and decreased cull yields. Broccoli yields were highest at 72,000 plants/ha and 224 kg N/ha.

MODELLING NUTRIENT NEED

A significant barrier to more efficient use of nitrogen fertilizer by *Brassica* growers is a lack of information on the seasonal, soil-related and cultural variations in the supply of mineral nitrogen from the soil and the requirements for nutrients by the crop. Considerable information has been accumulated for the WELL_N model in the UK in an attempt to remedy these deficiencies.

WELL_N is a user-friendly computer program, conceived by Warwick University-Horticulture Research International at Wellesbourne. It provides fertilizer recommendations and management advice on the use of nitrogen for a range of brassicas, and tailors recommendations for different weather conditions, soil factors and cultural practices at each site. The package has been compared with paper-based systems of recommendation such as the DEFRA (Department for Environment, Food and Rural Affairs, London) Reference Book 209 (*Fertilizer Recommendations for Agricultural and Horticultural Crops*) or those testing midrib sap (Gardener and Roth, 1989).

The electronic model allows users to select manually a particular model suited to their conditions. From that point, the program has the potential to set up automatically the optimum model for the user under particular field conditions. In field trials, a substantial database of measurements was built up from farm-scale experiments that were run over two seasons on 37 sites across the UK with a range of brassica crops. The database was constructed to store the descriptions and results of field trials and make it readily available for future research and development uses. Evaluation at the *Brassica* cropping sites indicated that the fertilizer recommendations from WELL_N and Reference Book 209 resulted in closely similar yields. WELL_N gave more accurate recommendations, however, than Reference Book 209 or farm practice, thereby reducing fertilizer costs and wastage to the potential benefit of the environment and the producer. Models similar to WELL_N have been developed in other European countries and in Canada.

IRRIGATION AND WATER USE

Efficient water management is a prerequisite to nitrogen management. Nitrate nitrogen leaching can be minimized by matching irrigation applications to evapo-transpiration (ET) need. In cauliflower, for example, water is needed throughout the crop life, but is most effective at the onset of curd formation (Salter, 1961; Wiebe, 1981). Improved product quality is greater on soils with higher water-holding capacity, but soil type has less effect than nitrogen fertilizer (Nilsson, 1980). Increasing nitrogen from 150 to 300 kg N/ha significantly increased yield. Yields increased up to 500 kg N/ha. Polish growers add nitrogen up to 500 kg N/ha with irrigation (probably accompanied by huge leaching losses into groundwaters and associated serious pollution), and this correlated with increasing nitrate nitrogen in the curds. At these levels of application, there was a linear increase in nitrate nitrogen in cauliflower leaves and curds (Kaniszewski and Rumpel, 1998).

Cabbage is intermediately susceptible to water stress, with the head formation stage more sensitive than the preceding growth periods (Smittle, 1994). Critical periods for water stress are in the 3–4 weeks before harvest. Yields of vegetable crops, including cabbage, are reduced when soil water tension is >25 kPa. Crops irrigated when the soil moisture tension is <25 kPa at 10 cm produced the highest total and marketable yields. This regime requires more water to be applied, but the water use efficiency rate is similar to that for cabbage irrigated at 50 and 75 kPa.

Several methods exist for measuring ET from climatic data. The modified Penman and the Jensen–Haise methods use combinations of solar radiation, temperature, humidity, wind velocity and vapour pressure measurement to estimate ET for a reference crop. This then requires a crop coefficient to adjust the values obtained for the reference crop to estimate the ET of the crop to be irrigated. The crop coefficient values (ET of the irrigated crop/ET of the reference crop) are multiplied by the ET values estimated by the specific method to estimate the ET of the irrigated crop.

Pan evaporation (E_p) incorporates the climatic factors influencing ET in a single measurement and has been used to schedule irrigation for several crops. The single crop factor value (ET/E_p) usually results in applications of excessive irrigation during some growth phases and water deficits in others. A generalized curve was developed to describe crop factor value changes during crop growth, but the generalized curve lacks precision. Smittle (1994) have developed regression equations to calculate the daily crop factor values during the growth of several vegetables and have incorporated these equations to estimate ET from E_p data into irrigation scheduling models.

Grieve *et al.* (2001) determined the effect of salinity and timing of water stress on leaf ion concentration in the field. Using Pak Choi (*B. rapa* L. Chinensis group), Tatsoi (*B. rapa* L. Narinosa group), kale (*B. oleracea* L. Acephala group), cooking greens (*B. rapa* L.) and mustard greens (*B. juncea* L., Czerniak), the experiment used saline solutions to simulate the high-sodium and high-sulphate drainage waters typical of the San Joaquin Valley of California, USA. The mineral ion concentrations in leaves were significantly affected by increasing salinity of irrigation water. The stage at which salinity was applied, however, had little effect. With increasing salinity, calcium ions and potassium ions decreased in the leaves of all species, whereas sodium ions and total sulphur significantly increased. Magnesium also rose in the leaves of brassicas with increasing salinity; there was also an increase in chloride ion content. The use of moderately saline irrigation water did not adversely affect crop quality as rated by colour, texture and the mineral nutrient content available to consumers.

In many areas of the world, water is now the most valuable and scarce resource, and this shortage is set to become more acute. Areas of intensive vegetable production include parts of: the USA, i.e. California and Florida; southern Europe, i.e. southern Spain, Portugal, Italy and Greece; and the Middle East, i.e. Israel and Egypt. In each of these areas, irrigation is an essential ingredient for *Brassica* crop production. The yield response gained by the use of irrigation exceeds 200%. Water used for irrigation on crops, parks and golf courses accounts for 80% of consumption in the USA for example. Irrigated vegetable production accounts for 1.9 Mha (7.5% of the irrigated area), and Arizona, California, Florida, Idaho, Nebraska, Oregon, Texas, Washington State and Wisconsin account for 80% of this production.

Originally, irrigation was supplied through surface and seepage systems, and these are currently employed on 45% of crops with a water use efficiency of 33%. Sprinkler or overhead systems were developed in the 1940s. At present, they are used on 50% of irrigated land with a water use efficiency of 75%. Since the late 1960s, microirrigation using drip or trickle methods has been developed. Currently these are used on only 5% of US field vegetables, but this is likely to increase substantially as their water use efficiency is between 90 and 95% (Locascio, 2005). As water becomes increasingly scarcer and more expensive, methods that increase the efficiency of its use will become ever more important.

Worldwide, growers using water for irrigation will be forced to confront the problems of decreased supply and increasing salinity in their irrigation water and the implications of this for the quality of vegetables made available to the consumer. The consumer is becoming ever more conscious of the health risks associated with nitrate nitrogen loading, especially in leafy vegetables of which brassicas comprise a major portion. One of the on-farm management options is the re-use of agricultural drainage effluents. This strategy is especially attractive because significant amounts of good quality water are preserved and also because the volumes of drainage water that ultimately need to be disposed of are substantially reduced (Sorenson, 2000).

In the suggested drainage water re-use system proposed for the San Joaquin Valley of California, USA, selected crops would be grown and irrigated in sequence, starting with the very salt-intolerant species (V. Rubatsky, personal communication, San Joaquin Valley Drainage Program, 1990). Drainage effluents from these crops would be used to irrigate crops of higher salt tolerance. At each step in the sequence, the drainage water becomes progressively more salty. The composition of the drainage effluent waters in this region is typically a mixture of salts with sodium > sulphate > chloride > magnesium > calcium predominating in that order.

Members of the Brassicaceae are relatively tolerant of salinity (see Chapter 1). Even with these crops, however, abnormally high levels of salinity severely limit plant growth. It is essential that brassicas are able to form large vigorous root systems. This requires sufficient oxygen in parallel with adequate water supply. Mylavarapu *et al.* (2005) identified the importance of subsoiling which allows adequate soil aeration for collard (*B. oleracea* var. *acephala*) in the southeastern states of the USA.

Leafy vegetables are the primary source of mineral nutrients for human diets. Numerous tables of food composition list the major constituents of vegetables and give values for predominant mineral ions (see Table 5.13). These values are only estimates in so far as the data are based on a limited number of samples and vary due to biological and environmental factors such as maturity, analytical procedures and processing. In addition, the availability, uptake and partitioning of mineral ions within the plant are controlled by numerous environmental factors including the concentration and composition of solutes in the soil solution. Under saline conditions, mineral ion interactions in the external growing medium may affect the internal requirements of elements essential for plant growth and development. These imbalances often influence the growth and nutrition of the crop, which in turn may affect crop quality in terms of colour, texture and nutritive value.

Calcium plays a vital nutritional and physiological role in plant metabolism (see Chapter 8). Under saline conditions, ion imbalances in the substrate or plant may adversely affect calcium nutrition. Substrate levels of calcium that are adequate for plant requirements under non-saline conditions may be nutritionally inadequate and growth limiting when the plant is salt stressed. The calcium status of the plant is strongly influenced by the ionic composition of the external medium in that the presence of other

Table 5.13. Water content and mineral ion concentration (mg/100 g edible portion) of selected *Brassica* crops irrigated with saline water (11 d/Sm) compared with non-saline conditions.[a]

Vegetable	Water (%)	Ca	Mg	Na	K	P	S	Cl	Reference
Tatsoi	93.7	123	26	236	323	24	63	208	1
Mustard greens									
Vitamin	95.0	106	20	220	262	23	48	490	1
Red Giant	94.1	116	24	61	419	23	52	118	1
	89.3	181	29	33	374	46			2
	92.6	138		18	220	32			3
Kale	89.0	306	91	222	476	53	175	246	1
	89.7	202	45	43	420	55			2
	91.2	134		43	221	46			3
Pak Choi	94.0	118	24	248	426	24	53	192	1
	93.8	118	19	71	234	39			2

[a]Data from crops grown with saline water are underlined.
1 = Grieve *et al.* (2001); 2 = Rubatsky and Yamaguchi (1997); 3 = Ensminger *et al.* (1995).

salinizing ions in the substrate may reduce calcium activity and limit the availability of calcium to the plant. Cations such as sodium and magnesium may disrupt calcium acquisition, uptake and transport. Leaf calcium concentrations in all vegetables tended to decrease as salinity increased despite added quantities being available in the soil. Salinity-induced calcium deficiency can result in physiological disorders in brassica vegetables such as internal tipburn, browning and necrosis of the inner leaves (Chapter 8).

Under conditions of salinity stress, maintenance of adequate levels of potassium is essential for plant survival. High levels of external sodium not only interfere with potassium acquisition through the roots but may also disrupt the integrity of root membranes and alter the selectivity of the root system for potassium as compared with sodium.

Total sulphur content in the leaves of all brassica vegetables increased as sulphate values rose. The brassicas are particularly active sulphur accumulators. Members of the Brassicaceae are among the 15 plant families which biosynthesize significant quantities of sulphur-rich glucosinolates. Hydrolysis of these compounds yields 'mustard oils' that impart the characteristic spicy tastes to these vegetables. Increases in total sulphur accumulation in response to irrigation with moderately saline, sulphur-dominated waters may enhance the flavour, provide benefits to human health and raise the acceptability of brassica vegetables to the consumer.

Competitive Ecology and Sustainable Production

INTRODUCTION

Until very recently, the agronomist's view of the relationship between crops and weeds was dominated by the desire to find the most effective herbicide with which to kill an offending competitor plant. This philosophy has changed radically over the past decade. Pressures coming largely from the final link in the chain from field to plate, i.e. those who eat *Brassica* vegetables, now require producers to sustain and protect the environment in addition to providing wholesome foods. Reducing or eliminating the use of herbicides demands new attitudes to weed control using wider ranges of methods that prevent competition from weeds. This encompasses the strategy of 'integrated crop management' (ICM), which links all other aspects of crop husbandry discussed in this book. In ICM, consideration is primarily directed at determining the costs and benefits of whether or not weeds cause sufficient crop losses to make it worthwhile eradicating them. If the answer to that question is 'yes', then consideration moves to the opportunities for cultural and husbandry weed-controlling methods in preference to spraying herbicidal poisons. This approach will become ever more dominant over the coming decades such that agronomists and ecologists will need to work as a team growing economically successful *Brassica* crops with very limited, if any, use of herbicidal chemicals. Consequently, this chapter treats the control of weed competition in *Brassica* crops specifically from this ecological (commonly termed 'sustainable') perspective and a similar approach is adopted in Chapter 7 dealing with pests and pathogens.

WHAT ARE 'WEEDS'?

An ecologist defines weeds as: 'plants growing entirely or predominantly in situations (that are) disturbed by man without being deliberately cultivated'

(Baker, 1965). This definition includes all plant types, not solely the flower-forming angiosperms, although these comprise the great majority of weed taxa. The world weed flora mainly contains plants from relatively few, highly advanced families (Table 6.1) (Hill, 1977) which thrive in the very fertile soils developed for crop production. Such plants are opportunistic and have evolved to exploit the crop environments created by man which differ markedly from natural ecosystems (Fogg, 1975).

THE EFFECTS OF WEEDS COMPETITION

The agronomist defines weeds as 'plants which interfere adversely with the production aims of the grower' (Spitters, 1990). Weeds affect crops by:

- reducing crop growth and yield, mainly due to competition that limits resources such as light, water and nutrients;
- reducing the financial value of the harvested product, either by contaminating the produce or by diminishing its quality; and
- hampering husbandry practices, especially harvesting operations, thereby increasing costs.

Weeds reduce the financial profits either directly by lowering output (kg yield per ha × price per kg), or indirectly either by reducing the market price achieved through the diminution of quality attributes or by increasing husbandry and harvesting costs. Each of these penalties is imposed on *Brassica* crops by weeds. Increasingly, it is their effects on crop quality that are of prime consideration. These effects and their interactions are summarized in Table 6.2.

Factors that interact influencing the degree of mutual interference (competition) between weeds and crops are as follows.

Table 6.1. List of angiosperm families whose species account for the majority of the 700 weeds introduced into eastern North America.

Family	Number of species
Compositae	122
Graminae	65
Cruciferae	62
Labiatae	60
Leguminosae	54
Caryophyllaceae	37
Scrophulariaceae	30

After Hill (1977).

Table 6.2. The effects of competition by an aggressor (A) species on a suppressed (B) species.

Effects	Competition for light only	Competition for nutrients only	Competition for both light and nutrients
Direct	(a) Intrusion of A into the light environment of B: reduced light supply for B	(c) Intrusion of A into the nutrient supply of B: reduced nutrient supply for B	B suffers: (a) reduced light supply (c) reduced nutrient supply
Indirect	(b) As a result of reduced light supply: B has reduced capacity to exploit its own nutrient supply	(d) As a result of reduced nutrient supply: B has reduced capacity to exploit its own light supply	(b) reduced capacity to exploit the nutrient supply (d) reduced capacity to exploit the light supply
Interactions	Interaction of (a) and (b)	Interactions of (c) and (d)	Interactions of ab, ac, bc, bd and cd plus any higher order interactions

Source: Donald (1958).

- *Crop factors:* type of crop, time of sowing and/or transplanting, time of thinning, and population density which is a function of seed rate, germination and thinning.
- *Weed factors:* species of weed(s), time of germination, start of growth and population density controlled by seed population in the soil, percentage germination and seed importation from other areas.

The level of competition resulting from these crop and weed factors will be modified further by weather conditions, soil type, soil fertility and cultivation practices.

IDEAL CHARACTERISTICS FOR A SUCCESSFUL WEED

The biological attributes favouring success as a weed are summarized in Table 6.3. An open growth habit combined with the early induction of flowering means that small plants reproduce rapidly by seed and continue growing as further flowering takes place. This ensures continuing outputs of seeds. Plants growing in disturbed (cultivated) habitats tend to allocate greater proportions of their photosynthetic output to seed production than the more stable types, as demonstrated in Table 6.4.

Table 6.3. Characteristics of a successful weed.

1. Germination requirements fulfilled by many environments
2. Discontinuous germination (internally controlled) and great longevity of seed
3. Rapid growth through the vegetative to the flowering phase
4. Continuous seed production during the entire growing season
5. Self-compatible breeding system but not completely autogamous or apomictic
6. Cross-pollinated by unspecialized visitors or wind
7. Very high seed output in favourable environmental circumstances
8. Produces at least modest seed yield in a wide range of adverse environments
9. Adapted for short- and long-distance seed dispersal
10. Where the life cycle is perennial, the plant has vigorous vegetative reproduction or regeneration from stem or root fragments
11. Perennials have brittle stems or roots that are not easily drawn out of the ground
12. Abilities to compete with other organisms by specialized means such as rosette habit, choking growth or the production of allelochemicals

Table 6.4. Average output of seeds or fruits for ten common weeds that compete with crops.

Weed species	Average output per plant	Seed (S) or fruit (F)
Groundsel (*Senecio vulgaris*)	1,000–1,200	F
Common chickweed (*Stellaria media*)	2,200–2,700	S
Shepherd's purse (*Capsella bursa-pastoris*)	3,500–4,000	S
Greater plantain (*Plantago major*)[a]	13,000–15,000	S
Common poppy (*Papaver rhoeas*)	14,000–19,500	S
Prickly sowthistle (*Sonchus asper*)	21,500–25,000	F
Perforate St John's Wort (*Hypericum perforatum*)[a]	26,000–34,000	S
Canadian fleabane (*Conyza canadensis*)[a]	38,000–60,000	F
Hard rush (*Juncus inflexus*)[a]	200,000–234,000	S

[a]Perennial. After E. Salisbury (1961).

The competitive ability of weeds results from a combination of high reproductive potential and rapid vegetative spread. This embraces physiological (efficient ion uptake and rapid root growth) and morphological factors (ability to climb, scramble over competitors for light; produces rosettes which spread out close to the ground and smother competitors; and large and vigorous habit). The *Brassica* crop, especially where it is transplanted, may even improve conditions for the establishment of weeds after their germination by offering protection from extremes of temperature and the drying effects of wind. This provides a competitive advantage for some weed species by promoting an early start to their growth.

CRITICAL PERIODS OF COMPETITION

There are critical periods of growth when weeds provide competition with crops and other times when they are not exerting depressing effects on yield (Nieto *et al.*, 1968). Soils used for *Brassica* production usually contain more than 2000 viable seeds of annual weeds/m^2 (20 million/ha). Only a small proportion of these germinate into seedlings at any one time, but this can still lead to populations of 100–300 weed plants/m^2 competing with the crop for resources. Yield losses vary according to the habit of the predominant weed species and the crop growth stage at which they begin competing with each other for resources. In drilled summer cabbage (*B. oleracea* var. *capitata*), for example, losses varied from 25 to 100% with a weed density of 100 plants/m^2. The scale of losses depended on the weed species present and the degree to which environmental conditions favoured either the crop or its competitors (Roberts *et al.*, 1976; Röhrig and Stützel, 2001).

In vegetable crops, the concept of 'critical periods' is highly developed to indicate times during which any competition from weeds will cause irreversible yield loss. A 'critical period' of weed competition is defined as 'the minimum time over which weeds must be suppressed to prevent yield loss'. There are two separate components to such critical periods: (i) the time a crop must be weed free after drilling or planting so that emerging weeds do not reduce yields; and (ii) the subsequent time when weeds which emerge within the crop can remain before they begin to interfere with crop growth and productivity.

These factors define the optimum timing for weed removal to prevent yield loss rather than stopping intense competition. Use of this concept is limited to timing of cultivation and other measures, and is dictated by crop tolerance and the stages susceptible to weed growth. The critical period varies with crop type and cultivar, location, spectrum of weed species and their density in any particular season (Weaver, 1984). In cabbage (*B. oleracea* var. *capitata*), the length of the weed-free period, weed density and ambient light reduction caused by competing weeds were important factors governing yield loss (Miller and Hopen, 1991). Maintaining the plants free from weed competition in the first 4 weeks after sowing was crucial for the establishment of yield and quality (Table 6.5).

The concept of critical periods can be criticized for failing to take into account the influence of the full range of ecological and agronomic factors specifically related to a particular site. These simulation models of competition tend to be generally empirical regression models based on one or more parameters such as weed density, relative leaf area or relative time of weed emergence. These parameters vary widely between sites and over seasons. Such models also tend to lack physiological values dependent on the interaction of the crop and its surrounding weeds competing for resources.

Table 6.5. Critical periods of weed competition in cabbage (*Brassica oleracea* var. *capitata*), related to yield.

Weeks weed-free after emergence[a]	Cabbage yield kg/m^2
14	10.39
6	10.19
4	10.24
2	3.16
1	3.30
0	0.86
LSD (0.05)	1.52

After Miller and Hopen (1991).
[a]Weed = velvetleaf (*Abutilon theophrasti*).
LSD = least significant difference.

This competition is process oriented and defined as the distribution of growth-limiting factors between species in a vegetation canopy and the efficiency with which each species uses resources for growth (Spitters, 1990).

The resources for which the crop and weed compete are radiation, nutrients and water. The rate of use of these resources depends on the structure and growth pattern of crop plants and interactions with the weeds. Cauliflower (*B. oleracea* var. *botrytis*), for instance, has a relatively constant resource allocation pattern throughout the vegetative and reproductive growth stages and does not respond to competition for radiation by significantly changing dry matter distribution. Since it is a relatively tall plant susceptible to damage from wind and heavy rainfall, it requires support from an appropriately sized stem which is responsible for plant stability. The lack of side shoots and branches in cauliflower ensures that the size of leaves primarily determines the lateral expansion of the plant.

Radiation is the most crucial resource for which there is competition between *Brassica* crops and weeds. Crops are usually planted in widely spaced rectangular patterns, thus developing ecologically as single plants rather than 'closed' or row canopies. The influence of weeds on light availability in direct-sown cabbage (*B. oleracea* var. *capitata*) is shown in Table 6.6.

Good husbandry ensures that ample irrigation and fertilizer are provided so that water and nutrients are not limiting either for the crop or for competing weeds. Hence the spatial development of competing plants and their morphological adaptation to unfavourable growing conditions strongly influence the distribution of radiation within the canopy.

The growth of spring- and summer-planted cauliflower as affected by the weed *Chenopodium album* (fat hen) was studied by Röhrig and Stützel (2001). It is evident that this weed had a much greater impact on crops planted later in the season. This is due to higher temperatures and the accelerating growth rate of the weed in summer.

Table 6.6. Effect of weed growth on light availability in direct-sown cabbage (*Brassica oleracea* var. *capitata*).

Treatment[a]	Cabbage yield kg/m^2	Light availability to cabbage as a percentage of full sunlight
Weed – free control	9.84	100.0
Weed – free 6 weeks	9.50	97.5
Weed – free 4 weeks	9.35	100.0
Weed – free 2 weeks	3.24	57.1
Weed – free 1 week	0.72	15.8
Unweeded 2 weeks	9.70	100.0
Unweeded 4 weeks	7.06	97.5
Unweeded 6 weeks	1.78	93.8
Unweeded 14 weeks	0	5.8
LSD (0.05)	1.35	14.1

After Miller and Hopen (1991).
[a]Weed = velvetleaf (*Abutilon theophrasti*).
LSD = least significant difference.

This encourages more vigorous growth of the weed and an increasing capacity to shade the cauliflower, thereby reducing radiation use efficiency in the crop. When this weed was growing at high density, it substantially reduced cauliflower total dry weight, leaf area index and curd diameter; crop height was reduced more in the late summer crops compared with earlier maturing types.

With spring and summer *Brassica* crops grown from direct seeding or using transplants, provided the weeds are removed effectively in a single weeding, then there are few adverse effects of competition. The crucial period for weeding *Brassica* crops is in the earliest stages of establishment and as subsequent growth develops. The crop canopy closes soon after weeding and prevents further competition. *Brassica* crops, however, show significant effects of row width and weed removal treatments on yield and quality. Yields tend to rise as row spacing is reduced, up to critical values, and thereafter they diminish. Weed removal becomes essential once competition between the individual crop plants and weed species commences after planting or drilling. Provided weeds are removed before this critical time, usually by a single weeding operation (using either mechanical or herbicidal methods), then yield and quality losses are prevented. This requires the use of herbicides applied pre- or post-planting, or mechanical methods to remove the developing competition. The period when *Brassica* crops are establishing is especially crucial in preventing the development of weed competition. This is because there are few, if any, environmentally acceptable herbicides that can be applied to established stands of *Brassica* crops. Single, thorough mechanical weed removal must normally occur between 2 and 3 weeks after transplanting, depending on the crop row spacing employed. Weed competition commences more quickly at narrow row spacings compared with wider ones.

Weeds have limited effects in *Brassica* crops grown during the autumn and winter provided they are removed before active crop growth recommences in spring. Those weeds remaining into the spring compete severely with the crop, resulting in smaller marketable cabbage (*B. oleracea* var. *capitata*) heads, reducing internal head quality, decreasing numbers of plants forming heads and in very severe cases weeds can cause death of the crop plants (Lawson, 1972). Particularly successful weed species in Lawson's northern European (Scottish) studies included chickweed (*Stellaria media*), annual meadow grass (*Poa annua*), shepherd's purse (*Capsella bursa-pastoris*) and knotgrass (*Polygonum* spp.). A major factor in the competitive fitness of these weeds was an ability to grow rapidly in spring, outpacing and overtopping the crop and causing shading effects. Weed competition was analogous to increasing the density of a crop planting up to a point where cabbage population densities became so high that normal development was prevented. Conversely, comparison of cropped and uncropped plots indicated that the crop could exert considerable competitive pressure on the growth and development of weeds, particularly where their capacity for expansion had been retarded by treatment with herbicides.

The management decision of whether or not weed control is economically worthwhile depends on a costs and benefits analysis. This identifies that at a given level of weed infestation, crop yield and/or quality are likely to be damaged if the weeds remain uncontrolled. Models of weed and crop competition are an essential part of both short- and long-term crop management planning.

The most favoured model that describes the relationship between weed infestation and crop yield loss is a rectangular hyperbolic curve (Cousens, 1985) (Fig. 6.1). Many discussions of the relationship between crop yield and weed densities have used either sigmoidal or quasi-sigmoidal models meaning that there is a threshold weed density below which there is no yield loss. In this relationship, there is virtually no competition between weed and crop at low weed density despite the plant size being at a maximum. This ignores the fact that the competitiveness of each individual weed plant is greatest when it has achieved maximum size and is exploiting nutrient and water resources from the soil. Support for the use of a sigmoidal relationship is based on the observation that at very low weed densities, statistically significant reductions of yield are difficult to demonstrate in crops such as cereals. In vegetables such as the brassicas, however, losses may result from the damage to product quality that is difficult to quantify by using field experiments. Also, experimental designs are often incapable of detecting differences in yield of the magnitude expected at low weed densities; this does not mean that such differences do not exist and, especially with *Brassica* vegetable crops, exert an indirect influence on their profitability. A hyperbolic model of the relationships between weed density and yield loss indicates, however, that wherever weeds are present in a crop there will be competition for resources such as nutrients and water. This occurs even when there is very limited or no shading effect reducing the light available to the crop plant. The hyperbolic model assumes that weeds are distributed at random in relation to crop plants. Consequently, as weed density increases, the mean distance from the weed to crop plant remains constant.

Several practical studies emphasize the influence of cultivar selection, row width and planting date on weed competition for *Brassica* crops. These vary substantially in husbandry systems using a wide range of crop spacings, population densities and methods of establishment (direct seeding versus transplanting). Efficient weed management is essential because of the high intrinsic values of the products, limited availability of herbicides and high labour costs.

EFFECTS OF WEED TYPE

The morphological habit of a weed will substantially affect its impact on the growth and yield of the crop. For tall-growing weeds such as *C. album* (fat hen), there is a linear relationship between the extent of yield reduction and weed

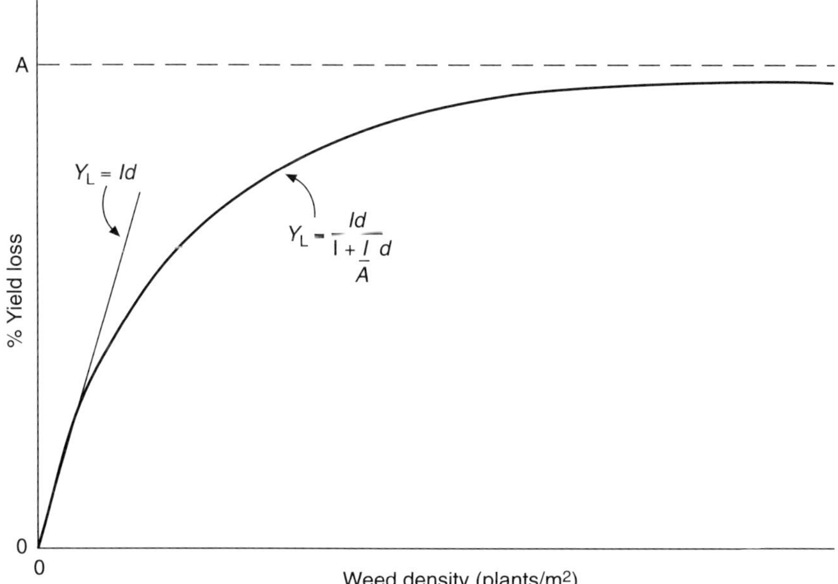

Fig. 6.1. The rectangular hyperbolic model for relating yield loss to weed density, illustrating its parameters A and I. Y_L = percentage of yield lost because of weed competition; d = weed density; I = percentage of yield loss per unit weed density as $d_{\rightarrow} 0$; A = percentage of yield loss as $d_{\rightarrow} \infty$. After Cousens (1985).

density; as few as 3 plants/m^2 were sufficient to cause statistically significant reductions to the yield of cabbage (*B. oleracea* var. *capitata*) crops. Low growing weeds such as *S. media* (chickweed), *Poa annua* (annual meadow grass) and *Urtica urens* (nettle) caused smaller reductions in cabbage yield. Detailed studies by Röhrig and Stützel (2001) demonstrate the effects of three levels of weed competition on the growth of cauliflower as determined by changes to aerial dry weight, leaf area index, height and curd diameter (Fig. 6.2).

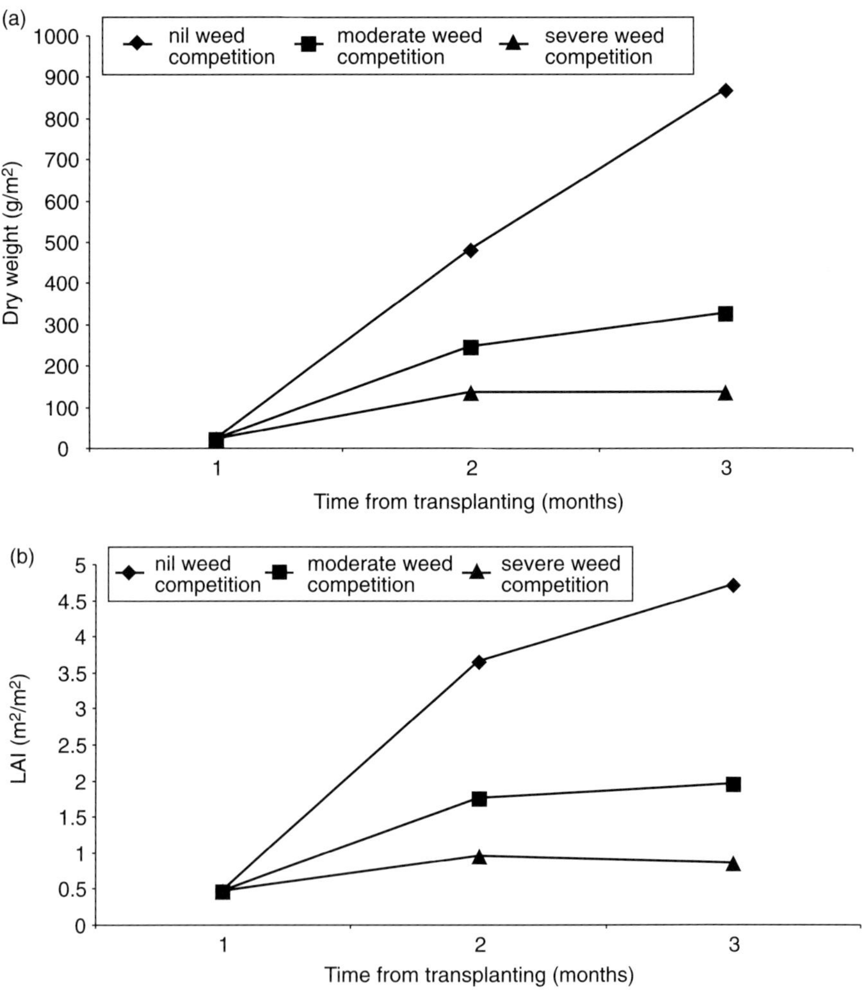

Fig. 6.2. The effect of competition from *Chenopodium album* (fat-hen) on the growth and curd size of spring- and summer-transplanted cauliflower (*Brassica oleracea* var. *botrytis*) cv. Fremont. Simulated (lines) and observed values (symbols) for (a) aerial dry weight, (b) leaf area index (LAI), (c) height and (d) curd diameter of cauliflower (after Röhrig and Stützel, 2001).

The presence or absence of a *Brassica* such as cabbage (*B. oleracea* var. *capitata*) does not normally affect the relative proportions of weed species. An exception to this seems to be shepherd's purse (*C. bursa-pastoris*), which when treated with the herbicide trifluralin (α,α,α-trifluro-2,6-dinitro-*N*,*N*-dipropyl-*p*-toluidine) was then suppressed by the crop. Freyman *et al.* (1992) studied the competitive effect of shepherd's purse. It is one of the most common and difficult weeds to control in vegetable cole crops, because of its botanical similarity to brassicas. Consequently, this is where the use of herbicides would be most likely to cause damage to the crop plant. Results indicated that intra-row competition from the weed could be reduced by use of closer spacing within the rows. Closer intra-row spacing is normally associated with the use

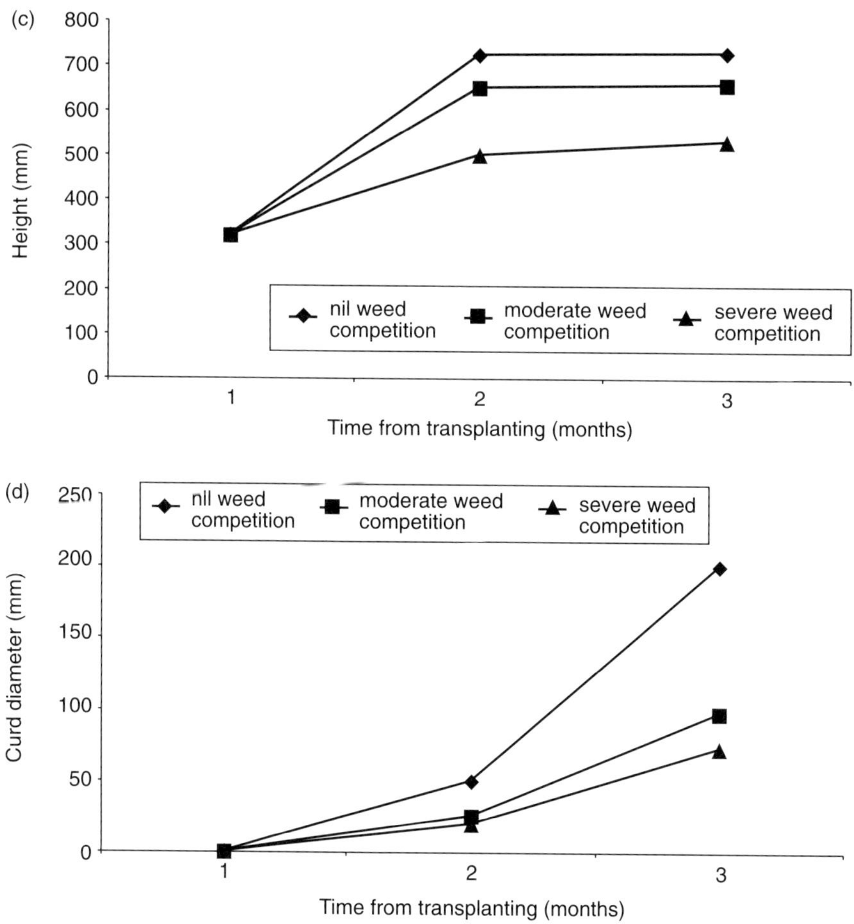

Fig. 6.2. *Continued.*

of wider inter-row spacing. Reducing the intra-row spacing diminished competition and made cultivation easier because of the increased distance between rows. Chickweed (*S. media*) was the main weed affecting winter cabbage in Lawson's (1972) studies; this was capable of surviving winter frost and then accelerating into rapid growth in the spring, becoming the dominant weed species and eventually shading the crop. Treatment with the herbicide propachlor (2-chloro-*N*-isopropylacetanilide) delayed the capacity of *S. media* to cause competition. Combining this with the earlier use of trifluralin controlled this weed and allowed the crop to dominate the competitive relationship. The dominance of weed species will change as a result of altering crop husbandry systems, including changing the spectrum of herbicides and other control techniques used and also resulting from several biological factors. This is what Lawson termed 'the ever moving target for weed control' (H. Lawson, personal communication).

In California (USA), for example, cultivated radish (*R. sativus*) and the weed (*R. raphanistrum*, i.e. wild radish) are both introductions from Europe. Continual interspecific hybridization since their arrival in America has converted cultivated radish into a weed and the climatic range has been enlarged continuously (Panetsos and Baker, 1986).

The outcome of competition is determined by the timing of emergence of weed seedlings relative to those of the crop *Brassica* and the extent to which environmental conditions during the early stages of growth favour either of the competitors. Difficulties arise where herbicides fail to be effective against the entire spectrum of weed species present in the crop. Under these circumstances, those weeds that are uncontrolled are given a selective advantage and become predominant, producing populations with strong competitive advantages in comparison with the crop. These will provide major problems during crop harvesting and can act as reservoirs for pests and pathogens. They will also return substantial numbers of viable weed seeds into the soil, causing an escalating problem for future years especially where the opportunities for crop rotation are limited (Roberts *et al.*, 1976).

The use of transplanted brassicas provides the crop with a considerable growth advantage since it is capable of competing with weed seedlings more quickly. The crop canopy closes rapidly, inhibiting the further growth of weed seedlings. This is usually linked to the use of a pre-planting soil-incorporated selective herbicide capable of destroying germinating weed seedlings. There is a defect in this strategy caused by the presence of cruciferous weeds within the soil flora that are unaffected by such selective herbicides. This leads to an upsurge in weeds such as shepherd's purse (*C. bursa-pastoris*) and of volunteer crop *Brassica* plants such as the increasingly problematical oilseed rape (*B. napus*) seedlings that are residual from preceding crops. There may also be other volunteer groundkeeping crops such as potatoes, which by virtue of their presence as resilient vegetative tubers have a growth advantage even when in competition with transplanted brassicas.

COMPETITION FOR NUTRIENTS

Anatomical and physiological differences between crop species and even between crop cultivars that influence their competitive abilities against weeds are beginning to be explored. So also are the capacities of weeds to compete with each other for resources. These interactions potentially offer means by which the crop and weed relationship can be manipulated to the advantage of the former through their respective nutritional demands. This is because crops and weeds may compete for nutrient resources at different levels of effectiveness. *Brassica* crops, for example, are better competitors for soil nutrients than many other field vegetables (Chapter 5). In turn, it is known that cabbage plants have a considerable effect on the growth of various aggressive weed species such as *C. album* (fat hen) and to a lesser extent with *Senecio vulgaris* (groundsel).

The presence of the crop plant markedly reduced the seed weight produced by *C. album* (Qasem and Hill, 1993). An aim of future plant breeding programmes may, therefore, be to develop cultivars that are sufficiently vigorous to be capable of out-competing weeds because of their increased efficiency in utilizing nutrients and water from the soil.

WEED MANAGEMENT

Integrated crop management (ICM) identifies the weed problems through regular crop inspections (so-called 'crop walking') (see also Chapter 7). In turn, this is combined with preventative, cultural, mechanical, biological and chemical control methods in a compatible manner to solve the problem. Integrated methods avoid relying solely on one management tool and help reduce the need for chemical weed control by combining a series of approaches to profitability, producing a marketable product while minimizing harm to the environment.

Weed management is an important aspect of integrated husbandry and illustrates many of the characteristics and problems also associated with pest and pathogen control (Chapter 7). Essentially, crop walking is no more than an application of the old Chinese adage that 'the best manure is the farmer's boot'. In other words, close and continuous attention to the crop ensures the best results. Crop walking demands visiting the fields regularly, taking quadrat samples in which the numbers of weeds, their differing species and locations within the crop are identified and recorded. Particular attention is paid to identifying the dominant species and also the occurrence of uncommon and perennial weeds that may pose especial problems for both the current and succeeding crops. It is an important aspect of integrated management that problems that might afflict future crops in the rotation are identified as early as possible, preferably before that crop is grown. Differentiating weed species by their life cycle is an essential aspect and is a key element that increases the success of integrated control.

Ephemeral and annual weeds reproduce primarily by seed, and hence husbandry controls aim to prevent them from seeding. Many annual weeds of *Brassica* crops have a periodicity of emergence and germinate in response to specific environmental cues. Cultivation strategies can be modified to provide 'stale seedbeds' for example, where weeds are encouraged to germinate and are then destroyed by subsequent tillage prior to sowing or transplanting with the *Brassica* crop.

Perennial plants reproduce both sexually by seed and asexually by vegetative organs such as rhizomes and tillers which can be broken and distributed by ill-timed cultivation, encouraging the spread and further propagation of the weed population. Most perennial weeds tend to occur sporadically and patchily within a field. Mapping the location of these concentrations allows control measures to be targeted to the problem areas.

This is a more efficient system of control and economizes on the resources used. Maintaining records of weed populations over a period of years provides an understanding of the manner by which weed species spread within fields and their response to control measures. An essential aspect of ICM is the prevention of the spread of weeds into cropped areas. Weeds may be introduced on cultivation equipment into new areas. Cleaning equipment before moving between fields is an essential first step to preventing the entry of new species.

Animals can spread weeds by attachment to their coats or hooves, and in their manure. Weeds are also spread in propagation media such as peat moss or shredded bark used to produce transplants, and in irrigation water. New infestations of weed species should be removed by chemical spot treatment while the colonies are small rather than waiting for a larger scale problem to develop before remedial action is taken. The effect of global climate change is encouraging plants to move into previously uncolonized regions and, because of the lack of predators, they become very successful weeds that are exceedingly difficult to eradicate.

EFFECTIVE CULTURAL PRACTICES

Cultural weed control aims to optimize sowing or planting dates, seed rates or transplant densities, spacing layouts, soil fertility, irrigation practices and cultivar selection to achieve rapidity of crop growth which is able to out-compete weeds for resources. The aim should be to ensure either that the crop plants emerge first or that transplants can establish ahead of weed development and close their canopy over the weeds, thereby smothering them.

Manipulating crop geometry can radically influence the grower's ability to control weeds. Thus seeds or transplants should be placed at uniform depths to produce even and regular crop growth. Use of high crop densities and narrow row spacing ensures that the crop canopy closes as quickly as possible. This may be aided by the use of crop covers (see Chapter 7) which increase the accumulation of units of soil heat and, therefore, encourage more rapid root and shoot growth. Integrated management requires that inputs such as nutrients and water are applied uniformly ensuring even growth without the development of stress within the crop. Similarly, biotic stresses caused by pests and pathogens should be minimized. Vigorous healthy crops have greater competitive abilities against weeds than stressed crops.

NON-TILLAGE AND TILLAGE CULTIVATION SYSTEMS

Non-tillage systems are a means of managing the disposal of crop residues with minimum cultivation (Unger and McCalla, 1980). Synonyms for this approach include conservation tillage, direct drilling, eco-fallowing, limited tillage, minimum tillage, no-tillage, reduced tillage and stubble mulching. The process resembles the use of organic mulches.

The aims are:

- leaving sufficient plant residues on the soil surface at all times to reduce wind and water erosion;
- reducing energy use;
- conserving soil and water.

The impact of this form of husbandry on yields varies depending on the region in which non-tillage is practised. Where there are heavy soils, wet climates and colder regions, yields may be depressed due to the following:

- lack of knowledge and/or equipment to manage the system;
- colder, wetter and less well aerated soils;
- weed, insect and pathogen problems;
- lower nitrogen availability, e.g. lower nitrate production;
- changes in the microbial status of the soil;
- production of phytotoxic substances.

In comparison, tillage systems help to control weeds by:

- killing emerging weed seedlings;
- burying weed seeds and delaying the growth of perennial weeds;
- leaving a rough surface that hinders weed seed germination;
- providing enough loose soil at the surface to permit effective further cultivation;
- leaving a clean uniform surface for subsequent efficient action by herbicides;
- incorporating herbicides where they are required.

Since the effects of tillage vary with soil type, so the degree of weed control will also alter.

Experiments with spring cabbage (*B.oleracea* var. *capitata*) by Knavel and Herron (1981) showed that with no-tillage culture, yield was less than that obtained with conventionally tilled crops when using the same nitrogen and spacing treatments. Yields obtained from non-tillage systems were increased by raising the density of plant populations and by applications of additional nitrogen, but head size remained smaller than with conventional systems. Large head size tended to correlate with the nitrogen and calcium content of the wrapper leaves.

It is only economical to use non-tillage systems for cabbage (*B. oleracea* var. *capitata*) where the soil is susceptible to erosion or drought. Non-tillage systems produce greater weed emergence compared with conventional tillage. This is probably due to increased weed seed mortality and seed burial resulting from the use of tillage implements. Tilling soil at three different depths significantly decreased weed emergence (Egley and Williams, 1990) compared with non-tillage. The effects varied, however, depending on which weed species was dominant within a particular field.

Fallowing land in the absence of crops allows the use of intensive non-selective weed control and can be used to eliminate tenacious plant species, particularly perennials, that are otherwise difficult to control. Land selection techniques can be linked with fallowing such that those parts of the holding infested with weeds which are difficult to control in *Brassica* crops are avoided. This technique is a form of crop rotation and can be used to exploit forms of husbandry or herbicidal control that totally eradicate weeds in advance of specific *Brassica* crops.

EFFECTS OF TILLAGE IMPLEMENTS

Vertical gradients in the soil microclimate such as water availability, temperature and light occur in the field (Heydecker, 1973; Fenner, 1985), so that one of the major factors influencing the success of weed germination and emergence is the position of a weed seed within the soil profile. Chancellor (1982) identified that weed seeds have mechanisms that respond to these gradients, thereby preventing germination at depths from which the seedling cannot reach the soil surface. Mechanisms such as the need for light to stimulate germination in scented mayweed (*Matricaria recutita*) ensure that germination only takes place close to the soil surface. Each weed species has a characteristic emergence response to depth of burial. The distributions of weed seeds in soil banks are spatially heterogeneous both horizontally and vertically. The major means of seed movement in soils are the implements used in cultivation themselves. As a consequence, cultivation practices can give rise to distinctive vertical distributions of seeds in the soil profile. While this distribution can be sampled and the products examined in controlled laboratory tests to establish the spectrum of weed seed within the soil profile, there is little evidence to determine the manner by which these seeds are placed in specific sectors of the profile in the first instance. Recent experiments have shown that spading and rotary cultivating tend to spread weed seeds throughout the soil profile. Spring tine and power harrows concentrate weed seeds in the top 6 cm of soil (Fig. 6.3). Investigations of the effects of implements on weed seed dispersal in soil compared rotavator, spring tine harrow, spader and power harrow implements (Grundy *et al.*, 1999). The rotavator caused a backward movement of seeds (as represented by beads), but neither the spring tine nor the spader had any significant effect on the horizontal displacement of seeds, while the power harrow had the greatest capacity to move seeds forward beyond 0.5 m in the soil.

Harrowing weeds before and after *Brassica* crop emergence considerably reduces the reliance on herbicides. Encouraging the use of harrows requires that the number of cultivations should be minimized – currently up to eight passes may be needed in a single weeding operation – and dependence on favourable weather conditions for soil working and weed killing should be reduced. Harrowing predominantly kills or suppresses small weeds in their early growth stages. A relatively smaller proportion of the weeds (up to 25%) are uprooted.

Most weeds are simply covered by a loose layer of soil. As weed plants mature, they become taller and less flexible and hence more resistant to being covered by a layer of soil from which they can regrow after the cultivation operation is over. Understanding and improving the effectiveness of harrowing demands knowledge of the relationships between: weed control and crop damage; weed density and crop yield; and crop damage and crop yield. Each of these relationships will be moderated by the prevailing weather and soil conditions at the time of cultivation. Harrowing depth, working speed and soil moisture content will have varying influences on the covering of different weeds, affecting the efficiency of weed control. The prime factor affecting the efficiency of weed control appears to be the depth to which weeds are buried by the harrowing operation (Fig. 6.4) (Kurstjens and Perdok, 2000).

The density of weed seedlings emerging in a crop following cultivation is related to:

- the ability of seedlings to emerge from various depths in the soil;
- the survival of seeds at different depths;
- the depth of seed burial in no-tillage, rotary tillage and plough tillage.

Where a great abundance of weed seeds is thoroughly mixed in the soil, probably the best approach is to use nil or minimum tillage and attempt to deplete the surface fraction of the weed seed pool. If this approach is to be successful, then reseeding must be prevented, for example, through the use of herbicides or by further shallow cultivation. Repeated surface cultivation has the added benefit of promoting the germination of seeds already in the pool

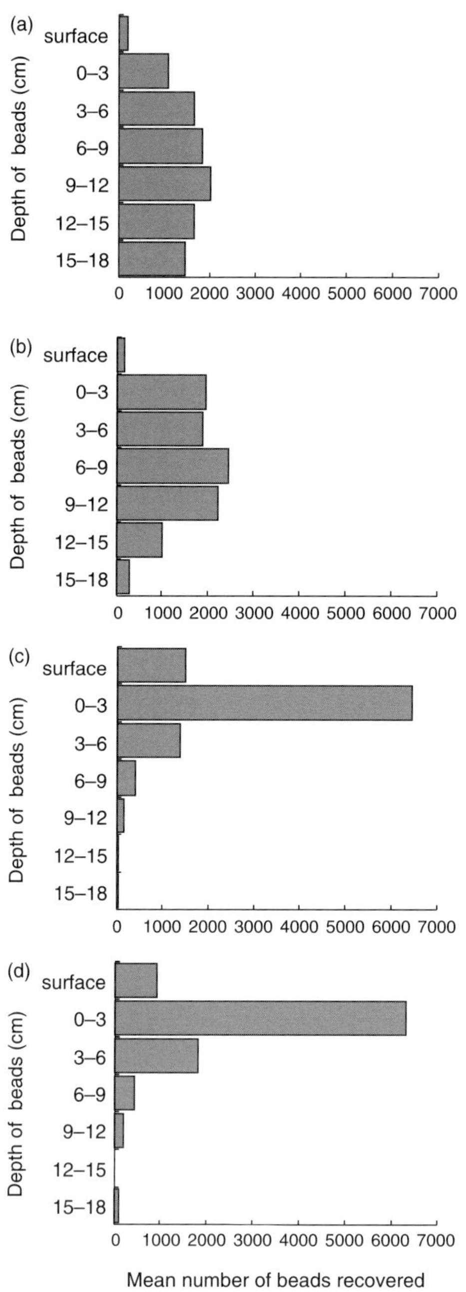

Fig. 6.3. Illustration of simulated vertical distribution of 10,000 beads initially sown on the soil surface following one pass of (a) spader, (b) rotavator, (c) spring tine and (d) power harrow (Grundy *et al.*, 1999).

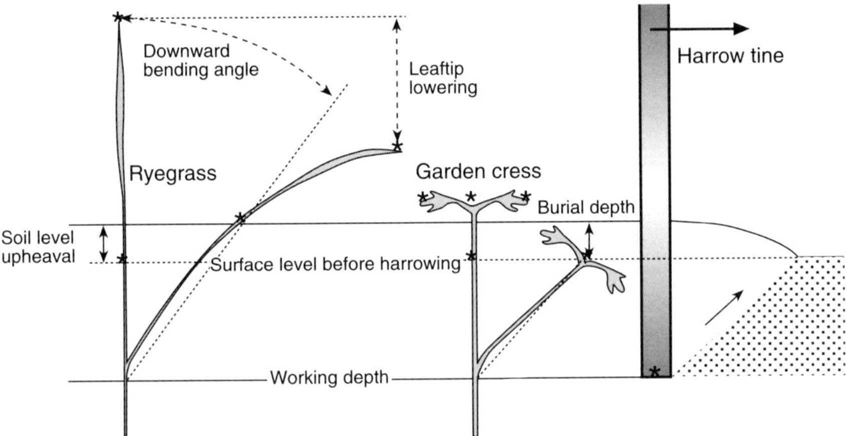

Fig. 6.4. Illustration of ryegrass and garden cress seedlings before and after harrowing. Digitized coordinates (marked *) of a harrow line and a ryegrass and garden cress seedling before and after harrowing, with an illustration of derived parameters (side view) (Kurstjens and Perdok, 2000).

and thereby speeding the decline in the number of seeds that are close enough to the surface for germination and emergence. When confronted by a year of weed control failure in which many seeds are shed on to the surface of an otherwise clean soil, the best strategy will be to plough as deeply as possible and then use minimal soil disturbance methods thereafter (Moss, 1985; Mohler, 1993).

EFFICACY OF TILLAGE WEED CONTROL FOR VEGETABLE BRASSICAS

Weed control is a major reason normally advocated for soil tillage. The use of primary cultivation, ploughing and secondary cultivation, discing and harrowing eliminates emerging annual weeds and suppresses perennial weeds. Ploughing buries about 80% of surface weed seeds and only returns about 40% to the surface. It also brings perennial rhizomes and tillers to the surface, allowing them to be killed by desiccation or freezing.

Conservation (limited) tillage or non-tillage systems have been advocated for oilseed brassicas, but are less frequently used for high-quality vegetable cole crops. The presence of crop residues on the surface tends to slow down the processes of soil warming, provides habitats for pests and pathogens, and may encourage the development of perennial and groundkeeper weed populations. Flame weeding has been used for selective control and the removal of crop residues applied prior to crop emergence in countries such as the USA. These systems are particularly suited to crops that emerge slowly. Hand weeding is largely discarded except as a labour-intensive and consequently expensive supplement to other strategies and when applied only as an emergency measure.

The side effects of mechanical weed control in *Brassica* production are that in addition to the intended effects on weeds, mechanical control causes some negative side effects such as physical damage to crops. A positive effect could be improved soil structure. Studies in cauliflower showed that where there was an overall treatment, this resulted in at least 4% loss of plants with ridging and tine harrowing. Yields of mechanically controlled crops ranged between 82 and 111%, with an average of 97% of untreated control (Laber and Stützel, 2000).

INTERCROPPING AND COVER CROPPING SYSTEMS

Conventional forms of tillage husbandry deplete soil organic matter and are associated with the use of substantial inputs of resources such as energy, soluble fertilizers, pesticides and water. Weed management for *Brassica* crops using solely agrochemicals is diminishing. This is because the registrations of many older pesticides have been cancelled or not renewed by the manufacturers as patent rights expired, and fewer new products are coming on to the market because of the expense involved in their testing compared with their profitability. This is because vegetable crops generally are seen as being very limited markets offering only small profits to the chemical companies. This attitude is reinforced by the need for rigorous safety testing of new agrochemicals to determine their potential risks to human health, environmental stability and natural biodiversity. Such testing is very expensive, upwards of US$500 million per molecule, and can only be justified economically by opportunities for sales for use on large area crops. Vegetable crops by virtue of their intensive culture do not offer sufficiently large land areas for the application of new chemical products and hence are not attractive to the agrochemical companies. Alternative methods of weed, pathogen, pest and fertility management are being developed as replacements for a reliance on agrochemicals.

Two systems that have formed an integral part of the 'organic' approach to *Brassica* production for many years are now receiving attention from researchers as potential components for integrated crop production strategies. These are intercropping and cover cropping systems. A stimulant to accepting inter- and cover cropping systems is that in addition to controlling weed competition, they also limit the impact of some pests on crops.

This approach is not new; studies of aphid populations on Brussels sprout (*B. oleracea* var. *gemmifera*) in the 1970s, for example, indicated that the presence of weed cover provided a useful means of reducing the size of aphid populations through increased natural predation (Smith, 1976a, b). Brussels sprouts grown in weed-free culture were more attractive to colonizing aphids compared with those where weeds remained. Clean weeding of sprouts provided ideal conditions for colonization by aphids, whitefly and certain Lepidoptera in the absence of their predators. These pests were attracted to the crop plants since they stood out against the bare background provided by a weed-free soil. In turn, colonization by predators of the pests of Brussels sprouts was related to plant density, height and contrast or colour of the background that affected the optical attractiveness of the crop, and encouraged by the presence of an understorey of non-crop plants. Variations in species composition of the weed flora affected the differential attractiveness of crop backgrounds to a range of predators of insect pests.

Intercropping

Increasing crop specialization and intensification in developed countries have virtually eliminated the use of rotations for vegetable *Brassica* production, resulting in undesirable side effects such as soil compaction, loss of soil structure and decreased organic matter content (Nicholson and Wien, 1983).

Research indicates that intercropping with rows of plants whose purpose is soil improvement, placed between the economic crop, restores soil structure with minimal deleterious effects on the efficiency of the cash crop. Intercropping is still a common cultural practice in tropical areas, but is much less frequently used in temperate regions (Theunissen *et al.*, 1995). Intercropping can be part of an ICM strategy contributing to ecologically and economically acceptable forms of sustainable horticulture. Favourable effects of intercropping particularly include suppressing or delaying pest population expansion and crop colonization.

Criteria for intercropping species are as follows.

- They must offer 'intercropping effects' in terms of pest population suppression.
- Competition with the main or cash crop must be minimal.
- They must not create weed problems for the following seasons.
- They must be predicable and manageable within the normal cropping pattern.
- The seed of the intercropping plant(s) must be commercially available in sufficient quantities.
- They must not generate or support pests or pathogens.
- Leaching of nitrogen must be prevented or reduced.

The requirement that intercropping reduces pest insect populations is not fulfilled in all cases. Some species of pest seem to be unaffected by either the presence or absence of intercrops or only react in specialized conditions, such as, for instance, the small white butterfly *Pieris rapae* (Risch, 1981).

Intercropping may affect the morphology and development of brassicas. For example, cabbage heads taken from intercropped plots were smaller but more compact, and this in turn can affect their sensitivity to pest attack. It is also suggested that stresses induced by polyculture may alter the physiology of the crop plant, rendering it less attractive or nutritious, or even more toxic to its pests. Results summarized in Table 6.7 compare the effects of mono- and polycropping on the yield and value of white cabbage grown in The Netherlands.

The successfulness of intercropping is increased where chemical suppression of the 'live mulch' prevents excessive competition. The most promising living mulches were the shorter and less vigorous turf grasses and clover. Chewing's fescue (*Festuca rubra* var. *commutata*), Kentucky bluegrass (smooth meadow grass) (*Poa pratensis*) and Kent wild white clover (*Trifolium repens*) did not affect the yield of cabbage. Some compromise is required between improving soil structure and limiting the competition with the *Brassica* crop when growth of the live mulch is not controlled. In principle, interseeded cover crops can be chosen to complement the main crop in resource use while directly or indirectly interfering with weed growth, thereby suppressing weeds but not the crop (Barnes and Putnam, 1983; van der Meer, 1989). In practice, however, success has been limited because interseeded crops often suppress neither the crop nor the weeds, or suppress both. Two basic designs for comparing the performance of intercropping with monoculture are advocated, the additive and substitutive formats.

Table 6.7. Yield and financial data from mono- and polycropping systems applied to white cabbage (*Brassica oleracea* var. *capitata*).

	% Marketable cabbages		Mean weight of marketable cabbage heads (kg)		Gross income (Dfl/ha)		
Treatment	1	2	1	2	1	2	3*
Monocrop	52	29	1.98a	1.97a	9,062	4,643	4,647
Polycrop							
T. repens	77	72	1.55b	1.68b	10,620	9,569	9,570
T. subterraneum	84	70	1.45b	1.71b	10,656	9,306	9,307

*Including marketable class 3; 1 = Experiment 1 (1990); 2 = Experiment 2 (1991).
Figures followed by a different letter differ significantly at $P = 0.05$.
After Theunissen *et al.* (1995); data are retained in Dutch guilders rather than conversion to euros.

Advantages are gained from intercropping where there is limited competition between the species or where one species provides a benefit to the other. A low level of competition occurs when two or more crops use different components of the ecosystem or use the same components in varying ways, or exploit alternative ecological niches. This is the Principle of Competitive Production as identified by van der Meer (1989). Low levels of competition can result from differing times of resource interception, rates of resource requirement, specificity for key resources and varying tolerances to biotic and abiotic stresses. These factors may interact with population densities of the components of the intercropping system. As yet, the impact of population density on the successfulness of intercropping has received only limited attention from researchers. In studies of broccoli and cauliflower intercropped together and with other crops such as potatoes and oats, Santos *et al.* (2002) concluded that successful combinations would exploit factors such as asynchrony of growth and maturity periods, and variations in canopy height and flexibility.

In seeking to use intercrops for transplanted broccoli production in New York State, USA, Brainard and Bellinder (2004) suggest that winter rye (*Secale cereale*) will be successful where it is sown at high densities, in localities or seasons with low initial temperatures and in combination with other weed management tools. This exploits the cold adaptation C_3 attributes of winter rye with germination requirements of 0–5°C compared with those of Powell's amaranth (green pigweed, *Amaranthus powelli*) which is a C_4 species adapted to warmer environments germinating at 10–15°C.

Cover cropping

Cover cropping resembles intercropping and exploits broadly similar ecological principles but involves mulching across the entire cropped area into which the cash crop itself may be inserted. The mulch may be mown or treated with a herbicide applied at reduced rates to diminish its competitive abilities prior to establishing the cash crop. Growers using the techniques of 'organic husbandry' have long recognized that non-tillage cover crops reduce pest populations and weed competition and improve soil quality by protecting the soil surface, decreasing soil, wind and water erosion, conserving soil moisture and reducing nutrient run-off. Non-tillage cover crops are sown in the autumn or very early spring and mown during late spring to form a mulch on the soil surface just before planting the main cash crop.

Cover crop mulch systems modify the microenvironment of the crop with subsequent effects on pest populations and crop yields. Plants in non-tillage systems tend to be smaller and paler than those with conventional tillage. This results from lower soil temperatures observed in non-tillage plots which decrease the rate of mineralization of soil nitrogen reducing its availability to

Brassica crops. If a legume, however, is included in the mulch, then nitrogen is added into the soil. Suitable plant species for use as cover crops are given in Table 6.8.

Indirect benefits obtained from cover crops

Some cover crops release chemicals that inhibit weed germination, a phenomenon termed allelopathy as initially defined by Molisch (1922) and discussed more recently by Rice (1984). Rye (*S. cereale*) residues release chemicals inhibitory to annual weeds, and suppression is greater where the rye residues are left on the soil surface. Hairy vetch (*Vicia villosa*) has weed-suppressive properties but to a lesser extent than rye. Cover crops may also affect the chemical constituents of soils. Cover crop residues may make nitrogen unavailable to vegetable crops due to assimilation or denitrification. Conversely, cover crops can also increase nitrogen available for brassicas either by the decomposition of legume residues or by preventing nitrogen losses during winter. Rye also increases the concentration of exchangeable potassium near the soil surface.

Key findings from a detailed study of cover cropping by Brandsæter (1996) and Brandsæter *et al.* (1998) in Norway identified the following effects.

1. Frost-sensitive cover crops are of no value for weed control in the subsequent spring – only those killed off in spring by herbicides (glyphosate-*N*-(phosphonomethyl) glycine) provided improved weed control.
2. Spring-sown cover crop (clover) provided better weed control when mown in early summer.
3. Competition and the failure of spring-sown clover to supply added nitrogen decreased the yield of white cabbage when compared with using monocultures.
4. No significant differences in yield of white cabbage between monoculture, late spring-sown subterranean clover, and rye and rye/vetch treatments were found.
5. Autumn-sown subclover (*Trifolium subterraneum*) did not suppress weeds effectively in spring, and mowing off the weeds that germinated in the spring or early summer was necessary.

Table 6.8. Plant species suitable as cover crops.

Grasses and cereals	Legumes
Rye – *Secale cereale*	Hairy vetch – *Vicia villosa*
Barley – *Hordeum vulgare*	Austrian winter pea – *Pisum sativum* subsp. *arvense*
Wheat – *Triticum aestivum*	Crimson clover – *Trifolium incarnatum*
Ryegrass – *Lolium multiflorum*	

6. Live mulch systems suppressed weeds more effectively than plant residues.
7. It is necessary to know how to manage the yield depression caused by competition, for example, by finding winter hardy annual legumes suited to local environmental conditions.
8. A considerable increase in cabbage yield was achieved both in subterranean clover (cv. Geraldton) and in white clover (cv. Pertina) treatments by rototilling between the rows 6 weeks after transplanting.
9. The cover crops did not reduce weed biomass or number of weeds early in the season compared with monoculture, but weed biomass in late summer became significantly lower in living mulch systems.
10. Rototilling was more effective for weed suppression than mowing.
11. Cover cropping combined with rototilling reduced weed biomass by 89% compared with untreated monoculture.
12. Subterranean clover used as a living mulch gave the earliest and most extensive ground cover and lowest cabbage yield, but it also reduced insect damage.
13. Both subterranean clover and white clover living mulches were associated with a greater number of marketable cabbage heads resulting from reduced insect damage.
14. White clover intercropped with cabbage gave significantly higher oat yields in the subsequent year compared with monocropping.
15. About half as many cabbage heads were marketable (rating 1 or 20) in monoculture plots (21%) as compared with either of the living mulch treatments (subterranean clover 42% and white clover 38%) based on caterpillar damage. This gave damage indices for cabbage heads of 72.2 for the monoculture, 63.5 for the white clover and 61.4 for the subterranean clover. The differences in the damage index were highly significant ($P = 0.0001$) for both clover species compared with the monoculture, but there were no significant differences ($P = 0.2$) between the clover species.
16. The number of turnip root fly eggs oviposited in the monoculture plots was significantly higher ($P = 0.05$) than in the plots with live mulch during peak oviposition of the turnip root fly. This effect was slightly more pronounced in the plots with subterranean clover, but there were no significant differences between subterranean clover and plots of white clover.
17. Overall there seems to be a reduction of insect damage, but also loss of yield, resulting from cover cropping. These two effects will balance out, especially if there is a saving in the amount of pesticide used. The effects of yield enhancement by rototilling at 6 weeks suggest that this reduces the regrowth of the legume and enhances the availability of nutrients and soil moisture to the crop.
18. Intercropping systems seem to reduce pest damage and are valuable for their green manuring effects on subsequent crops. Early establishment of the cover crop is important, but this must be balanced against yield depression that occurs as a result of maturing cover crops.

19. Trials of different overwintering legumes showed that the hardiest for Norwegian conditions was hairy vetch (*V. villosa*) since it provided the highest biomass production, the lowest regrowth ability after mowing and the best weed suppression.

20. Reducing competition between cover and cash crops has focused on the use of chemical or mechanical suppression systems. Alternatively, growth of the cover and cash crops could be synchronized to achieve maximum growth for the cash crop. One way to achieve this is to sow the cover crop in the middle of the growing period of the cash crop. Alternatively, the cover crop is sown first as a weed-smothering crop in advance of the cash crop. Here use is made of rapidly growing plants.

21. As a further alternative, use can be made of autumn-sown annual legumes such as subterranean clover or wild white clover which grow vegetatively during the autumn, become dormant or semi-dormant in winter and then resume growth in the spring. The most advantageous strategy would be to use species that flower, senesce and die in spring. This then allows the cash crop to be transplanted into the senescing mulch that provides additional soil moisture and nutrients as it decays. It is essential to use species or cultivars that have low canopy height and terminate their vegetative growth in late spring or early summer. An added valuable factor would be the ability to produce seed at senescence and then regrow in the autumn.

Nutritional benefits of cover cropping

Vegetable growers in the northeastern USA use cereal rye as a winter cover crop, but rye does not fix nitrogen. It ties up nitrogen after being incorporated into the soil (Schonbeck *et al.*, 1993). As an example of specialist studies of cover cropping, the effects on yield of broccoli and cabbage by cover crops of hairy vetch, grown alone or in combination with rye, were compared with rye alone and a no-cover control at four locations in New England. The effects of applied nitrogen were evaluated at two sites. Cover crops were grown until flowering and were either incorporated or mown and left on the surface. *Brassica* seedlings were transplanted into the plots, grown to maturity and assessed for yield, components of yield and foliar nutrients. Soil moisture was measured at two sites and inorganic nitrogen at one site.

Cover crops of vetch and vetch with rye consistently produced higher broccoli and cabbage yields than rye alone or with no cover crop. Measurements of foliar nitrogen contents indicated that nitrogen contribution was a major factor in the yield response, although cover crops sometimes affected concentrations of other nutrients. When rye was used alone, it reduced yields, probably through locking up soil nitrogen. At one location, broccoli yield responded dramatically to applied nitrogen (112 kg/ha as ammonium nitrate) regardless of cover crop or management system. Combinations of leguminous cover crop (cowpea, *Vigna unguiculata*; soybean,

Glycine max; velvet bean, *Mucuna pruriens*) and rates of applied nitrogen used for autumn broccoli production in coastal South Carolina, USA are reported by Harrison *et al.* (2004). The cover crops tended to promote higher and earlier yields of broccoli as compared with untreated plots. Other leguminous cover crops such as red clover (*Triflolium pratense*), biennial sweetclover (*Melilotus* spp.) and lucerne (*Medicago sativa*) will add nitrogen to the soil but take the land out of *Brassica* production for 12 months. Winter annual legumes such as hairy vetch (*V. villosa*), crimson clover (*Trifolium incarnatum*) and Austrian winter field pea (*Pisum sativum* subsp. *arvense*) can add nitrogen and organic matter without interrupting production for an entire season. Legumes when grown with a cereal grain or other grass (e.g. hairy vetch with rye) may produce more organic matter, protect the soil and suppress weeds better than either when grown alone.

Benefits of reduced tillage associated with cover crops

Traditionally cover crops are tilled into the soil before planting brassicas. Increasingly, however, growers adopt no-tillage systems to conserve soil and reduce fuel, machinery and labour costs. No-tillage cover crops reduced cabbage yields in Oklahoma, USA, but they also curbed soil erosion and reduced the number of pesticide applications required. Such crops may be killed with paraquat (1,1′-dimethyl-4,4′ bipyridinum ion) herbicide sprays or by mowing close to the ground once flowering has commenced.

Manipulating competition from the cover crop

The main objection to cover cropping is the yield depression due to competition with the living cover crop. This may be mitigated by the following:

- Using less competitive cover crop cultivars or choosing a combination of cover crop and sowing times to reduce mutual competition. This may be achieved by use of winter annual legumes whereby weeds are suppressed in the critical period for their development early in the cover crop life history and which synchronizes with the onset of maximum cash crop growth. This demands a species of legume suited to the local climate of the crop.
- Suppressing the live mulch to reduce interference with crop growth by mechanical cultivation systems. Alternatively, tilled strips may be used in a similar manner to intercropping for the main (cash) crop rows; or timely overseeding of the main crop into the cover crop and chemical methods for cover crop suppression.

There were reduced yields of spring cabbage (*B. oleracea* var. *capitata*) in non-tillage crops compared with conventional tillage (Moore and Seward, 1986). By contrast, yields of autumn cabbage (*B. oleracea* var. *capitata*) increased with non-tillage systems. Differences in planting date influenced this response. Using legumes in non-tillage systems is a logical means of adding nitrogen. Hairy vetch (*V. villosa*) is the most efficient legume in this respect, adding 90–100 kg N/ha annually. Many experiments fail to mitigate the effects of competition between the cash and cover crops and hence find the crop growth component is less in living mulch treatments compared with bare soil control treatments. Hence, the biomass of cauliflower (*B. oleracea* var. *botrytis*) was reduced by 62% when grown in a vetch living mulch compared with clean cultivation (Altieri *et al.*, 1985).

The yields of cauliflower and Brussels sprout (*B. oleracea* var. *gemmifera*) were reduced by 42 and 61%, respectively, when overseeded with white clover (Dempster and Coaker, 1974; O'Donnell and Coaker, 1975). These authors planted Brussels sprouts in to an established stand of white clover (*T. repens*) at differing levels of soil cover (25, 50, 75 and 100%) and yields were reduced by 38, 40, 56 and 81%, respectively.

Andow *et al.* (1986) found that compared with clean cultivation cabbage (*B. oleracea* var. *capitata*), head size was smaller in live mulch combinations consisting of cv. Idaho clover compared with the dwarf cv. Kent. Further, Ryan *et al.* (1980) found that when cabbage was interplanted with clover in every third row (33% cover), average head weight was 15% higher compared with the 'no cover' treatment. The use of sublethal doses of herbicide has been frequently attempted as a means of cover crop suppression; or the use of non-chemical suppression that has involved mowing or partial tillage of the cover crop. It was found that when a white clover cover crop was kept mown to 15 cm high, then the components of yield of broccoli (*B. oleracea* var. *italica*) were statistically equivalent between 'no cover' and 'living mulch' treatments. The combined effects of cropping, tillage and herbicide treatments are shown in Table 6.9.

There is evidence that water stress reduces the main (cash) crop yield in a living mulch system, hence there is need for added irrigation to adjust this deficiency. The amount of water required to produce a given cabbage yield increased when perennial ryegrass was added into the system as live mulch (Graham and Crabtree, 1987). Mechanical suppression of the ryegrass by mowing twice per season did not increase cabbage yield. Use of herbicides (fluazifop-*p*-butyl; (*R*)-2 [4-(5-trifluoromethyl-2-pyridyloxy) phenoxy] propanoic acid) decreased competition for water, and at high irrigation rates cabbage yields were similar to those of conventional husbandry systems. Cover cropping rests on the acceptability of the ability of the cover crop to suppress weeds and tolerating the resultant competition with the main cash crop.

Cover cropping is seen as a particularly appropriate technique for the integrated management of brassicas. This is because the fields are finely tilled before sowing or planting and the soil is frequently wet during harvesting, which may involve the use of mechanical or mechanized harvesting such as gantry systems causing serious damage to the soil structure (Stivers-Young,

Table 6.9. Yield of broccoli (calabrese, *Brassica oleracea* var. *italica*) in relation to cover cropping, tillage and herbicide treatment.

Treatment	Head fresh weight (Mg/ha)	Maturity date[a]	Plant fresh weight (Mg/ha)
Rye			
Till – herbicide	7.61	10.8	22.11
Till + herbicide	7.16	9.9	27.34
Mow – herbicide	7.32	21.8	20.58
Mow + herbicide	7.19	21.5	24.04
Rye + vetch			
Till – herbicide	8.28	3.6	28.15
Till + herbicide	9.27	1.7	33.55
Mow – herbicide	10.19	7.8	35.26
Mow + herbicide	8.37	8.4	33.32
No cover			
Till – herbicide	7.78	6.3	24.69
Till + herbicide	7.92	2.5	25.74
Mow – herbicide	0.05	43.0	1.38
Mow + herbicide	8.05	19.5	20.12
Factorial T Test[b]			
Cover (C)	**	**	**
Tillage (T)	*	**	**
Herbicide (H)	**	**	**
C × T	**	**	**
C × H	**	**	**
T × H	*	NS	**
C × T × H	**	**	**

After Mangan *et al.* (1995).
[a]Days after 18 July.
NS, i.e. $P > 0.05$; *$P < 0.05$; ** $P < 0.01$.
Site = South Deerfield, Massachusetts, USA; 3 × 2 × 2 factorial experiment, cover crops = rye (*Secale cereale*) + vetch (*Vicia villosa*); rye alone; no cover crop; tillage = conventional = cover crop tilled in and no-till = cover crop mowed; herbicide = DCPA (dimethyl tetrachloroterephthalate).

1998). Many brassicas are relatively inefficient users of nutrients; the rates of fertilizer applied frequently exceed crop demand, and excess nutrients, especially nitrogen, are lost through leaching into the soil water and eventually into free-flowing water courses such as streams and rivers. Nitrogen loss may also take place through denitrification and volatilization. *Brassica* crops tend to return relatively little organic matter to the soil and leave little surface residue that will protect the soil from wind and water erosion.

Growers need cover crops that have a relatively short growing season, accumulate leachable nitrogen and suppress autumn-growing weeds and are killed during the winter or easily destroyed by use of herbicides such as glyphosate (*N*-(phosphonomethyl) glycine). For efficient, profitable *Brassica* husbandry, however, it is essential that the use of an autumn cover crop does not delay spring cultivation and subsequent seeding or planting. In the temperate northeastern USA, *Brassica* cover crops such as oilseed radish (*Raphanus sativus*), white senf mustard and yellow mustard (*Sinapis alba*) are seen as well suited to autumn sowing as cover crops. These grow quickly, accumulate significant amounts of biomass, deplete soil inorganic nitrogen concentrations and suppress weeds. Agricultural brassicas such as oilseed rape (*B. napus*) or yellow mustard (*S. alba*) have been advocated as cover crops for horticultural brassicas to exploit the weed-suppressive effect of their secondary products such as isothiocyanates, ionic cyanates and epinitriles. Results obtained by Haramoto and Gallandt (2005a, b), however, suggest that weed suppression may not be any greater than that achieved by other methods.

Cover crops tend to decrease water evaporation from soil and increase infiltration, resulting in greater moisture content. This can lower soil temperature and delay the growth of early season brassicas. If, however, a rye crop grows excessively in the early spring, then soil moisture will be depleted. It was concluded that autumn cabbage and broccoli could be produced effectively in non-tillage soil management systems. Hairy vetch (*V. villosa*) and Austrian winter pea (*P. sativum* subsp. *arvense*) are more efficient cover crops than rye (*S. cereale*). Possibly this results from these legumes releasing nitrogen through mineralization.

In studies of the integrated use of cover crops and herbicides, transplanted cabbage were grown in conventional tillage (100% cultivated) and strip tillage (25% cultivated and 75% residue). Cabbage (*B. oleracea* var. *capitata*) yielded similar head weights irrespective of tillage treatment. Herbicides tested were: propanil (3′,4′ dichloropropionanilide), napropamide ((*RS*)-*N*,*N*-diethyl-2-(1-naphthyloxy) propionamide), oxyfluorfen (2-chloro-α,α,α-trifluoro-*p*-tolyl 3-ethoxy-4 nitrophenyl ether) and napropamide + oxyfluorfen (Hoyt *et al.*, 1996).

Cabbage produced on sloping land prone to erosion should be grown using some form of conservation tillage to reduce water and soil erosion and to maintain soil productivity. Weed management programmes for conservation tillage are critical for successful economic yields.

Reduced tillage practices offer advantages of soil and water conservation, reduced erosion potential, less energy and time (labour) spent producing the crop, and more efficient land use. Grass and cereal cover cropping with cabbages tends to depress yields in comparison with systems using legumes or where adequate nitrogen is applied artificially. A highly significant yield response was obtained from broccoli treated with cover crop, tillage and herbicide treatments. The vetch plus rye treatment gave higher yields than rye alone and non-cover crop. The vetch plus rye treatment was also associated with earlier harvesting dates for the broccoli crop. The lowest yielding treatment was the non-cover crop, non-tillage and no herbicide. Incorporation of each cover crop treatment and herbicide use in tilled treatments hastened the maturity of the broccoli crop.

ECOLOGY OF CROP–PEST INTERACTIONS IN INTERCROPPING AND COVER CROPPING

The interaction of weed species, crop type and the presence of insect pests has been summarized by Schellhorn and Sork (1997). Where the weeds are closely related botanically to the crop type, for example cruciferous weeds in a *Brassica* crop, these will encourage specialist feeding insects such as flea beetles. Where the flora is a mixture of botanical types, this depresses the numbers of specialist feeding insects and encourages more generalist feeders. Botanically mixed flora also encouraged natural insect predators such as coccinellids, carabids and staphylinids; in turn, these could be responsible for the reduction in the presence of specialist feeding pests such as imported cabbage worm (small cabbage white butterfly; *P. rapae*) and diamond back moth larvae (*Plutella xylostella*). Polycultures of cash, inter-row and cover crops often support fewer insect pests at lower densities compared with those found in monocultures (Risch, 1981). Various biotic, structural and microclimatic factors in multispecies plant communities work synergistically in producing pest control. Root (1973) suggested two ways by which this could be achieved. The first is the 'Enemies Hypothesis' which predicts that the increased abundance of insect predators and parasites in species-rich associations can better control herbivore populations. Richer plant associations supposedly supply a more favourable environment for predators and parasites, reducing the probability that they will leave or become locally extinct.

These conditions include: (i) greater temporal and spatial distribution of nectar and pollen sources – both of these attract natural enemies and increase their reproductive potential; (ii) increased ground cover provided by a more diverse environment, particularly valuable to nocturnal predators; and (iii) improved herbivore richness, providing alternative hosts and prey when other hosts and prey are scarce or at inappropriate stages of their life cycle.

The second hypothesis is the 'Resources Concentration Hypothesis' (Root, 1973) and involves changes in the behaviour of the herbivorous insects themselves. Visual and chemical stimuli from the host and non-host plant affect both the rate at which herbivores colonize habitats and their behaviour in those habitats. The total strength of attractive stimuli for any particular herbivore species determines what Root (1973) called 'resource concentration' and it is the result of the following interacting effects: (i) number of host species present and the relative preference of the herbivores for each; (ii) the absolute density and spatial arrangement of each host species; and (iii) interference effects from the non-host plants.

An herbivorous insect will have greater difficulty locating a host plant when the relative resource concentration is low. Relative resource concentration may also influence the probability of the herbivore staying in a habitat once it has arrived. For instance, a herbivore may tend to fly sooner, farther or straighter after landing on a non-host plant than a host plant, resulting in a more rapid exit from those habitats with lower resource concentrations. Finally, reproductive behaviour can be affected, for example, when a herbivore tends to lay fewer eggs on host plants in an environment of lower resource concentration.

The results of Risch's experiments suggest that the 'Resource Concentration Hypothesis' is correct as opposed to the 'Enemies Hypothesis'. The factors of importance were differences in levels of beetle colonization and residency time.

The evidence for this was as follows:

- There were no differences in parasitism or predation of beetles among treatments – hence no 'enemy' effects.
- There was higher beetle emergence from monocultures than polycultures in only one instance (*Acalymma thiemei* (Baly) from squash), and even in this case the difference was not large enough to account for the observed difference in adult abundance.
- The number of beetles per host was lower in the polyculture than the monoculture only when there was a non-host present in the polyculture. When two host plants were present in the polyculture and no non-host plant, the numbers of beetles per host plant were in fact higher in the polycultures than in the monocultures. This pattern of abundance could be predicted only if beetle movements, and not mortality due to natural enemies, primarily determined beetle abundance.
- Field measurements of beetle colonization and experimental studies directly showed that differences in resource concentration between monocultures and polycultures affected patterns of beetle distribution.

The impact of ground cover mulches on yield and quality is illustrated in Table 6.10.

In the particular case of brassicas, cover crop mulches affect insects by interfering visually or olfactorily with host plant selection, thus reducing pest dispersal, reproduction and colonization of *Brassica* crops. Beneficial insects such as ground beetles are favoured by reduced tillage. It was concluded that rye mulch, for example, offers significant levels of weed suppression for the most critical stages of cabbage and diminishes the populations of several important insect pests. These improvements, however, are at the expense of yield losses due to difficulties in crop management. The lower populations of diamond back moth (*P. xylostella*), imported cabbage worm (small cabbage

Table 6.10. Effects of ground covers on the quality of cabbage (*Brassica oleracea* var. *capitata*) heads.

Ground cover	Total number heads/plot	% Heads marketable	% Heads unmarketable		
			Tip burn	Worm[1]	Other[2]
Bare ground	19.6a	87.7ab	1.9a	4.9a	5.3b
Vetch	19.1a	83.0b	0.5ab	2.2a	14.3a
Rye	21.0a	92.0a	0.3b	1.9a	4.4b

[1]Worm = small white butterfly (*Pieris rapae*) caterpillars.
[2]Includes heads damaged by thrips, black rot and other pests and pathogens.
Means separation by Duncan's multiple range test; means followed by the same letter are not significantly different at $P = 0.05$.
After Roberts and Cartwright (1991).

white butterfly, *P. rapae*) and aphids in rye mulch may have been related to the much smaller size and lower head weights of the crop plants. Cabbage planted in rye mulch and treated with Bt-insecticide (*Bacillus thuringiensis*) had the lowest insect damage ratings of any of the treatments, but yields were still less than those obtained by conventional tillage. A major yield constraint in the rye residue treatments was probably initial soil compaction and later competition with rye and red clover. Soil compaction was caused by equipment movement on wet soil necessary to mow the cover crop prior to planting of the cash crop.

Cabbage aphid (*Brevicoryne brassicae*) is a major pest of broccoli (*B. oleracea* var. *italica*) (Chapter 7); it colonizes the developing florets rendering them unmarketable. The effects of living mulch on aphid abundance are directly proportional to the amount of inter-row vegetation present; the aphids colonize more heavily plants surrounded by bare soil compared with those planted in vegetation (A'Brook, 1964, 1968; Gonzales and Rawlins, 1968; Costello, 1994), as illustrated in Table 6.11.

Flea beetle populations are generally lower on brassicas in weedy habitats compared with bare ground monocultures. This is possibly related to their movement and host-finding behaviour. Flea beetles are extremely mobile and their host-finding ability is impeded by non-host odours (Tahvanainen and Root, 1972). Non-host foliage may inhibit movement, resulting in faster leaving rates and lower colonization rates in living mulch plots. The response of the aphid *B. brassicae* to mixtures of host and non-host has been even more consistent than that of *Phyllotreta cruciferae*. Compared with monocultures, aphid populations were lower in cole crops with weeds in experiments in both the USA and the UK.

The response of *P. rapae* to mixtures of host and non-host plants has been variable in both the USA and UK experiments. The specific relationship between the physical and chemical structure of the cropping system and the precise host-

Table 6.11. Effect of living mulches and bare ground culture on insect population densities in cabbage (*Brassica oleracea* var. *capitata*) cv. Excel.

		Cropping system – living mulches		
Insect and stages	Bare ground	Creeping bentgrass	Kentucky bluegrass	Kent wild white clover
Phyllotreta cruciferae				
Adults[a]	70.0 (4.3)	36.1 (7.7)	31.5 (5.4)	55.0 (7.7)
Pieris rapae				
Eggs[b]	7.45 (1.45)	9.05 (2.16)	9.00 (1.14)	5.35 (1.59)
Larvae[c]	1.65 (0.45)	2.65 (0.17)	2.85 (0.35)	2.25 (0.43)

Population density = numbers per plant; data are the means of four replicates with standard errors in parentheses.
[a]Population density on cabbage in bare ground was significantly greater than on the living mulches ($P = <0.0001$).
[b]Population density on cabbage on bare ground was not significantly different from that on the living mulches; the density on cabbage in clover was significantly lower than on cabbage in grass ($P = 0.05$).
[c]Not significant.
After Andow *et al.* (1986).

searching behaviour of *P. rapae* may be critical, since the oviposition behaviour of *P. rapae* is sensitive to plant size and development, plant water content and plant dispersion. Clover inhibited oviposition in the late summer generation of *P. rapae* but had no effect on oviposition in the mid-summer generation.

Cabbages grown in a polycropped system showed less infection by thrips (*Thrips tabaci*) in terms of pest incidence and reduced population size and, in consequence, lower levels of physical damage were sustained. Limitation of damage by cabbage root fly (*Delia radicum*) and caterpillar (mainly *Mamestra brassicae* with smaller populations of *P. rapae* and *P. xylostella* and the occurrence of *Plusia gamma* in one experiment only) was quite substantial. The rates of infestation of heads by cabbage gall midges (*Confarinia nasturtii*) were low and evenly distributed across treatments. Feeding damage from flea beetles (*Phyllotreta* spp) was low and concentrated in the monocropped plots. These effects are illustrated in Table 6.12.

Living mulches compete with cabbage plants for resources within 2 weeks of transplanting. Living mulches could not control flea beetle populations below economic thresholds by themselves. Commonly used early season chemical treatments for flea beetles might be eliminated when living mulches are used. The lower incidence of insect pests, however, may be offset by cabbage yield reductions from competition between cabbage and living mulch. The market may demand smaller heads, however, and the increased quality noted where Kent wild white clover was used could be an added incentive to using these husbandry systems (Table 6.13).

Table 6.12. Causes of non-marketability in white cabbage (*Brassica oleracea* var. *capitata*) grown in monocropping and polycropping husbandry systems.

Cause	Monocrop	Polycrop	
		T. repens	*T. subterraneum*
Cabbage root fly	45a	23ab	19b
Thrips	56c	3d	12d
Caterpillars	17a	2b	4b
Cabbage gall midge	6	4	5
Flea beetles	6	0	0
Cabbage aphid	1	0	1

Figures with different letters in rows are significantly different at $P<0.05$; small percentages have not been taken into account.
After Theunissen *et al.* (1995).

Table 6.13. Effect of cropping systems using living mulches on cabbage (*Brassica oleracea* var. *capitata*) cv. Excel yields and quality.

Yield and quality	Cropping system – living mulch				
	Bare ground	Creeping bentgrass	Red fescue	Kent wild white clover	Idaho clover
Marketable head size (kg/head)	1.59 (0.07)	1.21[b] (0.12)	1.37 (0.04)	1.43 (0.12)	1.15[b] (0.08)
Harvest date[a]	22 September	6 October	14 October	6 October	14 October
% Marketable heads	92.5 (4.8)	75.0 (14.4)	87.5 (9.5)	100 (0.0)	55.0 (11.9)
Mulch dry wt (g/m^2)	Nil	674 (134)	467[b]	314 (78)[b]	407 (90)[b]

[a]Planting date = 2 July
[b]Significantly different from bare ground by Duncan's procedure ($P<0.05$).
After Andow *et al.* (1986).

Intercropping cabbage and beans reduced oviposition by brassica root flies (*Delia radicum* and *D. floralis*) by 29% compared with monocultures (Hofsvang, 1991). There was also a reduction in oviposition when the crops were mixed with 'weeds', where reductions of 63 and 40% in eggs per cabbage plant were recorded in two seasons. Such effects are summarized in Table 6.14.

In summary, reduced tillage in combination with cover crop mulch systems can conserve beneficial insects. For example, predatory wasps nest in the ground and tillage interferes with their reproduction. Cover crop mulches may also reduce pest dispersal, reproduction and colonization of host plants. Plant compounds released by cover crop residues may influence host plant

Table 6.14. Studies on the effect of plant diversity (intercropping/undersowing/weeds) on oviposition by *Delia radicum.*

Plant diversity	Reduction in oviposition compared with monoculture (%)	Observations	References
Brussels sprout/cauliflower/ clover	29	Eggs	Demster and Coaker (1974)
Brussels sprout/clover	60	Eggs	O'Donnell and Coaker (1975)
Brassicas/beans, spinach, clover, grass	53–77	Eggs	Coaker (1980)
Cabbage/clover	26–65	Eggs	Ryan *et al.* (1980)
Brussels sprout/spurry	30–99	Infestation	Theunissen and Den Ouden (1980)
Cabbage/spinach	36–44	Eggs	Tukahirwa and Coaker (1982)
Rape/clover, weeds	64–89	Infestation	Coaker (1988)

After Hofsvang (1991).

selection for oviposition and larval feeding. Cover crop mulches can confuse pests visually or olfactorily, reducing colonization of *Brassica* crops. Important visual cues for insects are leaf colour, area and visual prominence of the hosts. Twice as many cabbage root fly eggs were found on green and yellow models compared with red or blue ones. Cabbage root fly landings increased linearly with host leaf area. Diamond back moth has a strong preference for egg laying on dark green hosts.

Tillage had a significant effect on cabbage maggot and diamond back moth incidence. Larger numbers of both pests were associated with the tillage treatments compared with non-tillage treatments. Rye or hairy vetch can, however, reduce cabbage yields. Rye residues have a high carbon to nitrogen ratio and their decomposition could immobilize soil nitrogen, thereby reducing cabbage yields.

In several studies, particular soil properties were improved by cover cropping, and weed and insect pest populations were reduced but yields also fell. Yield reductions could have resulted from the immobilization of soil nitrogen, lower soil temperatures or allelopathy. Strip tillage, which cultivates the row where the brassicas are planted leaving residues between crop rows intact, may overcome reduced *Brassica* vegetable yields while combining with the advantages of both conventional tillage and cover crop mulch systems as identified by Mangan *et al.* (1995) (Table 6.15).

This information can be used to predict how mixtures of plants can be used to reduce pest infestation and damage. So far, it has been difficult to

Table 6.15. Incidence of cabbage maggot eggs and diamond back moth larvae in relation to tillage and cover cropping.

Treatment	Numbers per plant	
	Cabbage maggot eggs (*Delia radicum*)	Diamond back moth larvae (*Plutella xylostella*)
Rye		
Tillage	0.13	0.44
Mowing	0.07	0.00
Rye + vetch		
Tillage	0.34	0.47
Mowing	0.03	0.00
No cover		
Tillage	0.40	0.20
Mowing	0.20	0.12
Factorial F test[a]		
Cover (C)	NS	NS
Tillage (T)	*	**
C × T	NS	NS

[a] NS, i.e. $P > 0.05$; * $P < 0.05$; ** $P < 0.01$.
Site = South Deerfield, Massachusetts, USA; 3 × 2 × 2 factorial experiment., cover crops = rye (*Secale cereale*) + vetch (*Vicia villosa*); rye alone; no cover crop; tillage = conventional = cover crop tilled in and no till = cover crop mowed; herbicide = DCPA (dimethyl tetrachloroterephthalate).
After Mangan *et al.* (1995).

make such predictions in relation to the manipulation of *Brassica* habitats. If animal behaviour is the key to such predictions, however, then it may be possible to identify herbivore abundance in novel horticultural systems. This requires a basic understanding of the pest and predator's natural history (habitat preference, diet and general behaviour). This information can be used to suggest the taxonomic and structural plant diversity needed to achieve a degree of resistance developed by placing *Brassica* crops in association with other plants.

OVERSEEDING

Overseeding is a technique used after the crop plants have been established to obtain a uniform stand of living mulch and suppress emerging weeds. It is particularly effective in a non-tillage system where the soil surface is left undisturbed before and after transplanting. Transplanted crops are suitable for overseeding because the transplants have a growth advantage over the post-transplant emergence and growth of weeds and overseeded species. In addition to controlling weeds, overseeding when used to form living mulches can minimize erosion, decrease soil temperature, improve the water infiltration rate, improve soil structure, favour microbial activity and increase crop yield.

In most non-tillage systems, all live vegetation is desiccated with contact herbicide before planting to achieve a weed-free, stale seedbed environment. The stale seedbed is a form of limited tillage normally applied to plough and disc harrowing systems in which a flush of new weed seedlings germinating after tillage is killed by chemical or mechanical means before planting the cash crop. Often the desiccated residues rapidly decompose after planting, exposing the soil to potential erosion and soil moisture deficits. Overseeding can maintain the integrity of non-tillage systems by replacing the dead decaying mulch with a non-competitive living mulch. After harvesting the cash crop and chopping the remaining crop residues to encourage decomposition, the overseeded living mulch becomes a valuable overwintering cover crop. It is claimed to provide a non-tillage system in which there is no competition with the cash crop for resources such as water, nutrients, space and light. The results of overseeding were found neither to increase nor to decrease the yield of broccoli.

CHEMICAL MANAGEMENT OF WEEDS

Chemical management of weeds includes the use of herbicides and soil fumigants. In the past, fumigants have been used on land where intensive brassicas are grown. In particular, this involved the use of methyl-bromide to destroy soil-borne pathogens and weed seeds. Use of this chemical is being discontinued internationally to prevent further damage to the atmospheric ozone layer, and safer alternatives are being sought.

Herbicides may be selective or non-selective in their mode of action. Selective herbicides destroy specific plant species and may thus be used to eliminate them from populations of other species. Probably the most widely used selective herbicide for *Brassica* crops is trifluralin (α,α,α-trifluoro-2, 6-dinitro-*N*,*N*-dipropyl-*p*-toluidine). Non-selective herbicides such as glyphosate (*N*-(phosphonomethyl) glycine) and paraquat (1,1′-dimethyl-4,4′-bipyridinium) will destroy all green tissues that they come into contact with. Thus they are used, for example, to destroy weeds that germinate in stale seedbeds prior to drilling or transplanting the cash crop. These two chemicals are inactivated once they come into contact with soil.

Herbicides can also be classified according to the timing of their application in relation to crop growth stage. These are broadly the pre-emergence and post-emergence categories. Those with pre-emergence characteristics are applied prior to or after sowing the seeds of *Brassica* crops and prior to their emergence. Post-emergence herbicides are applied once the crop has emerged, to destroy competing weed species.

Well-established pre-emergence herbicides for the control of annual weeds in drilled and transplanted *Brassica* crops in the UK include trifluralin (α,α,α-trifluoro-2,6-dinitro-*N*,*N*-dipropyl-*p*-toluidine) and propachlor (2-chloro-*N*-isopropylacetanilide). Neither kills all the weed species commonly present in these crops so their spectra of activity are largely complementary and they can be applied in combination (Roberts and Bond, 1975). Initially this involved two operations because trifluralin needs to be incorporated into the soil before drilling or transplanting, while it is recommended that propachlor is applied to the soil surface after drilling. In attempts to avoid this level of complexity in crop management, experiments studied the effects of combined applications. Trifluralin can be applied as a granular formulation at between 0.56 and 1.12 kg/ha and propachlor as wettable powder at between 2.18 and 4.37 kg/ha. Applying mixtures and incorporating propachlor into the soil had little effect on the efficacy of propachlor, and cabbage yields were not significantly affected. Incorporation of both chemicals in a combined application using granular formulations appears to be another cost-effective strategy. There may be a need to increase the dose rate of propachlor to compensate for the effects of incorporation. There are very few if any recommended post-emergence selective herbicides that may safely be applied to vegetable *Brassica* crops. Generally, growers attempt to ensure that crops are established either from seed or by transplanting without weed competition. Subsequently, the crops normally grow speedily enough to close their canopies before weeds become established.

LIMITED AVAILABILITY OF HERBICIDES FOR *BRASSICA* CROPS

In the USA, a collaborative project was established in 1963, known as the IR-4 Project (Inter-regional Research Project Number Four). Collaboration has been established between state agricultural research stations, the US Department of Agriculture-Cooperative States Research Service (USDA-CSRS), chemical manufacturers and growers organizations to identify and register herbicides for use with speciality crops such as the brassicas (Hopen, 1995; Baron *et al.*, 2002). This initiative was driven by the lack of suitable chemicals being placed on the market by the chemical industry for these small area crops. The project (IR-4), with headquarters at Rutgers University, New Jersey, aimed to assist in gaining registration and label recommendations of agrochemicals for use on highly intensive vegetable crops (Baron *et al.*, 2004). Labels are of national, regional, state need and state emergency categories. The programme succeeded in expanding the availability of herbicides for *Brassica* crops. In particular, the participants attempted to find molecules that were safe and effective for use as selective herbicides applied to established crops.

Weeds such as hairy galinsoga (*Galinsoga ciliata*), wild proso millet (*Panicum miliaceum*), velvet leaf (*Abutilon theophrasti*) and yellow nutsedge (*Cyperus esculentus*) are difficult to control in standing crops and cause considerable loss of yield and quality. Comparable schemes are under development in Canada and Australia.

A project with similar aims is established in the UK, operated by the Horticultural Development Council (HDC) and known as the Specific Off-label Approval Scheme (SOLA). The HDC assembles a dossier of information for a particular chemical that might be valuable for growers and submits this for consideration for SOLA approval to the Pesticide Safety Directorate (PSD). Regrettably, the agrochemical industry in Europe is not prepared to contribute financially towards the success of the SOLA programme. Similar types of scheme are emerging in many countries to support vegetable growers. The European Union (EU) is currently considering the use of a scheme similar to SOLA across the Community. At the same time, opportunities for the use of chemicals in vegetable crop production are being severely limited by national governments and international organizations such as the EU. The latter, for example, is undertaking a re-registration process for herbicides and other crop protection chemicals, as is the American Environmental Protection Agency (EPA). Agrochemical companies are being required to re-submit applications for older chemicals with data brought up to current standards. Frequently, the companies are reluctant to do this for crops grown on limited areas of land because the expense involved is not justified by the income gained from sales. In consequence, many of the vegetable and especially *Brassica* crops are now devoid of herbicides for specific and specialized aspects of weed, pest and pathogen control. Although a few herbicides have been used in *Brassica* production for many years, these are exceptional. New molecules with apparent potential value for these crops are frequently suggested, but few achieve regulatory acceptance and even if this happens they may disappear from the market very quickly for a range of commercial reasons. Because of this transient situation which applies to all agrochemicals (herbicides, insecticides, fungicides and growth regulators), no lists of chemicals are included in this book, but these are available in Anon (2006).

WEED CONTROL IN *BRASSICA* SEED CROPS

Brassica seed crops suffer from weed competition in an analogous manner to ware or cash crops. Weed competition may reduce seed yield dramatically in cabbage by >50% (Al-Khatib and Libbey, 1992; Al-Khatib *et al.*, 1995). The presence of weeds at seed harvest increases mechanical damage to cabbage seed and reduces harvest efficiency; weeds also reduce seed quality by interfering with the processing operations. Weeds such as bedstraw (or cleavers) (*Galium aparine*), charlock (*S. arvensis*) and wild mustard (*B. kaber*) are of similar size to cabbage seed, making separation during cleaning operations difficult. Crop seed contaminated by weeds will be rejected during seed testing and certification as regulated by the International Seed Testing Association (ISTA) and controlled by national and international statutory legislation.

Weed control in seed crops is difficult because of the extended growing seasons required. The seed production cycle is at least 12 months in extent and in some crops even longer. Herbicides are available for the control of grass weeds (sethoxydim, (+/–)-(*EZ*)-2-(1-ethoxyiminobutyl)-5-[2-(ethylthio) propyl]-3-hydroxycyclohex-2-enone; and fluaziflop, (R)-2-[4-(5-trifluromethyl-2-pyridyloxy) phenoxy] propionic acid), but the control of broadleaf weeds is limited by the lack of effective registered chemicals.

Triflualin (α,α,α-trifluoro-2,6-dinitro-*N,N*-dipropyl-*p*-toluidine), oxyfluorfen (2-chloro-α,α,α-trifluoro-*p*-tolyl-3-ethoxy-4-nitrophenyl ether), clomazone (2-(2-chlorobenzyl)-4,4-dimethyl-l,2-oxazolidin-3-one), napropamide ((*RS*)-*N,N*-diethyl-2-(1-naphthyloxy)proprionamide), clopyralid (3,6-dichloropyridine-2-carboxylic acid) and pyridate (6-chloro-3-phenylpyridazin-4 yl) have been used to control broadleaf weeds in both seed crops and brassicas grown for processing into sauerkraut.

Withdrawl of nitrofen (2,4-dichlorophenyl *p*-nitrophenyl ester) in 1980 made the control of broadleaved weeds in all cole crops difficult (Bhowmik and McGlew, 1986). Oxyfluorfen provides good to excellent control in broccoli (*B. oleracea* var. *italica*), Brussels sprout (*B. oleracea* var. *gemmifera*), cauliflower (*B. oleracea* var. *botrytis*) and cabbage (*B. oleracea* var. *capitata*) grown for seed production. Several reports indicate varying degrees of crop tolerance, especially with liquid formulations applied post-transplanting. Pre-transplanting treatments of oxyfluorfen alone or followed by sequential post-emergence grass herbicides could be safe at rates that provided good weed control without any harmful effects on cabbage seed yield and quality. The recommended rate for oxyfluorfen for acceptable weed control is 0.43 kg/ha. The safe use of oxyfluorfen at 0.43 kg/ha as a post-transplanting treatment in cabbage production would increase the choice of herbicides for broadleaved weeds.

7

Pests and Pathogens

INTRODUCTION

Pests and pathogens of brassicas are considered in this chapter mainly from the perspective of their control. Details of host–pest and host–pathogen interactions, biology and epidemiology are widely available elsewhere in numerous textbooks. As with weeds (Chapter 6), the strategies underlying methods of control are shifting very substantially away from reliance on single 'magic bullet chemicals'. Previous decades saw an explosion in the use of synthetic chemicals that frequently acted by disrupting specific sites in enzyme-guided metabolic pathways of both pests and pathogens. These chemicals had modes of action which broadly included eradicants, protectants, soil sterilants, seed dressings and systemic compounds. Many were used in *Brassica* crops destined for human consumption albeit following stringent evaluation and testing to determine the extent of their interaction with the environment, human health and other non-target organisms. Although such chemicals do provide remarkably effective pest and pathogen control, they have many disadvantages. These chemicals exert intense selection pressures on the populations of the target organisms, leading to the rapid appearance of tolerant strains in the well-established 'boom and bust' cycle (Dixon, 1981). They adversely affect surrounding populations of non-target organisms, diminishing forms of natural biological control, and may lead to undesirable residues and by-products reaching human and other food chains. Public recognition of these disadvantages has led to social pressures demanding the drastic limitation, and in many cases complete banning, of many pesticides.

The use of synthetic chemicals in *Brassica* crop production is diminishing especially in Europe, and also increasingly elsewhere wherever the quality assurance processes demanded by Western supermarkets are made mandatory. One consequence is that most agrochemical companies are very reluctant to develop new molecules for use specifically in *Brassica* crops. These are no longer seen as a profitable market sector and, additionally, many traditional pesticides have disappeared because the companies find it uneconomical to support them as a result of the ever more stringent statutory requirements for data concerning their toxicological and environmental impact. It is predicable that brassicas grown worldwide for human consumption will be produced with only very minimal use, if any, of synthetic pesticides within the next two decades. Consequently, specific references to individual pesticides and their uses for the control of pests and pathogens of *Brassica* crops are avoided in this book except where there is relevance to a general biological or agronomic principle concerned with control. Specific references to the use of insecticides, fungicides and other agrochemicals are largely limited to the latter parts of this chapter where particular examples of pests and pathogens are provided.

Currently, developing strategies for pest and pathogen control utilize combinations of control measures in integrated systems (Theunissen, 1984). A basic component of integrated control systems is the use of forms of genetic resistance. *Brassica* crops, because of their genetic flexibility, lend themselves admirably to this approach and, even in the era of chemical control, pest- and pathogen-resistant cultivars played a major role in assisting crop production. Examples of the development of resistant cultivars are described for several of the pests and pathogens discussed later in this chapter.

The techniques used for pest and pathogen control on brassicas, like other horticultural crops, are typified by their diversity, as advocated in Dixon (1984). These include the use of biological methods of control, a range of cultural techniques that minimize pest and pathogen attack, and certified healthy seed and planting material. The recent emergence of a market for seed grown by organic methods which excludes the use of any synthetic chemicals has added impetus for the development of diversity in the components used in integrated control strategies. The control of pests and pathogens in *Brassica* crops now demands an open-minded and flexible approach that accepts novel and unusual techniques. This typifies the changing attitude overall to pest and pathogen control that now seeks to alter the edaphic and aerial environments in favour of the crop and to the disadvantage of the pest or pathogen. Successful use of this strategy demands a detailed understanding of host and pest or pathogen biology combined with an understanding of the surrounding physical, chemical and biological environments.

A diverse range of factors governs the extent of damage caused by predatory organisms. Factors such as host plant chemistry, the geographical region in which the crop is grown, the availability of alternative hosts and environmental temperature, rainfall and soil structure affect the outcome of encounters between pests, pathogens and their hosts. *Brassica* crops are vulnerable to very large numbers of pests and pathogens. Regrettably, only a few of those with major significance worldwide can be included here. The selection of damaging pests and pathogens discussed in this chapter solely highlights general principles related to those that are of greatest significance for both ware and seed vegetable *Brassica* crops (Dixon, 1981, 1984; Jones and Jones, 1984; Finch and Thompson, 1992).

FORMS OF CROP LOSS CAUSED BY PESTS AND PATHOGENS

Land capital value

Land areas devoted to individual vegetable *Brassica* crops tend to be relatively small, although in specific geographical regions they may appear as substantial concentrations by virtue of favourable soil types, climatic conditions, cultural traditions and, consequently, the availability of skilled labour. Concentration of production especially where several *Brassica* crops are grown on one area of land in successive seasons almost inevitably leads to rampant infestation with soil-borne pests and pathogens. These diminish crop yield and may well render the land completely unfit for economic production of *Brassica* crops. In consequence, the longer term asset value of the land drops. Classic examples of pathogen and pest infestations that diminish land value include clubroot disease caused by *Plasmodiophora brassicae*, white rot caused by *Sclerotinia sclerotiorum* and the *Brassica* cyst nematode, *Heterodera cruciferae*.

Direct yield loss

Single vegetable *Brassica* plants have high individual cash values. Pest and pathogen damage to even small percentages of a crop can have devastating effects on yields and economic returns. Indeed, even the presence of only slight blemishes may make produce unacceptable to the market and hence totally valueless. Consequently, the impact of damage is not directly correlated with the incidence and intensity of pathogen or pest population as is often the case with extensive agricultural crops. A range of forms of loss may arise in *Brassica* crops throughout the supply chain from the field to the point of retail sale, for example:

- failure of growth and reproduction by individual plants in the field;
- damage to the saleable portions of the plant in the field;
- distortion of the harvesting sequence with consequent failure to reach the desired market at an optimal time and maximum value;
- rejection during harvesting, grading or display;
- development of blemishes in transit, during storage or at points of sale;
- downgrading during the marketing chain;
- total rejection at any point from field to retail sale.

The consumer increasingly fails to recognize the connection between demands for reduced use of pesticides in food production and the likelihood of increasing the frequency of blemishes developing. Consumers are not prepared to accept *Brassica* products where there are signs of pest activity, eggs, larvae or frass, nor areas of disease symptoms caused by pathogen infection.

These demands expressed through supermarket buyers, for blemish- and pesticide-free produce, are driving strenuous research and development programmes which aim to find alternative approaches to pest and pathogen control.

CROP MONITORING, FORECASTING AND DISEASE MANAGEMENT

The philosophy of 'forewarned is forearmed' whereby pest and pathogen invasion is anticipated and crops monitored for the first signs of their arrival is not new, but the application of new technologies makes it more efficient and effective. Monitoring *Brassica* crops helps to ensure that control measures, particularly where chemical methods remain available, are applied swiftly and effectively, and as part of the integrated strategy. This contrasts with the previous practices whereby chemical control measures were either applied at specified and regular intervals during the growing season irrespective of the presence or absence of a particular pest or pathogen or they were used as 'fire brigade' tactics once an outbreak had reached visibly damaging proportions.

In the former case, pesticides were often used in excess of their requirement, resulting in financially costly, wasted applications and environmental pollution as, for example, with powdery mildew (*Erysiphe cruciferarum*) on Brussels sprouts where little is known about the threshold values for spray control and chemical application is advocated as soon as symptoms appear in August or early September. With the 'fire brigade' approach to control, applications were made when damage had reached the level of visual symptoms and the attendant reduction to quality of the product was probably beyond repair. Again expensive chemicals were wasted accompanied by increased environmental pollution. The essence of integrated control lies in the combined use of genetically resistant cultivars, husbandry and nutritional strategies (Chapters 2, 4 and 5) that decrease the likelihood of pest and pathogen outbreaks and minimal use of pesticides applied on the basis that an outbreak is predictable in its occurrence. Components that can form elements in the integrated control strategies for either pests or pathogens were summarized by S. Finch (personal communication) under two headings: (i) *well established methods* such as: insecticides, forecasting, supervised control, cultural control and physical control; and (ii) *experimental systems still being developed* such as: undersowing, parasitoids, predators, parasitic viruses and bacteria, parasitic fungi and nematodes, and behaviour-modifying chemicals.

Each of these components is considered at various points in the chapters of this book.

Environmental monitoring

Prediction and monitoring programmes require the regular availability of current weather information that may be compared with past seasonal records and data as to those environmental factors that encourage pest and pathogen establishment and reproduction. Mathematical models help to describe the interaction between pests or pathogens, their hosts and the environment. They incorporate data for factors such as:

- dew formation and its timing;
- minimum and maximum temperatures and duration;
- sunshine hours;
- rainfall;
- relative humidity;
- the succession or periodicity of conducive or suppressive weather such as numbers of days with warm, dry or wet conditions;
- cloudiness and the length of periods of overcast skies;
- leaf wetness periods.

The cabbage aphid (*Brevicoryne brassicae*), for example, arrives on newly transplanted crops in Europe at any time between May and August, with normally a natural dramatic decline in population numbers in late August followed by an upsurge during early autumn. Routinely programmed spraying is unlikely to provide adequate control and may well encourage the development of pesticide tolerance in aphid populations. This is especially likely as the number of effective pesticides has been reduced recently and hence it becomes increasingly more difficult to use combinations of chemicals of differing molecular structures and modes of action. A link has been established between the rate of aphid development and ambient temperature. A mathematical model is used to link seasonal weather and the rate of aphid development, thereby predicting when damaging insect populations will expand and are most susceptible to control.

Models are in development that could help control of pathogens in brassicas such as *Alternaria brassicae* (dark leaf spot), *Mycosphaerella brassicicola* (ringspot) and *Albugo candida* (white blister). These relate weather conditions such as the duration of leaf wetness and ambient temperatures to spore formation, release, dispersal and survival, with re-infection and colonization and subsequent disease development.

Electronic systems for weather monitoring in crops have become increasingly available (Fig. 7.1).

Fig. 7.1. Meteorological monitoring station (Skye Instruments).

Data gathered from meteorological stations set up within *Brassica* crops are compared with the mathematical models providing prediction programmes for individual pests and pathogens.

Pest and pathogen monitoring

Air-borne spread of pests and pathogens needs to be monitored in order to identify incoming population loads, allowing the assessment and prediction of the likely speed with which an epidemic may develop.

Knowledge of the effects on spore dispersal of air movements and the resultant distances over which pathogens may be dispersed are vital information in constructing models of epidemic development. Climatic factors such as long-term rainfall, short-term intermittent rain patterns, relative humidity and temperature are all important in determining the success or otherwise of parasite epidemics. Some of these factors will also influence the success of soil-borne pests and pathogens, but the amount of background knowledge with which to interpret the data is limited for these organisms. Prediction programmes are based on years of accumulated knowledge of the epidemiology of specific pests and pathogens and their attendant symptoms. A further prerequisite for gathering such information is knowledge of the form and structure of symptoms caused by particular parasites.

Field monitoring

Regular crop walking and inspection are essential for the successful prediction of pest and pathogen epidemic development. The crop is inspected at regular intervals to determine whether pest and pathogen problems are emerging. Quantitative evaluations of a developing pest or pathogen problem are gained visually through the manual use of pests and disease assessment keys that quantify developing signs and symptoms. Examples are shown in Figs 7.2 and 7.3 for powdery mildew (*E. cruciferarum*) on the buds and leaves of Brussels sprout (*B. oleracea* var. *gemmifera*) (Dixon, 1976), respectively.

Quantifying a developing pest infestation can utilize basic biological processes of pests to advantage for their control. The first such process is that of their ability to detect odours (smelling) (Cobb, 1999). Insects may be drawn to odours emitted by the crops.

The cabbage seed weevil (*Ceutorhynchus assimilis*), for example, responds positively to extracts of oilseed rape (*B. napus*) and to the colour of the crop plants. The active attractant molecules are α-farnesene (component of flowers) and 3-butenyl and 4-pentenyl isothiocyanates (analogous to components of foliage) which are components of *Brassica* (Evans and Allen-Williams, 1998).

The insect sex drive can be utilized in monitoring through attractive chemicals that females emit to attract males which encourages them into what are termed pheromone traps that can be used to assess insect populations and their likely potential to cause damage to crops. Monitoring the size of pest infestations using traps treated with sex pheromone chemicals to attract males is a well established technique in integrated pest management (Ridgway *et al.*, 1990). The timing of subsequent applications of insecticides in order to gain control depends on the establishment of reliable threshold values with which to predict the onset of damage.

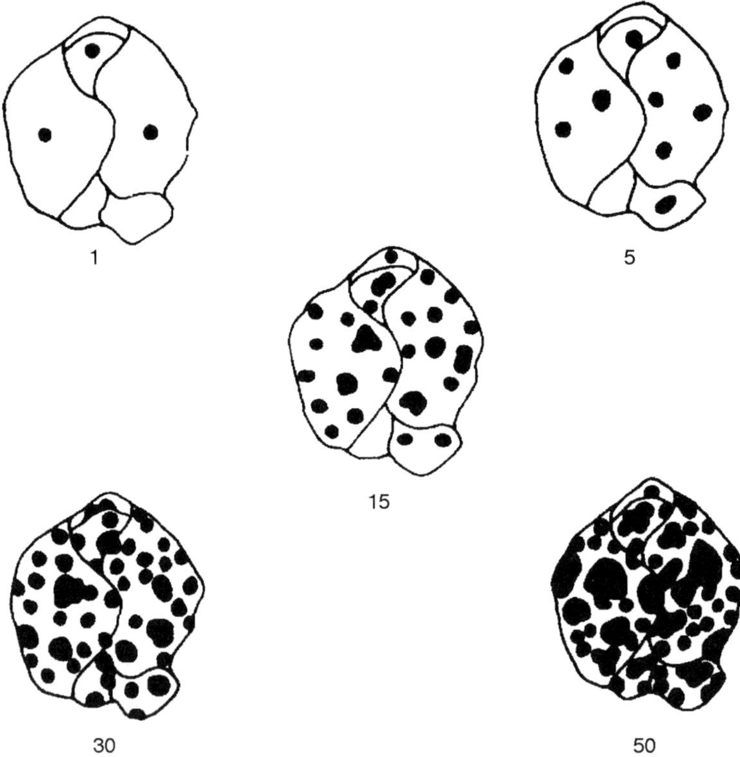

Fig. 7.2. Brussels sprout bud powdery mildew (*Erysiphe cruciferarum*) disease assessment key.

There is frequent use of natural pheromones and their synthetic analogues attractive to male diamond back moth (*Plutella xylostella*). Factors such as type of trap, trap height, duration of pheromone activity and diurnal patterns of attraction need to be optimized by field studies allowing the establishment of ecologically efficient thresholds for integrated pest control. This approach for diamond back moth (*P. xylostella*) in crops of cabbage (*B. oleracea* var. *capitata*), cauliflower (*B. oleracea* var. *botrytis*) and knol khol (*B. oleracea* var. *gongylodes*) has been compared with visual monitoring where insecticides are applied as soon as a threshold value of one hole per leaf is reached. Reddy and Guerrero (2001) showed that when eight, 12 and 16 male moths were caught per night in cabbage, cauliflower and knol khol, respectively, this was indicative of the optimum time to spray (Table 7.1). This replaced a routine spraying system of making applications every 7–15 days depending on the risk of pest damage. The use of pheromone-mediated monitoring replaces more costly and often more cumbersome sampling techniques. This work has established a good correlation between numbers of insect larvae, levels of infestation and crop yields with the numbers of adult males attracted to and caught by the pheromone traps.

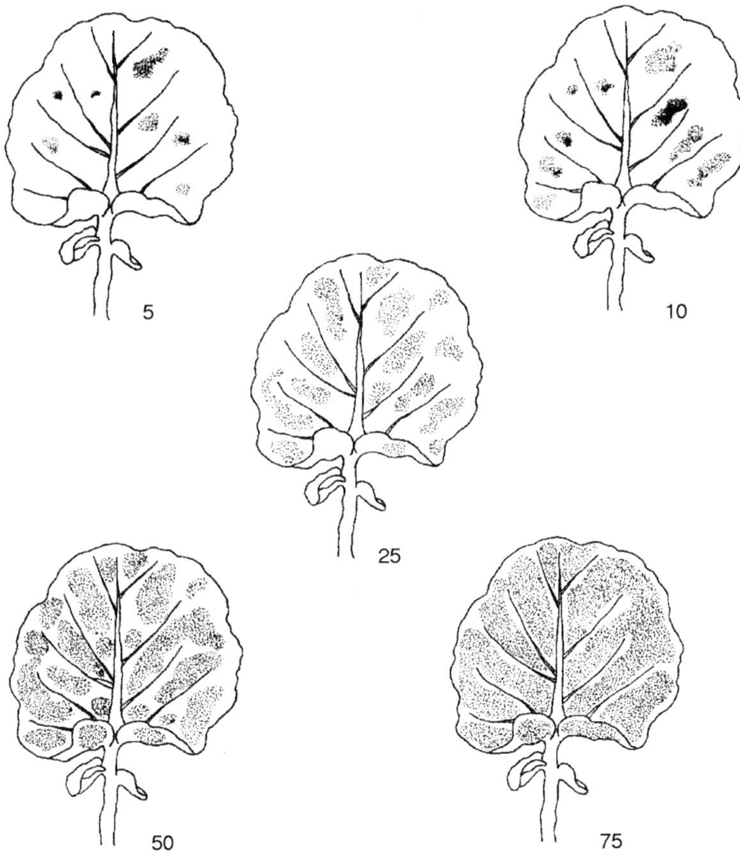

Fig. 7.3. Brussels sprout leaf powdery mildew (*Erysiphe cruciferarum*) disease assessment key.

Laboratory systems

Increasingly, visual assessment of the onset of pathogen damage is supplemented by automatic sampling for fungal spores that can be identified subsequently in the laboratory. Monitoring the ingress of fungal spores identifies the early stages in the development of a disease epidemic and adds significantly to the precision with which decisions on the use of fungicidal sprays are made. Progress in establishing criteria for evaluating population expansion by leaf and pod spot (*Alternaria* species), light leaf spot (*Pyrenopeziza brassicae*), ringspot (*M. brassicicola*), powdery mildew (*E. cruciferarum*) and downy mildew (*Peronospora parasitica*) is substantial. Monitoring incoming air-borne spores offers a direct measure of the risk of crop infection. For example, concentrations of the air-borne ascospores of *S. sclerotiorum* have been related

Table 7.1. Comparison of integrated pest control of diamond back moth (*Plutella xylostella*) with pheromone trapping and routine insecticide spraying on marketable yield of three *Brassica* crops.

	Site 1 – Hebbal			Site 2 – Devanahally		
Treatment	Cabbage	Cauliflower	Knol khol	Cabbage	Cauliflower	Knol khol
Pheromone (4 males/trap/night)	19.3a	15.5a	17.5a	18.5a	14.0a	16.5a
Pheromone (8 males/trap/night)	16.5a	14.0a	16.2a	16.0a	12.5a	16.0a
Pheromone (12 males/trap/night)	14.5b	12.0a	15.4a	14.0b	10.0a	14.5a
Pheromone (16 males/trap/night)	8.0d	7.5c	16.5a	8.5d	7.0c	14.5a
Pheromone (20 males/trap/night)	4.9e	4.8d	9.2c	5.4e	3.9d	8.8c
Routine spraying 7 DAT	12.5c	8.5b	13.0b	10.8c	8.5b	12.4b
Routine spraying 9 DAT	13.0c	8.3b	12.7b	10.5c	9.0b	12.2b
Routine spraying 12 DAT	12.8c	8.7b	12.8b	10.7c	8.4b	12.4b
Routine spraying 15 DAT	12.5c	8.5b	12.5b	10.5c	8.6b	12.0b
Control – no pest control	3.6f	2.2e	4.2d	2.2f	1.8e	3.8d

The sites studied were in Karnataka State, India.
Marketable yield = t/ha.
DAT = days after transplanting.
Means within each column that are followed by a similar letter do not differ significantly when tested with Duncan's multiple range test, $P = 0.05$.
After Reddy and Guerrero (2001).

to subsequent disease development in oilseed rape (*B. napus*) crops. Traditional methods of detection of air-borne inoculum are time-consuming, labour intensive and subjective, since they rely on the identification of spores by microscopy or aseptic cultural techniques. The ascospores of *S. sclerotiorum*, for example, are small, lack unique characteristics and are similar to those of other fungal species, and are therefore extremely difficult to identify morphologically with any degree of certainty.

Although the pathogen can be cultured, it is slow growing and it can take at least several days to obtain results. There is a need for new, more accurate and easy to use field-based methods. The ELISA (enzyme-linked immunosorbent assay) method is used successfully to detect virus pathogens especially in potatoes, and could be applied to pests and pathogens of *Brassica* vegetables. Recently, the potential of other molecular methods for the detection of air-borne microbes has been recognized. DNA-based methods are used to detect several species of air-borne bacteria. At present, DNA-based methods have only been developed for a few fungi. A PCR (polymerase chain reaction) assay for the detection of *S. sclerotiorum* involves specific primers designed to use nuclear ribosomal DNA (rDNA) internal transcribed spacer sequences (ITS). The ITS regions of nuclear rDNA evolve relatively rapidly and thus are highly variable, showing differences between closely related species and sometimes within species.

It is anticipated that molecular field test kits sensitive to unique genetic fragments of the pathogen will be marketed in the next few years. These will allow field detection of incoming fungal spores, leading to quicker prediction of disease development. The time lag between the arrival of spores and the expression of visual disease symptoms provided by early and rapid identification could increase the practical window during which control measures are applied. This increases very substantially the precision with which agrochemicals may be utilized, reducing their cost and potential environmental impact.

Economic thresholds

Where prediction and forecasting systems indicate the development of an epidemic caused by a pest or pathogen, ultimately a decision is needed as to whether or not it is economically sensible to apply control measures. Frequently, in vegetable brassicas, this decision turns on whether or not the saleable portion of the crop will be affected by the pest or pathogen. For example, high levels of leaf injury by caterpillars can be tolerated in Brussels sprouts during their early stages of growth. A threshold level for chemical control must consider both quantitative and qualitative damage to the axillary buds. The relationship between artificial defoliation and yield on Brussels sprouts was used to mimic the quantitative damage induced by leaf-eating insects. From the beginning of September onwards, the yield of Brussels sprout plants was insensitive to defoliation of up to 60% of leaf area. Similarly, it is unlikely that powdery mildew will cause quantitative damage since infection by *E. cruciferarum* does not start until late August to early September; the major effect is likely to be on the quality of buds produced rather than on the growth potential of the plant. The leaf injury by powdery mildew that can be tolerated from late August onwards may be determined by the following equation:

$$T = 5 + 0.42 \times (\text{Julian date} - 235)$$

where T = tolerable leaf injury in percentage leaf area covered.

When sampling the crop to assess the powdery mildew infection, it is necessary that leaves are taken from all parts of the stem as the top leaves tend to be much less heavily infested compared with those older ones lower down. This threshold level must be considered as preliminary and has to be validated under higher disease pressures with differing levels of quantitative damage.

PRACTICAL REQUIREMENTS FOR FORECASTING

The following conditions need to be met for satisfactory forecasting:

- The pest or pathogen causes economically significant damage by reducing the quantity or quality of yield in the region or area monitored.
- Arrival of the pest or pathogen can be identified prior to damage reaching economically measurable proportions.
- There is seasonal variation in the impact of the pest or pathogen in terms of time of onset, speed at which the epidemic builds up and the ultimate upper level of disease intensity.
- An appreciable part of the variation in impact of the pest or pathogen may be attributed either directly or indirectly to environmental factors.
- Control measures, in either curative or preventative forms, are available, acceptable to the consumer and at an economically bearable cost to the producer.
- Information is available from field and laboratory studies on the nature of the interactions between the pest or pathogen and weather conditions.

THE CHANGING SPECTRA OF CROPS, PATHOGENS AND PESTS

Recently in Europe and parts of Asia, greater production of agricultural oilseed brassicas (mainly *B. napus*) has increased the incidence of some *Brassica* pests and pathogens on horticultural crops (Lamb, 1989). This situation is exacerbated as the economic threshold for damage caused by pests and pathogens on oilseed rape is higher than for horticultural brassicas, and therefore less control is used. Consequently, large pest or pathogen populations can develop in oilseed rape in the absence of control and then move *en masse* on to horticultural crops with devastating consequences. Overwintering oilseed rape (*B. napus*) provides a substantial 'green bridge' for light leaf spot (*P. brassicae*) and in consequence it is the major foliar disease of that crop. In turn, rape provides a reservoir of infection that transfers on to vegetable brassicas and can cause appreciable losses of quality to leafy types.

Similar effects have been noted with blackleg (*Leptosphaeria maculans*) and dark leaf and pod spot (*Alternaria* spp.) where the production of large areas of agricultural brassicas adjacent to vegetable crops poses considerable threats due to their release of vast quantities of inoculum.

Changes to the spectrum of vegetable brassicas grown are a result of a widening of diversity in consumer demand. Increasingly, European and North American growers are producing 'novelty' *Brassica* crops, especially those of Asian origin. This provides omnivorous pests with increased feeding grounds and limits those of specialist feeders. This means that certain pests are becoming more widely spread around the world and there are increasing populations of insects ready and able to destroy crop profitability. The increased worldwide genetic homogeneity of *Brassica* crops with cultivars of similar genetic backgrounds being used in Asia, Europe and North America means that pests are presented with an acceptable, uniform diet wherever they spread round the globe. The value of heterogeneity of flora in field cropping has been described in Chapter 6 in relation to weed control. The impact of global warming on the distribution and frequency of plants and animals is now established (Root *et al.*, 2003). The relative importance of individual organisms is also changing. The diamond back moth (*P. xylostella*), for example, appears to be increasing in importance as a pest in parts of northern Europe where previously it was viewed as solely an entomological novelty. In a similar manner, black rot caused by the bacterium *Xanthomonas campestris* pv. *campestris* has become a major disease of northern European *Brassica* crops.

The feeding, growth and reproductive activity of insects and other ectothermic animals is strongly dependent on their body temperature. Prevailing weather conditions determine the temperature microclimate within the crops on which the insects live. There is strong selection pressure for rapid growth and development to decrease the time during which caterpillars of, for example, small cabbage white butterfly (*Pieris rapae*) are exposed to enemies. Field experiments with this insect showed that a slower growth was correlated with higher mortality due to the parasitoid *Cotesia glomerata* (Benrey and Denno, 1997). Consumption of collard leaves (*B. oleracea* var. *acephala*) by small cabbage white butterflies and their growth rates increased by six- to eightfold as temperature rose from 10 to 35°C (Fig. 7.4) (Kingsolver, 2000) and declined rapidly at higher temperatures. Herbivores such as *P. rapae* are able to sustain growth over the wide and fluctuating range of body temperatures that they routinely experience in the field. This insect thrives in a wide range of weather and climatic conditions throughout its larger geographical range (Gilbert and Raworth, 1996).

STRATEGIES FOR PEST AND PATHOGEN CONTROL

Host resistance

Genetic resistance to pests and pathogens is the principal means of control that is both economically and environmentally acceptable. Plant resistance to insects can result from: (i) antixenosis or the dislike of the plant or non-preference (Kogan and Ortman, 1978); (ii) antibiosis or antagonistic reaction resulting from the presence of harmful compounds; or (iii) tolerance and combinations of these attributes.

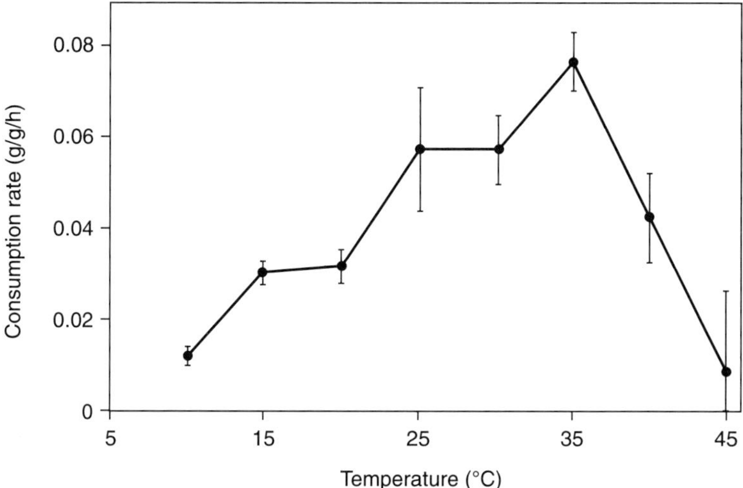

Fig. 7.4. Food consumption by cabbage white butterfly (*Pieris rapae*) larvae of the fourth instar (g/g/h) as related to temperature (°C) over 24 h (Kingsolver, 2000).

The mechanisms of resistance have a direct influence on the durability and ultimate success of a pest- or pathogen-resistant cultivar. For example, the integration of low to moderate levels of antixenosis, antibiosis and tolerance can be an effective strategy in controlling a resident pest population that invades early in the development of a crop and increases gradually during the growing season (Kennedy *et al.*, 1987). Also, combinations of both antixenosis and antibiosis can decrease the likelihood that the pest population will overcome that resistance, compared with either antibiosis or antixenosis alone, as long as the preferred host plant is available.

Breeding resistance to pathogen-induced diseases was developed rapidly in the second half of the 20th century using single or a few genes of major dominant effect. This practice was particularly apparent in the breeding of cereal and other agricultural crop cultivars. This form of resistance was frequently eroded by the development of populations of pathogen physiological races compatible with these major genes. The disadvantage of using monogenic, dominant resistance especially against air-borne pathogens is that as the land area devoted to a resistant cultivar increases, this provides an opportunity for growth and reproduction by the pathogen to develop physiological races that are unaffected by the resistance. These tolerant races form an increasing proportion of the pathogen population and eventually the monogenic resistance is rendered obsolete (this is the so called 'boom and bust cycle', Fig. 7.5).

In the latter part of the 20th century, plant breeders sought to use more durable or long-lasting forms of resistance. This involves using combinations of genes, some of which will have only minor impact on the pathogen when

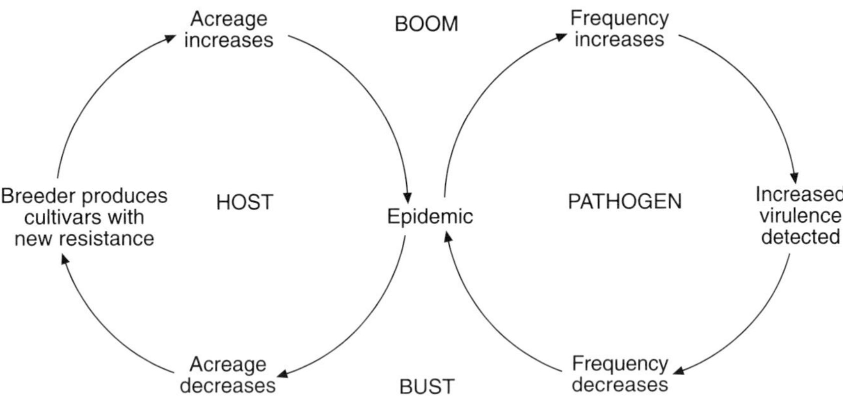

Fig. 7.5. The boom and bust cycles of host and pathogen expansion and collapse.

used alone. By this technique, combinations of resistance may be developed that only have an effect on the pathogen in the field and on adult plants.

Regrettably, breeding cultivars that carry traits for field resistance to crop pathogens is a longer and more difficult process that is less easily used by commercial seed companies. In field vegetables, there have been some very notably successful uses of durable disease resistance, for example the development of resistance to the soil-borne organism *Fusarium oxysporum* f.sp. *conglutinans*, the causal agent of cabbage yellows, in Wisconsin, USA. This resistance retained its effectiveness for many decades until a new pathogenic strain of the fungus emerged in California.

As knowledge of the genetic controls of resistance has been unravelled by molecular studies over the past 20 years, a gradual understanding of mechanisms of pathogen resistance has emerged. Saliently, the understanding of interactions between the pathogen and its host are elucidated, chemicals functioning as messengers, such as calcium, that stimulate resistance responses are being identified and the nature of the response itself unravelled. Much progress has been made with the model brassica *Arabidopsis thaliana*, especially with foliar pathogens such as downy mildew (*P. parasitica*), white rust (*A. candida*) and light leaf spot (*P. brassicae*). This research holds out the prospect of more robust resistance being developed in crop plants. Reviews of the basic underlying science are given by Keen *et al.* (2001) and Leong *et al.* (2002).

Reductions in the availability of pesticides especially for the control of cabbage root fly (*Delia radicum*) have stimulated intense interest in searching for sources of durable resistance. Some cultivated brassicas have moderate levels of resistance to cabbage root fly, for example, notably red cabbage, glossy brassicas, kale and swedes. None of the levels of resistance is considered sufficiently effective on its own to form the basis of a breeding programme aiming to produce cabbage root fly-resistant cultivars. Consequently, there has been a search for resistance sources in allied cruciferous species. To be most easily exploited, such sources of resistance should belong to the same genome as the cultivated crop. Ellis *et al.* (1998) found very high levels of cabbage root fly resistance in *B. incana*, *B. insularis* and *B. villosa*, and moderate resistance in *B. macrocarpa*. Each of these species is part of the CC genome grouping (see Chapter 1) in which the *B. oleracea* horticultural brassicas are included ($2n = 18$). Very high levels of antibiosis resistance have also been identified in *B. fruticulosa* and *B. spinescens* which belong to the FF genome ($2n = 16$) which is closely related to the B genome of *B. nigra* (Mizushima, 1980).

Research indicates that *Sinapis alba* may be resistant to *D. radicum* (cabbage root fly or cabbage maggot) as a result of reduced oviposition, diminished weights and survival of larvae, pupae and adults, and lessened damage to host plants. *Sinapis alba* (white mustard) also possesses tolerance to flea beetle (*Phyllotreta crucifera*), weevil (*Ceutorhynchus assimilis*), cabbage aphid (*B. brassicae*) and diamond back moth (*P. xylostella*) (Jyoti *et al.*, 2001).

These effects may be due to the structural properties of the root and stem of *S. alba*; thus the breeding accession Cornell Alt 543 tends to have a more profuse root structure and harder main stem than others tested. Reduced body weight of *D. radicum* could be the result of antibiosis coming from the rapid growth of *S. alba* Cornell Alt 543.

In vegetable brassicas, it is most important to prevent damage by root fly during the seedling period or soon after the plants are transplanted. The *Brassica* species possessing resistance to *D. radicum* were also found to resist the cabbage aphid, *B. brassicae*, and cabbage whitefly, *Aleyrodes proletella*. Multiple resistance of this nature is especially valuable. Cultivars of swede with moderate levels of resistance to *D. radicum* have been developed in Scotland and Scandinavia. These combine this resistance with that to turnip fly (*D. floralis*). To achieve commercially acceptable control, however, these cultivars need to be used in an integrated programme which still includes the use of some insecticides.

Antixenosis can provide a promising form of resistance for use against flea beetles such as *P. cruciferae* (crucifer flea beetle) and *P. striolata* (striped flea beetle) which are oligophagous herbivores feeding principally on brassicas. Related species such as *S. alba*, *Thaspis arvense* and *Lunaria annua* appear unacceptable to *P. stiolata* because of the compounds they possess, while *Camellina sativa* fails to offer the pest the necessary cues that initiate feeding, and the leaves of *B. villosa* subsp. *drepanensis* have high densities of hairs ($> 2172/cm^2$) and these act as a physical barrier to flea beetle feeding by preventing the insects from settling firmly on the leaf surface and initiating feeding (Gavloski *et al.*, 2000). Insect feeding preferences may change with leaf type or growth stage as well as host species. Thus the true leaves of *B. oleracea* were less preferred by *P. stiolata* compared with those of *Sinapis arvensis*, whereas the opposite preference was observed for cotyledons (Palaniswany and Lamb, 1992).

One avenue for producing resistance to crop pests is the development of transgenic plants which express endotoxins from the bacterium *Bacillus thuringiensis* (Bt). These effects come from insecticidal crystal (cry) proteins, δ-endotoxins, in the bacterium. They are highly toxic to lepidopteran, dipteran and coleopteran insects. They have been used widely as selective pesticides that are acceptable in both organic and integrated pest management (IPM) forms of husbandry (Entwistle *et al.*, 1993). Various Bt-transgenic *B. oleracea* vegetables and *B. napus* oilseeds have been produced (Earle and Knauf, 1999), mostly containing the *cry1A* gene. Other related genes have also been advocated for use in transgenic plants including a synthetic *cry1C* which has been incorporated into Chinese cabbage (Cho *et al.*, 2001) and the gene *cry1A(b)* used by Viswakarma *et al.* (2004) to develop transgenic broccoli (calabrese) in India. Resistance to diamond back moth, cabbage looper and imported cabbage worm resulted from this breeding work. The use of several genes makes the resistance more stable and potentially longer lasting.

Two problems are perceived with using transgenic plants. First, insect resistance to the Bt genes may come to dominate the pest population, thereby eroding the usefulness of the resistance. Development of resistance to Bt toxins in transgenic plants can be slowed by the planting of adjacent refuges containing non-transformed cultivars that either are treated with conventional pesticides or receive no chemical treatment at all. Secondly, antibiotic markers are used during the early laboratory stages of gene transfer, and these, if not removed in subsequent steps, may pose an unacceptable hazard to human health.

Much of the resistance that has been developed in vegetable brassicas utilizes one or possibly more genes of large effect. This effect is frequently expressed throughout the life of the resistant plant and can be identified by testing seedlings exposed to the pathogen when they are infected artificially. The drawback to this form of plant breeding is the 'boom and bust' cycle referred to earlier where a resistant cultivar becomes used in large proportions and then, as the size of the tolerant pathogen population increases concomitantly, the host resistance is eroded.

One route suggested to overcome this phenomenon is the development of field resistance (Kocks and Ruissen, 1996). Field resistance is defined as 'any resistance that effects epidemics in the field but which is not immediately apparent in laboratory or glasshouse tests' (Robinson, 1969). For example, plant breeders seeking resistance to black rot (*X. campestris* pv. *campestris*) concluded that major gene (hypersensitive) resistance could be overcome and that incorporation of additional levels of field resistance might be developed as supplements. Black rot juvenile (major gene) and field (multigenic) resistance are controlled by separate genetic systems (Camargo *et al.*, 1995). Evaluating and combining these into acceptable commercial cultivars could be expected to require a complex breeding process using tests applied to seedling and mature progeny lines grown in greenhouse and field trials, respectively. Comparisons of wound and spray inoculation with the pathogen applied to juvenile and mature plants, however, showed that an analysis of symptoms that developed following the spraying of juvenile leaves could be equated with mature plant-field resistance (Griffiths and Roe, 2005; see Table 7.2). In consequence, it should be possible for plant breeders to evaluate progeny for black rot resistance at an early growth stage with reasonable expectations that both seedling and mature plant systems are present and being evaluated.

MANIPULATION OF HUSBANDRY SYSTEMS

Integrated control strategies have the initial benefit of being less open to erosion by the development of tolerant strains of the pest or pathogen. Consequently, the life expectancy of such control strategies should be extended. These include the use of husbandry systems as a central factor in achieving control. The crop is manipulated by sowing and maturity date so as to avoid those calendar periods when the pest is most active; this can be supported by changing the nutritional regime used for the crop.

Rotation

The manipulation of husbandry systems to lessen the impact of pests and pathogens is the major avenue for the control that is directly placed in the hands of the grower. Such an approach is particularly appropriate with soil-inhabiting organisms. One of the first elements of husbandry control is the

Table 7.2. Comparison of black rot disease severity ratings (*Xanthomonas campestris* pv. *campestris*) following the inoculation of seedling and mature plants.

Comparison of inoculation method	Correlation[a]
Juvenile wounding versus juvenile spraying	0.75
Juvenile wounding versus mature wounding	0.58
Juvenile wounding versus mature spraying	0.51
Juvenile spraying versus mature wounding	0.89
Juvenile spraying versus mature spraying	0.86
Mature spraying versus mature wounding	0.89

[a] Spearman's ranking of correlations.
Wounding = piercing the leaves either side of the midrib with two needles carrying *X. campestris* pv. *campestris*; spraying = small or large volume sprays with bacterial slurries under controlled or field conditions for juvenile or mature plants, respectively.
After Griffiths and Roe (2005).

use of crop rotation. Appropriate crop rotations lengthen the time between susceptible crop types so that pathogen populations have time to decline as a result of degradation by other naturally suppressive organisms. For some diseases, field rotations using non-hosts for a sufficient period of time allows the decomposition of infested crop residues and/or a reduction in the viability of pathogen survival structures, eliminating one of the sources of future primary inoculum. By increasing the diversity of crops grown within a rotation, the pathogens residing in the soil or on residues from the previous crop may fail to infect the subsequent *Brassica* crop. Crop rotation can be one of the most effective biological control tools available to growers. Knowledge of the impact of crop residues on survival and carry-over are important in determining the efficacy of rotations.

In particular, it is necessary to know if the pathogen can survive as a saprophyte as well as a pathogen. Additionally, the roles of weeds and volunteer crop plants as hosts and reservoirs of infection should be established. In using rotation as a tool to limit or reduce pathogen survival and spread, knowledge of the extent of the host range is of major significance. Pathogens with a wide host range and robust survival mechanisms such as *S. sclerotiorum* (white rot) are unaffected or may even be encouraged by the use of rotations. Regrettably, in many instances, the pressure of market forces now means that growers cannot afford to use rotations for pest and pathogen control. In consequence, the incidence and severity of several *Brassica* pathogens are increasing dramatically.

Crop density

Stand density and crop plant architecture will affect the risk of disease infection and its subsequent impact. Stand density affects air movement, shading and moisture retention within the crop. Root (1973) proposed the 'resource concentration hypothesis' whereby 'any herbivores, especially those with a narrow host range are more likely to find hosts that are concentrated'. This hypothesis predicts that the density of herbivores per host plant is higher in dense stands of their host plants. The density of herbivores per host plant in the field is determined by: (i) the number of eggs laid by the individuals that enter from the surrounding area; and (ii) the survival rate of individuals growing inside the field.

Yamamura and Yano (1999) argue that in small areas, the density of herbivores per plant is dominated by the frequency of entry but in large areas of crop the density will be dominated by the survival rate of individuals inside the plot or crop (Chapter 6).

Host nutrition

Plant nutrition affects the rate at which the signs and symptoms of pest and pathogen invasion develop. Balanced fertility is more easily maintained in a diverse cropping rotation. Each crop species has varying nutrient requirements for optimal growth, development and reproduction, and so utilize varying fertilizer elements at different rates. The suitability of plants as insect food influences the ovipositional responses, for example *P. rapae* females can detect the physiological status of plants (Myers, 1985). The behavioural events prior to oviposition by *P. rapae* include searching flight, landing and contact evaluation on a potential host plant.

These factors may be influenced by the presence of non-host plants. Chlorophyll levels and percentage composition of nitrogen, sulphur and other macronutrients could affect host selection (Hooks and Johnson, 2001). Plant water content may also be an important indicator of nutritional status of plants to ovipositing butterflies.

Female small cabbage white butterflies (imported or European cabbage worm; *P. rapae*), for instance, assess the nutritional value of plants and tend to lay more eggs on plants that are richer in nutrients. This insect will also respond to other host characteristics such as size, leaf water content, host colour (green and blue-greens which relate to chlorophyll content) and plant growth stage.

Changing the nutritional regime used for the crop could offer an avenue for pest control. It is well recognized that use of some fertilizers is associated with increased pest incidence, while others tend to have the opposite effect of diminishing pest impact. Integral to this approach is the use of crop walking and assessment. Excessive use of nitrogen fertilizers, especially where this results in an imbalance between ammonium and nitrate ions in the soil, causes stress which then allows pests and pathogens to cause damage more easily. The use of environmentally benign nitrogen sources such as calcium cyanamide or calcium nitrate is associated with diminished crop stress and, in consequence, less disease; this topic is discussed by Engelhard (1996) and Hall (1996). Appreciation that some fertilizers and other compounds are associated with the diminution of pest and pathogen impact has led to the recent development of the concept of 'bio-stimulation'. This identifies that in addition to their direct nutritional properties, some compounds are capable of stimulating benign microbial populations associated with plant roots and the diminution of populations of pathogenic organisms.

The crop is inspected at regular intervals to determine whether pest problems are emerging. If there is a developing infestation, then growers are advised to apply chemical remedies that may be used at a reduced application rate. Kurppa and Ollula (1993) suggest that limiting the quantities of nitrogen applied to *Brassica* crops could result in fewer herbivores affecting them; their work was done with summer turnip rape and the blossom beetle (*Meligethes aenus*).

Physical barriers

Crop covering involves laying transparent plastic sheets over drilled or transplanted crops. The growing plants support the cover as they develop underneath. Alternatively, opaque plastic sheets are placed on the soil surface and on the crop. Then the crop either grows through holes in the sheet or is transplanted through them. Both covers and mulches offer barriers between the developing crop and pests. For *Brassica* crops, covers have more universal value in the control of pests.

The original crop covers were simply polythene sheets. These were often perforated to allow ventilation and ease of handling. Tougher polypropylene products are displacing perforated polythene. Polypropylene tends to heat up less quickly on hot, sunny days and hence can be left on the crops longer than was the case with polythene sheeting. The sheets are either 'woven' or 'non-woven'. Fibres in the non-woven forms are bonded together by heat – a process described as 'thermal spun bonding'. These fabrics are lighter and cheaper than the woven forms and are often described as 'fleeces'. They are permeable to air and water. Woven polypropylene is sold as netting with extremely tiny gaps between the warp and weft fibres. These products are heavier and longer lasting compared with fleeces.

Crop covers exclude many major pests from *Brassica* crops and are becoming important elements in pest control; for example, as the insecticide chlorfenvinphos is withdrawn in Europe, crop covers have become the most reliable way of producing swede crops free from damage caused by cabbage root fly (*D. radicum*). Mesh size in woven netting governs the species of insect that are excluded. Mesh of 0.6 mm^2 excludes leaf miner and whitefly, 0.36 mm^2 excludes aphids and 0.06 mm^2 excludes thrips. *Brassica* pests that have been controlled satisfactorily with crop covers include cabbage root fly (*D. radicum*), and cabbage aphid (*B. brassicae*) caterpillars on brassicas. The efficacy of this control technique was demonstrated by Dutch research where Chinese cabbage was grown under covers without the use of pesticides and compared with plants receiving chlorpyrifos (Table 7.3).

BIOLOGICAL CONTROL

Single organisms

Biological control of pests and pathogens has been sought for many years as an alternative to the use of synthetic pesticides. Biological control of pests using single predatory species is particularly successful with protected glasshouse crops. Similar success could be anticipated with field crops where they are covered with protective plastic sheets and netting. For pathogens, the initial approach was to find individual microbes that would inhibit their

Table 7.3. Protection of Chinese cabbage with crop cover netting.

Treatment	% Plants infested			% Marketable yield	Weight per head (g)
	Cabbage root fly	Midge	Leaf miner		
Untreated	98	10	40	56	800
Insecticide	2	0	93	82	1030
Net, mesh 1.35 mm, removed 2 weeks pre-harvest	0	0	3	98	900
Net, mesh 0.8 mm, removed 2 weeks pre-harvest	0	1	6	93	1000
Net, mesh 1.35 mm, removed at harvest	0	0	2	94	970
Net, mesh 0.8 mm, removed at harvest	0	2	7	94	1010

After Ester *et al.* (1994)

growth and reproduction and then market these in a manner analogous to conventional pesticides.

Most fungi used in biocontrol are Hyphomycetes, especially the genera *Trichoderma*, *Penicillium* and *Gliocladium*, and the bacteria *Pseudomonas* and *Bacillus* (Bello *et al.*, 2002). *Penicillium oxalicum* has been successfully employed against *F. oxysporum* in the laboratory. Mingling hyphae of *P. oxalicum* with *F. oxysporum* f.sp. *campestris* caused hyphae of the pathogen to swell and become disrupted. So far, however, the use of single beneficial microbes in the biological control of field pathogens has not gained acceptance because of the variability of success, probably as a result of changing physical environments.

Multiple organisms

An alternative recent approach is to use populations of microbes manipulating the husbandry environment to their advantage. This is substantially a means of developing artificially suppressive soils that mimic natural processes in the field. The beneficial impact of benign microbes on pests and pathogens can substantially reduce their invasion, colonization, disease, and damage initiation and progress. Slight environmental changes resulting from changed husbandry practices will alter the biological balance between benign antagonists and pests and pathogens. Many microbes are natural antagonists to parasitic organisms; only small proportions of the natural soil flora attack crop plants while the majority are beneficial to crops. The relative numbers of microbes in the top 15 cm are 10^9 bacteria, 10^8 actinomycetes, 10^6 fungi and 10^5 algae/g of soil. Their roles in soil include decomposition of organic matter, mineralization and immobilization of nutrients for plant growth, nitrogen fixation, environmental remediation, plant growth promotion, biological control for crop protection and only to a small extent pathogenic activities leading to crop disease.

Substantial efforts are being made to understand how cropping practices may be designed to encourage benign organisms and discourage pests and pathogens. The most basic husbandry practice is primary cultivation by ploughing. It has disadvantages, however, since it can increase water loss due to runoff and evaporation by leaving the soil exposed to heat and wind, and increases the loss of organic matter which is required for an active benign microbial population. Reducing tillage (see Chapter 6) under American conditions enhances microbial species diversity, with larger populations in the upper soil horizons, whereas tillage redistributes microbes throughout the upper and lower soil horizons.

Several higher fungi produce ergosterol that can be utilized as a biological marker to measure the activity of soil microbial populations. The efficiencies of extraction of ergosterol are relatively high, and it is an exactly defined compound that can be determined using chromatographic methods. Few other simple methods are available and reliable apart from counting microbial colonies grown in asceptic culture, which is laborious, expensive and long winded. Ergosterol content was greater in non-tillage conditions (Monreal *et al.*, 2000; Krupinsky *et al.*, 2002), indicating increased microbial activity. Conversely, the advantages of tillage lay in the burial of crop residues, levelling, consolidation and warming of seedbeds in spring, reducing surface compaction, breaking up soil pans, incorporation of pesticides and fertilizers, and reducing weed competition.

In those geographical regions with high ambient temperatures, use can be made of solarization to diminish soil-borne pest and pathogen populations. Solarization exploits the heating effects gained by placing plastic sheeting over soil. Population counts of *F. oxysporum* f.sp. *conglutinans* were greatly reduced, and cabbage yellows disease was undetected in plots treated with solar heating and cabbage amendments (1% w/w). Dried cabbage was mixed with soil and covered with translucent polyethylene sheeting (tarp) for 4–6 weeks; the use of solar heating or cabbage residues alone was ineffective. The sheeting traps fungitoxic gases derived from the cabbage residues that are concentrated; these together with high temperatures are responsible for the diminution in the viability of pathogen propagules. Ramirez-Villapudua and Munnecke (1987) postulated that these effects result from changes to the biodiversity of soil microflora, diminished soil fungistasis and release of nutrients that may induce chlamydospore germination leading to biological control of *F. oxysporum* f.sp. *conglutinans*. There may also be direct toxicity from the sulphurous compounds present in cabbage residues as a similar fungistatic effect of brassica residues was reported by Kocks *et al.* (1998) for black rot (*X. campestris* pv. *campestris*). Epidemics of this pathogen build up over years (polyetic) with a carry-over of inoculum from one vegetative period to the next. However, where black rot-infested plants were ploughed back in after harvest and the plants broken up before incorporation, presumably releasing sulphurous compounds, then the risk of infection in the following season was diminished.

Enhanced host resistance

Recently it has become apparent that selected strains of non-pathogenic, root-colonizing rhizobacteria are capable of inducing disease resistance in crop plants. This is referred to as rhizobacteria-mediated induced systemic resistance (ISR). Rhizobacteria are present in large numbers on the root surface where plant exudates and lysates provide nutrients. Besides inducing resistance, rhizobacterial strains are reported to antagonize soil-borne pathogens directly and to stimulate plant growth.

When appropriately stimulated, plants systemically enhance their defensive capacity against pathogen attack. This induced resistance is generally characterized by a restriction of pathogen growth and a reduction in disease severity. The development of a broad-spectrum systemic acquired resistance (SAR) after primary pathogen infection is characterized by an early increase in endogenously synthesized salicylic acid (SA). This is an essential compound in the SAR signalling pathway, because transgenic plants that are unable to accumulate SA are incapable of developing SAR. Further, SAR is associated with systemic activation of SAR genes. These include genes that encode pathogenesis-related (PR) proteins. Some of these have *in vivo* antifungal activity and may contribute to the resistance state of SAR (Ton *et al.*, 2001). Induced systematic resistance has been demonstrated in different plant species operating against a broad spectrum of pathogens. The mechanisms by which rhizobacteria induce resistance vary. Some rhizobacteria trigger the SA-dependent resistance pathway by producing this compound at the root surface; others use alternative pathways.

There may be opportunities to load seed with biological control organisms, a process termed bacterization (Boruagh and Kumar, 2002a, b). Seed bacterization improved germination, shoot height, root length, and fresh and dry mass, and enhanced yield and chlorophyll content of leaves in the field. Plant growth promotion resulted from the presence of siderophores and disease suppression from the presence of an antibiotic. Siderophores are high-affinity iron (Fe^{3+})-chelating molecules that enhance the acquisition of iron in a deficient environment, thereby making the element available to the crop plants but unavailable to pathogens. Identified strains of *Pseudomonas fluorescens* are capable of enhancing plant growth and diminishing the effects of fungal and bacterial plant pathogens. Attempts are being made to utilize this approach with *Bacillus polymyxa* (Pichard and Thouvenot, 1999) isolated from cauliflower seeds in order to control black rot and damping-off of cauliflower.

Natural products

As an alternative to conventional synthetic pesticides, recently interest has focused on evaluating plant extracts as potential insecticides on the premise that these materials are less specific in their mode of action. Hence insects will require longer lasting resistance in order to circumvent them. In addition, plant extracts are mostly biodegradable, which suggests that their application would be more environmentally acceptable and compatible with IPM programmes. Numerous plant extracts or plant-derived compounds can potentially be incorporated into alternative and novel strategies aiming to control various insects such as diamond back moth (*P. xylostella*).

Extracts from tropical plants have been considered, such as those from the rhizomes of *Alpinia galanga* (Zingiberaceae), fruits of *Amomum cardamomum* (Zingiberaceae), tubers of *Cyperus rotundus* (Cyperaceae) and seeds of *Gomphrena globosa* (Amaranthaceae) (Ohsawa and Ohsawa, 2001). Active compounds in *C. rotundus* and *A. galanga* have been identified as α-cyperone and 1′-acetoxychavicol, respectively; both reduced the larval density of *P. xylostella* at least as effectively as the pyrethroid insecticide deltamethrin. Most recently, Sureshgouda and Kalidhar (2005) reported that extracts of karanj (*Pongamia pinnata*) seed when sprayed on to cabbage leaves reduced the feeding ability of the first instar of the diamond back moth (*P. xylostella*).

STATUTORY CONTROL

The international network of legislation designed to offer growers reliable seed mainly provides information covering the determination of germination, purity and trueness to type (see Chapter 3). There are also requirements for freedom from pests and pathogens where these are seed borne. Relatively few *Brassica* pests and pathogens fall into this category, with the prime exception of *X. campestris* pv. *campestris*, the causal agent of black rot. Normally, batches of *Brassica* seed are tested rigorously for the presence of this bacterium. The International Seed Testing Association (ISTA) has evaluated methods for testing for seed-borne organisms. These tests are sufficiently robust to satisfy the needs of direct-drilled crops where the seed is going straight into its cropping stations. Much of the European and Asian cauliflower and calabrese crops are now transplanted and, in consequence, even minute traces of black rot can cause devastating epidemics during propagation and when the plants are subsequently spaced out in the field. The seedlings are raised in warm, moist, protected glasshouse environments that are ideal for the reproduction and spread of this bacterial pathogen.

The rate of spread can reach a 1% increase per day. This position is complicated by the capacity of *X. campestris* pv. *campestris* for symptomless spread (Shigaki *et al.*, 2000). As a consequence of these abilities, this pathogen has become a major source of loss for the many of world vegetable *Brassica* propagators who supply transplants for field production. Legislation normally prohibits the sale of diseased transplants. This applies to bacterial diseases such as black rot and, in some countries, also to plants infected with *P. brassicae*, the cause of clubroot disease. While this pathogen is not transmitted by seed, it can be carried in propagation media, especially those that are peat based. There have been several recent cases where peat infested with *P. brassicae* has been used for the propagation of *Brassica* transplants with devastating consequences for propagators and growers.

A SURVEY OF PESTS AND PATHOGENS DAMAGING *BRASSICA* CROPS

Aphids

Brevicoryne brassicae L. – cabbage aphid and *Myzus persicae* Sulzer – green peach aphid, peach-potato aphid (Fig. 7.6)

The intensity of aphid epidemics is regulated by the prevailing weather. In warm dry conditions, there is rapid and extensive colony growth, while cool damp weather inhibits population expansion (Blackman and Eastop, 1984; Minks and Harrewijn, 1987). Initial spring invasions rely on the proximity of overwintered crops to new plantations; consequently field hygiene by ploughing in or rotavating residues is of paramount significance to deprive these pests of sources of cover for hibernation. Spring-planted vegetable brassicas should not be sited close to overwintered oilseed rape crops. Attraction of new crops may be diminished by the use of inter-row mulches of transparent or blue plastic or straw.

Breeding for resistance to cabbage aphid is made difficult because of the large number of insect biotypes. There has been extensive selection and some breeding work in several parts of the world, notably in England and in France. It is suggested that good dominant resistance with some maternal effects is present in several cabbages. Some F_1 hybrids between susceptible and resistant lines were susceptible (Metz *et al.*, 1995), whereas Ellis *et al.* (1996) reported partial levels of antixenosis in red coloured accessions of cvs Ruby Ball and Yates Giant Red Brussels sprout cv. Rubine and a strain of 'Italian Red Kale'. Glossy accessions of cabbage and cauliflower possessed antixenosis and antibiosis resistance that lasted throughout the growing season of the crop. In kale, Ellis *et al.* (1998) found they were more likely than other *B. oleracea* to be resistant, while broccoli was a poor source of resistance. Several Portuguese kales carry resistance to the green peach aphid (*Myzus persicae*), especially selections of 'Crista-de-Galinha', followed by 'Joenes' and 'Roxa' (Leite *et al.*, 1996). Seven *Brassica* species were tested by Ellis *et al.* (2000) who found that *B. fruticulosa*, *B. spinescens*, *B. incana* and *B. villosa* had high levels of antibiosis;

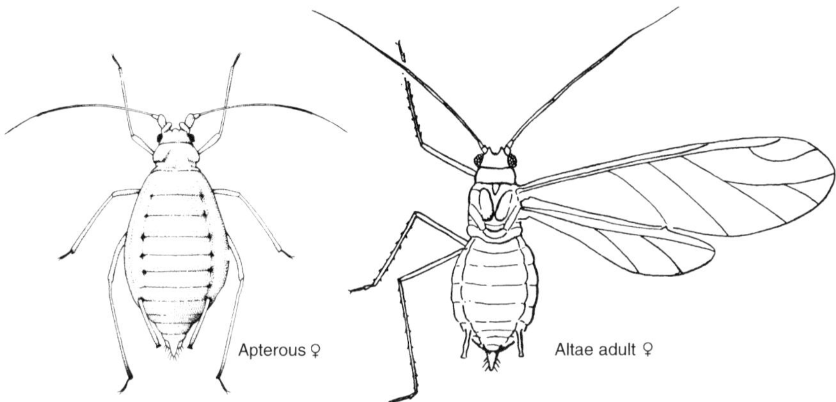

Fig. 7.6. Line diagram of the green peach aphid, peach-potato aphid (*Myzus persicae*).

these were the same species reported to possess resistance to cabbage root maggot (cabbage root fly, *D. radicum*). There is a suggestion that light coloured foliage is associated with resistance as in some of the Australian cauliflower cultivars. Possibly this resistance results from disrupting the inability of these aphids to locate such coloured crops.

Early season control has often relied on the use of granular formulations of soil-incorporated insecticides. Later in the season, foliar sprays are used as the efficacy of the soil-applied materials declines. Strategies for the use of insecticides are dominated by factors such as the permitted interval between application of a particular chemical and harvesting, the declining availability of carbamate and organophosphorus formulations and the relationship between insecticide toxicity and its effect on natural predators.

There are subsidiary considerations such as the failure of systemic granular and foliage-applied formulations to provide control in periods of drought when translocation is slower within the plants. Irrigation may then be used to improve translocation by increasing the availability of soil moisture. Minimizing nitrogen fertilizers and the use of mixed cropping have also been advocated as means of diminishing the impact of these pests. Similar principles and problems apply to the control of *M. persicae* when using pesticides as for *B. brassicae*. The development of strains of these pests resistant to insecticides is a major problem, and with *M. persicae* is made more extreme by the wide host range and consequently increased exposure of aphid populations to sprays applied to many host crops. Integrated control systems using combinations of husbandry, and chemical and genetic host resistance offer the most environmentally attractive means for control, but this demands regular crop monitoring in order to evaluate the build up of pest populations (Finch, 1987). Crop monitoring is essential in order to target the use of insecticides as aphid colonies are established and before significant injury to quality or crop growth has developed. The concept of 'control thresholds' has been used whereby the build up of aphid colonies is monitored and applications varied according to crop growth stage. Thus, with Brussels sprout, fewer aphids are required to trigger the application of remedial sprays in the early stages of growth after transplanting than in the main growing season, while acceptable tolerance levels are reduced again as the buds form and zero tolerance is permitted close to harvest (Theunissen, 1984). Aphids are generally controlled in nature by parasites, pathogens and predators. These may provide opportunities for the development of biological control systems in the future, but so far none have been successfully used in vegetable brassicas.

Whitefly

Aleyrodes proletella L. – cabbage whitefly

Once established in a crop, whitefly are amongst the most difficult pests to eradicate. This pest inhabits the undersurface of leaves; consequently control is only effective when insecticides are directed at high volume on to this target with suitable drop leg spray booms (Mound and Halsey, 1978; Bryne and Bellows, 1991). Control is essential in vegetable brassicas to achieve clean, high-quality produce acceptable to supermarkets. Pyrethroid formulations will kill all stages, whereas others are only effective against the adults.

Regular applications of organophosphorus products every 3–5 days will destroy adults, preventing egg laying and subsequent new generations. A crop inspection and application routine has been developed in Germany where sprays are used when 20 adults or 50 larvae per plant are found. This substantially reduces the number of chemical applications required for crops such as savoy and red cabbage. There may be opportunities for biological control (Hulden, 1986).

Beetles

Meligethes spp. – blossom beetles (Fig. 7.7)

The biology of beetles is discussed by Booth *et al.* (1990). Trap crops may be employed to divert these pests from summer cauliflower; success has been achieved at control using this technique with flowering crops of Chinese cabbage, broccoli (calabrese), sunflower, marigold and oilseed rape. The disadvantage of this method is that a substantial area of land is diverted from the cash crop to the trap crop.

Spray applications may be based on the population density of adult beetles and for autumn crops are 15–20 beetles per plant and in the spring

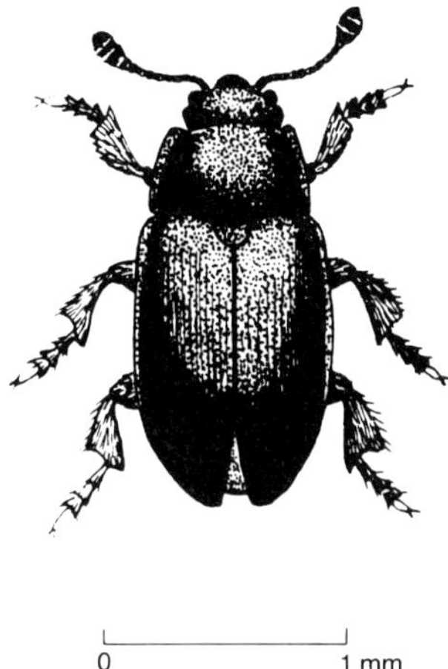

Fig. 7.7. Line diagram of the blossom beetle (*Meligethes* spp.).

three beetles per plant. Once crops have reached the flowering stage, most of the damage has been done and spraying is likely to be counterproductive. Computer-based forecasting systems have been developed for this and other similar pests.

Phyllotreta spp. – flea beetles and *Psylliodes chrysocephala* – cabbage stem flea beetle (Fig. 7.8)

Dry, warm spring weather favours these insects and, where seedlings are under water stress, increases crop losses. Sowing or mulching of *Brassica* crops diminishes their attractiveness for flea beetles. There is some suggestion that the use of organic as opposed to soluble fertilizers also discourages flea beetles. Unsuccessful attempts at biological control have involved the use of nematodes pathogenic to flea beetles. Some researchers have suggested that brassicas such as Chinese cabbage for which flea beetles show markedly preferential feeding might be used as trap crops for these pests and could be planted mixed with or adjacent to crops such as white head cabbage (Trdan *et al.*, 2005). Regrettably, populations of insects on Chinese cabbage become so high that there is no benefit for the cash crop white cabbage.

Insecticides applied to the seed, at drilling or post-emergence, will reduce pest impact. Overwintered crops are sprayed in early autumn, reducing the populations of egg-laying adults, and again after about 4 weeks to eliminate larval survivors. Crop monitoring aids the effectiveness of chemical control; those materials used for caterpillar control may also be effective against flea beetles.

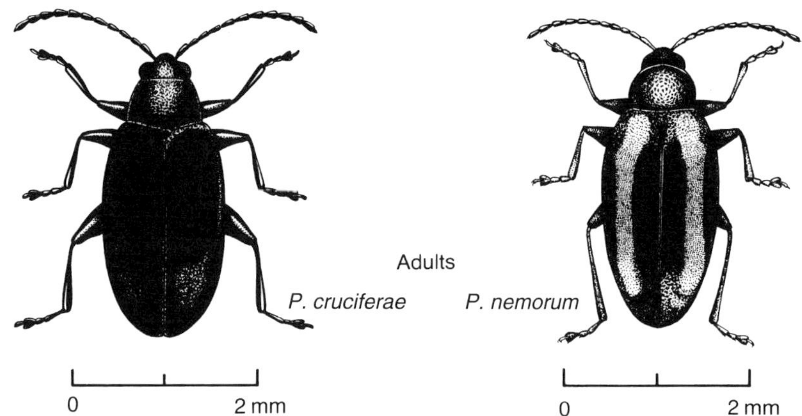

Fig. 7.8. Line diagram of the flea beetle (*Phyllotreta* spp.).

Ceutorhynchus quadridens – cabbage stem weevil and *C. assimilis* – cabbage seed weevil

Husbandry control measures include avoiding growing susceptible horticultural brassicas adjacent to oilseed rape crops. Chemical treatments applied to control pollen beetle (*Meligethes aeneus*) will also reduce populations of seed weevil (*C. assimilis*). Chemicals used against cabbage root fly (*D. radicum*) may give some control of stem weevil (*C. quadridens*).

Butterflies and moths

Pieris brassicae – large cabbage white butterfly and *P. rapae* small cabbage white butterfly (Fig. 7.9) (also known as the imported or European cabbage worm in North America)

The biology of butterflies is discussed by Vane-Wright and Ackery (1984). Natural control of cabbage white caterpillars comes from birds, carabid beetles, the parasitoid *Apanteles glomeratus* and a baculovirus. Several predators such as wasp parasitiods (*Apanteles* spp., *Pteromalus puparum* and *Meteorus versicolor*) and nematodes (*Heterorhabditis*, *Poinax* and *Neoaplectana* spp.) have been tested as potentially commercial biological control agents (Feltwell, 1982; Carter, 1984).

Commercial preparations of *B. thuringiensis* provide biological control using spores and a toxin preparation. Several regular applications are needed at 14 day intervals to sustain crop quality. The butterfly caterpillars colonize deeply into the host canopy; consequently, applications of insecticide sprays require accurate timing in order to coincide with the emergence of adults and subsequent larval growth, and several applications may be needed. As the crops mature and canopies become dense, it is essential to direct sprays

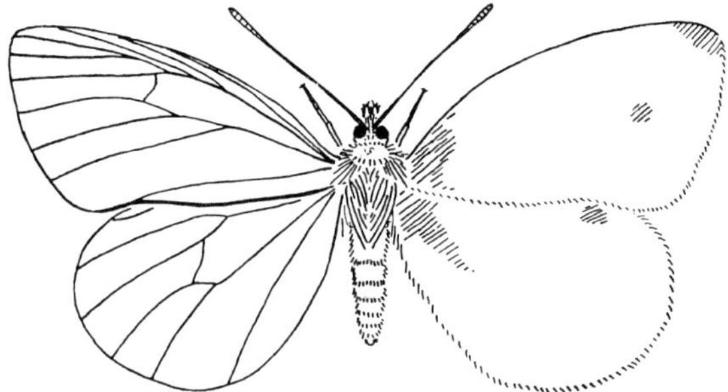

Fig. 7.9. Line diagram of the small cabbage white butterfly also known as the imported or European cabbage worm (*Pieris rapa*).

beneath the foliage and into the developing stem apex. Crop monitoring for egg laying and timing the application of sprays before the caterpillars become mature is essential. Currently, several chemical pesticides are available to kill the larvae.

Mamestra brassicae – cabbage moth

Several wasps colonize the eggs (*Trichogramma evanescens*), larvae (*Apanteles glomeratus* and *Hyposoter ebenius*) and pupae (*Ptetomalus puparum* and *Pimpla instigator*); additionally the flies *Phryxe vulgaris* and *Compsilura concinnata* are parasites of *M. brassicae*. Studies indicate that species of fungal and viral parasites offer potential biological control systems; the larvae are not susceptible to current commercial strains of *B. thuringiensis* unlike other lepidopteran pests (e.g. *Pieris* spp.). Reputedly, there is resistance in the cauliflower accession line PI234599; this was the first line associated with glossy leaves and resistance to leidopteran pests. Insecticide treatments are most effective when applied before the emergence of the migratory fourth instar. The timing of chemical applications should be supported by use of pheromone traps. There are no effective cultural methods of control.

Plutella xylostella – diamond back moth (Fig. 7.10)

Several wasp predators reportedly attack the larvae of *P. xylostella*. The larvae are generally susceptible to the biological control agent *B. thuringiensis*, but this may cost more than using insecticides (Talekar *et al.*, 1985; Talekar, 1986; Talekar and Shelton, 1993). Genes from two strains of *B. thuringiensis* have been transferred into broccoli by *Agrobacterium tumefaciens*-mediated transformation, which has effectively controlled the diamond back moth (Cao *et al.*, 1999). Plants with high levels of Cry1c protein cause rapid and complete mortality of three types of diamond back moth larvae, with no defoliation. Because of the rapid ability of diamond back moth to mutate and overcome *B. thuringiensis*, Shelton *et al.* (2000) have shown that it is

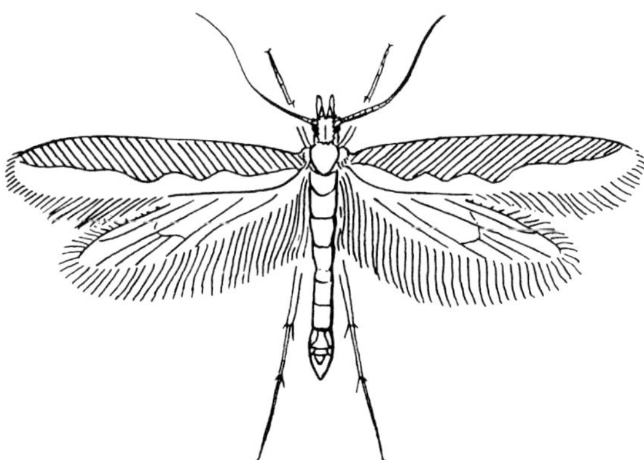

Fig. 7.10. Line diagram of the diamond back moth (*Plutella xylostella*).

necessary to provide refuges to conserve susceptible larvae and thus reduce the numbers of homozygous offspring. At the same time, refuges with high levels of *B. thuringiensis* toxic plants resulted in a reduction of predators (Riggins and Gould, 1997).

Considerable effort has been invested in developing resistance to the diamond back moth and small cabbage white butterfly in *B. oleracea*. Some glossy-leafed lines have shown good consistent resistance (Lin *et al.*, 1983) in many parts of the world. The glossy leaf is, however, unacceptable horticulturally and reduces yield and delays maturity, and so far no adequate resistance has been found in non-glossy lines. Resistance was found in cauliflower PI 234599, cabbage G8329 and lines of Chinese cabbage.

Some normal leaf lines developed by selection had moderate resistance, but this was probably insufficient to be economically successful. The mechanism of host resistance appears to operate by preventing penetration of the first instar into the leaf tissue. There are, however, reports of good resistance from a commercial breeding programme in South Korea. Predictive control based on counts of the numbers of plants infested allowed the application rate of insecticides to be reduced.

Pheromone traps may aid chemical treatment timing. Single spray applications timed correctly may be sufficient to prevent yield losses from this pest in northern regions; in more southerly locations where there is intense population pressure and frequent reproduction, sprays at weekly intervals or even more often are necessary. Some strains of this pest, particularly in Asia, have resistance to many chemical pesticides because of overuse. There are no methods of cultural control.

Evergestis forficalis – garden pebble moth

Intercropping especially in Brussels sprout reduces the amount of insect damage, but yields are adversely affected because the population density of the main crop is diminished. Spraying with insecticides is the only currently feasible means of commercial control.

Agrotis spp., especially the turnip moth – *Agrotis segetum* (cutworms) (Fig. 7.11)

This pest is of greater significance in seed compared with ware crops. A spatial barrier of 500 m between new and old season seed crops is frequently sufficient to prevent infestation because the adults are weak fliers. *Brassica* crops vary in their attractiveness to this pest, with fewer eggs being laid in mustard crops such as *B. nigra* (black mustard), *B. juncea* (leaf mustard) and *B. carinata* (Abyssinian mustard) compared with forms of *B. rapa* and *B. napus*. Forecasting using computerized prediction systems allows this and other pests

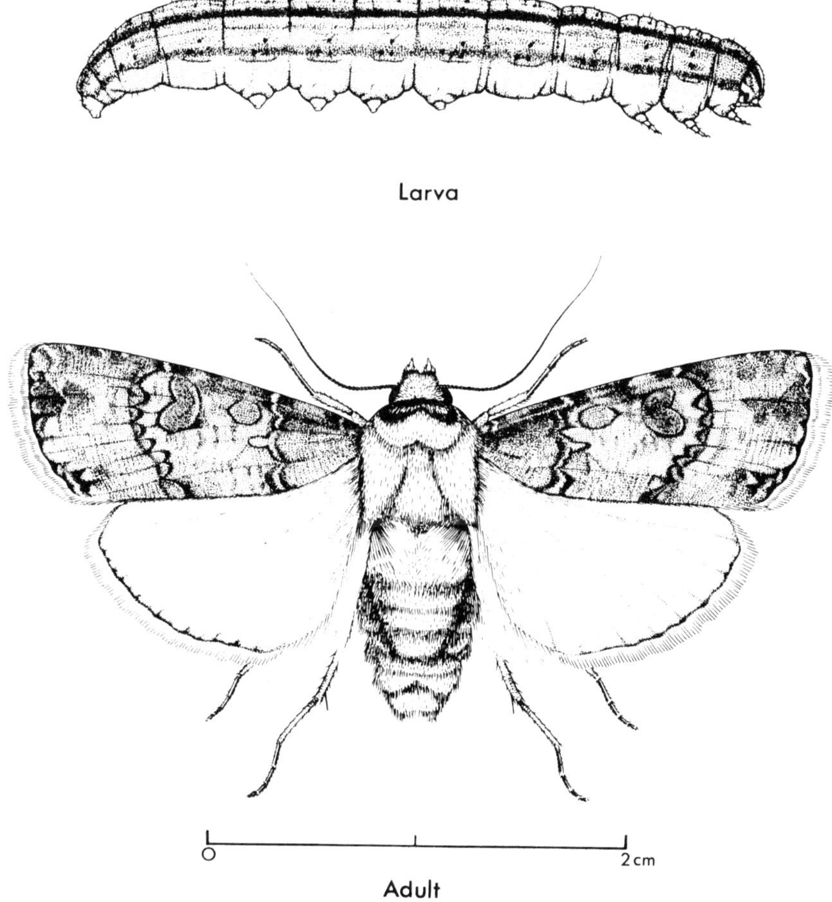

Fig. 7.11. Line diagram of the turnip moth (cutworm) (*Agrotis segetum*).

to be controlled by targeted applications of insecticide based on the build up of population numbers relative to prevailing and predicted weather conditions. This pest only causes measurable damage when *C. assimilis* damages the host pods in advance and provides entry to them.

Flies

Phytomyza rufipes – cabbage leaf miner (Fig. 7.12 *Phytomyza* spp.) and *Continaria nasturtii* – swede midge

Damage caused by cabbage leaf miner could be dismissed as cosmetic but, with current demands for blemish-free horticultural produce, this pest can be exceedingly damaging. Severe outbreaks will ruin what superficially appear to be very profitable crops with great rapidity.

Despite its name, swede midge will colonize most brassicas, radish, watercress and several weed species (Lole, 2005). Feeding by midge larvae induces physiological changes causing swollen petioles and flower buds, blindness, brown corky scarring in the growing apex and on petioles and inflorescence stalks, crinkling, puckering and distortion of the leaves, and the formation of multiple low quality heads. Damage becomes apparent only after the larvae have been feeding for several days. It can be confused with abiotic problems such as mechanical or weather injury, herbicide toxicity or nutrient imbalances. Rotation of the land used for *Brassica* crops and efficient weed control will help to defeat the pest. Chemical control of the larval stages is difficult because they are well hidden by foliage and inflorescences. Control may be effective using forms of carbamates, synthetic pyrethroids, nicotine and organophosphates where these are approved for use.

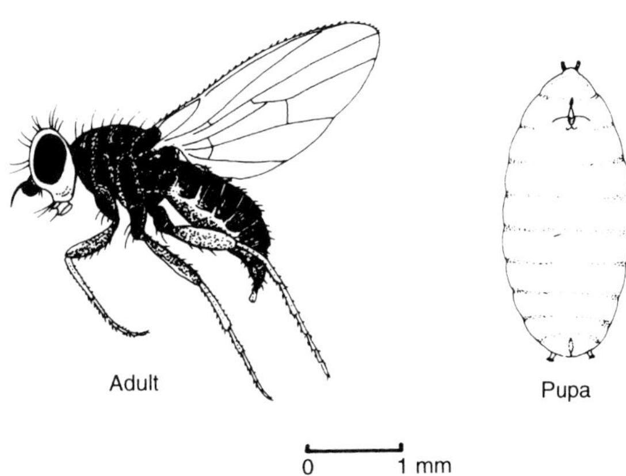

Fig. 7.12. Line diagram of the leaf miner (*Phytomyza* spp.).

Delia radicum – cabbage root fly, also known as cabbage root maggot in North America, and *Delia floralis* – turnip root fly

Cabbage root fly (*D. radicum*) is subject to predation by several flies, mites, ants, beetles and fungi at various stages in its life cycle. So far, no commercially acceptable system of biological control has been developed. Several beetles will consume the eggs of turnip root fly (*D. floralis*), while the fungus *Strongwellsea castrans* sterilizes the females but only after egg laying has commenced. Neither the beetles nor the fungus have yet been formulated into a satisfactory commercial biocontrol system.

Husbandry may be manipulated to avoid infestation by cabbage root fly through crop rotation, the destruction of infected hosts and the avoidance of autumn sowing which provides overwintering sites. Spatially separating ware production of vegetable brassicas from seed crops is of major importance. Forming a physical barrier by mulching with finely netted or non-woven covers or placing collars around the stem bases of transplants prevents the adult flies from placing their eggs adjacent to host plants.

Chemical control, where permitted, is achieved by incorporating granular compounds into soil used as nursery beds for raising transplants or using seed treatments. Further applications are essential by direct incorporation before transplanting either as liquid drenches or as granular formulations directed in bands along the crop rows. Applications at drilling or transplanting are normally sufficient to protect rapidly maturing crops such as calabrese (green broccoli). Longer season crops may require repeated applications to combat second and further generations of flies. There is relatively rapid mutation by root flies, producing tolerance to new insecticide formulations. Granular insecticides are most frequently used for the commercial control of turnip root fly. So far, no commercial resistant cultivars have been produced. Screening for resistance in root brassicas and cauliflower has been a main breeding strategy, and some tolerant strains have been identified. There appear, however, to be differences in response depending on which side of the Atlantic the screening for resistance occurs. Resistance has been reported to be additive. High levels of antibiosis resistance are recorded in *B. fruticulosa*, *B. incana*, *B. villosa* and *B. spinescens* to cabbage root fly (*D. radicum*). All *Brassica* species were attractive for egg laying by *D. radicum* and therefore lack antixenosis resistance. Freuler *et al.* (2000) found resistance to *D. radicum* in the cauliflower cvs Imperator Nouveau, Panda, Asterix, Belot, Talbot and Lara by counting eggs laid around the plant stem and estimating root damage. A.M. Shelton and M.H. Dickson (unpublished information) identified resistance to *D. brassicae* in cauliflower by response to egg placement at the stem base in the greenhouse, and in the field by damage assessment following baiting with offal. Dodsall *et al.* (2000) identified that *S. alba* had the greatest resistance to infestation by cabbage root fly maggots (*D. radicum*), and in crosses of *S. alba* × *B. napus* some hybrids had high levels of resistance similar to that of the *S. alba* parent which persisted from year to year.

Scottish-bred swede cvs Angus and Melfort apparently have some resistance to turnip root fly (*D. florialis*). Resistance in a range of *Brassica* types has been described by Alborn *et al.* (1985).

Thrip

Thrips tabaci – thrip (Fig. 7.13)

Probably the most successful breeding programmes for resistance to insects have been against thrips (*T. tabaci*) for which quantitative and recessive resistance was found, and many new hybrids have been tested for pest tolerance (Stoner *et al.*, 1989). Evaluation can best be done in the field under natural conditions.

Growing the lines to be tested adjacent to and downwind of wheat or oats enhances the chances of a severe test, since when the grass matures the thrips become airborne and drift over the plants to be tested (Shelton, 1995). Later as the heads mature, they can be evaluated. A sector of the head is cut and both the number of leaves and the severity of damage are scored on several heads of each cultivar. The results compared with known controls give a good estimate of resistance to thrips. An estimate of the number of thrips per plant is not an indication of resistance, in fact it is the opposite. On resistant plants, many thrips may rest on the outer leaves, but not enter the inner leaves of the head. Thus a resistant plant could superficially have a higher insect population than a susceptible one. Good economic resistance has been identified and selected using this procedure. Cultural control is possible when the crop maturity is scheduled to avoid periods during which thrip populations are at their peak.

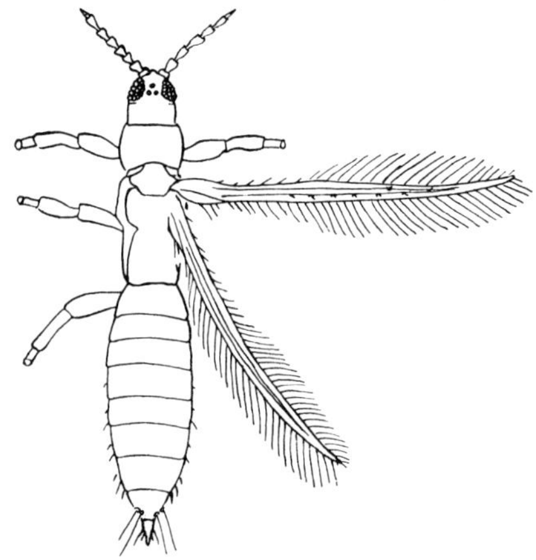

Fig. 7.13. Line diagram of thrip (*Thrips tabaci*).

Pest monitoring using yellow coloured sticky traps allows spray applications to be geared with periods of highest risk. The lacewing predator (*Chrysopa* spp.) has been used to provide biological control in India and has been integrated with pesticide applications.

Nematode

Heterodera cruciferae – brassica cyst nematode

Control is largely obtained by crop rotation, avoiding *Brassica* overcropping; if a rotation of brassicas (including oilseed rape) of no greater frequency than 1 year in 5 is maintained, the nematode should not become a problem. Some carbamate products used against cabbage root fly (*D. radicum*) offer some control of cyst nematodes.

Mollusc

Derocerus spp., _Arion hortensis_ and _Milax_ spp. – slugs

Where possible, avoid growing crops on soils that are infested heavily with slugs. Baits such as metaldehyde and methiocarb have been applied as chemical controls. For specialist situations, a parasitic nematode is commercially available as a biological control agent for slugs in the UK, but the product is currently too expensive for widespread field use. The use of some natural fertilizers, such as calcium cyanamide, appears to be associated with reductions in slug populations and is widely used in Germany for the reduction of slug populations in oilseed rape crops. Other natural products such as extracts of garlic have proved effective at controlling slug populations, but sprays require the addition of 'sticker' compounds to prevent them being washed away by rain. Molluscs as crop pests are reviewed by Barker (2002).

Vertebrates

Columba palumbus palumbus – woodpigeon (ring dove), _Oryctolagus cuniculus_ – rabbits, _Lepus europaeus occidentalis_ – hares, deer (e.g. _Capreolus capreolus_ – roe deer, _Dama dama_ – fallow deer, _Sika sika_ – Japanese deer, _Muntiacus muntjac_ – Indian muntjac) and rats (_Rattus norvegicus_ – brown rat, _Rattus rattus_ – black rat)

Preventing damage caused by vertebrates requires a combination of methods. Their use tends to be regulated by a mixture of environmental and conservation legislation and social pressures. Shooting has short-term benefit, but will only slightly reduce the populations of larger mammals such as deer, hares and rabbits. Scaring by various methods such as static carbide guns, vibrating tapes or simple scarecrows can be effective, but these need constant readjustment and changed positioning, as pigeons, for example, learn quickly to ignore anything that presents no actual danger to them. The use of toxic baits has now been largely abandoned because of their potential to harm non-target animals especially domestic pets. Transplants raised in seedbeds may be protected most effectively by netting or the use of vibrating tapes. Mammals can be diverted from young *Brassica* crops by the presence of adjacent sowings of more attractive food sources such as clover (*Trifolium* spp.) and grasses. The loss of poisonous baits is resulting in increasing problems with rats, in particular, attacking maturing crops of autumn brassicas. The presence of rats in crops raises significant potential human health risks both to those harvesting the crops and to the ultimate consumer from contamination by disease-causing microbes carried in rat excreta and urine.

Air-borne microbes

Albugo candida – white blister or white rust (Fig. 7.14)

Control can be achieved by preventative and eradicant fungicidal sprays against foliar and systemic infections. Race-specific resistance governed by single dominant genes has been identified in several *Brassica* and *Raphanus* spp.; quantitative inheritance of disease reaction type was demonstrated in *B. rapa*.

Resistance to race 2 of *A. candida* is controlled by a single dominant gene, and resistance was associated with leaf pubescence, which is also governed by a single dominant allele (Kole *et al.*, 1996). The resistance locus *ACA1* and pubescence locus *PUB1* were mapped using restriction fragment length polymorphisms (RFLPs) to linkage group 4. Portuguese land races of *B. oleracea* were screened for white rust resistance, and 'Couve Agarvia', 'Couve Glória de Portugal' and 'Couve Portuguesa' were the most resistant (Santos *et al.*, 1996). Some cultivars of *B. carinata* have resistance to both *A. candida* and *P. parasitica*, while dominant resistance is present in *B. juncea* and in *B. napus*. Resistance to *A. candida* is present in interspecific crosses between *B. carinata* and *B. juncea*. Mass selection and selection between and within half-sib families were effective methods of accumulating minor genes conferring reduced pathogen sporulation in rapid cycling populations of *B. rapa* (Edwards and Williams, 1987). Similar genes have been identified and mapped by Kole *et al.* (2002) in *A. thaliana*.

Alternaria spp. – dark leaf and pod spot

Host resistance is dominant and quantitative to this pathogen. Selections from species such as *B. tournifortii*, *Camelina sativa*, *S. alba*, *Capsella bursa-pastoris* and *B. carinata* appear to be highly resistant. Successful fusion of *C. sativa* with *B. oleracea* transferred a high level of resistance to *B. oleracea* (Sigareva and Earle, 1995). Subsequently, Hansen and Earle (1997) fused *S. alba* with *B.*

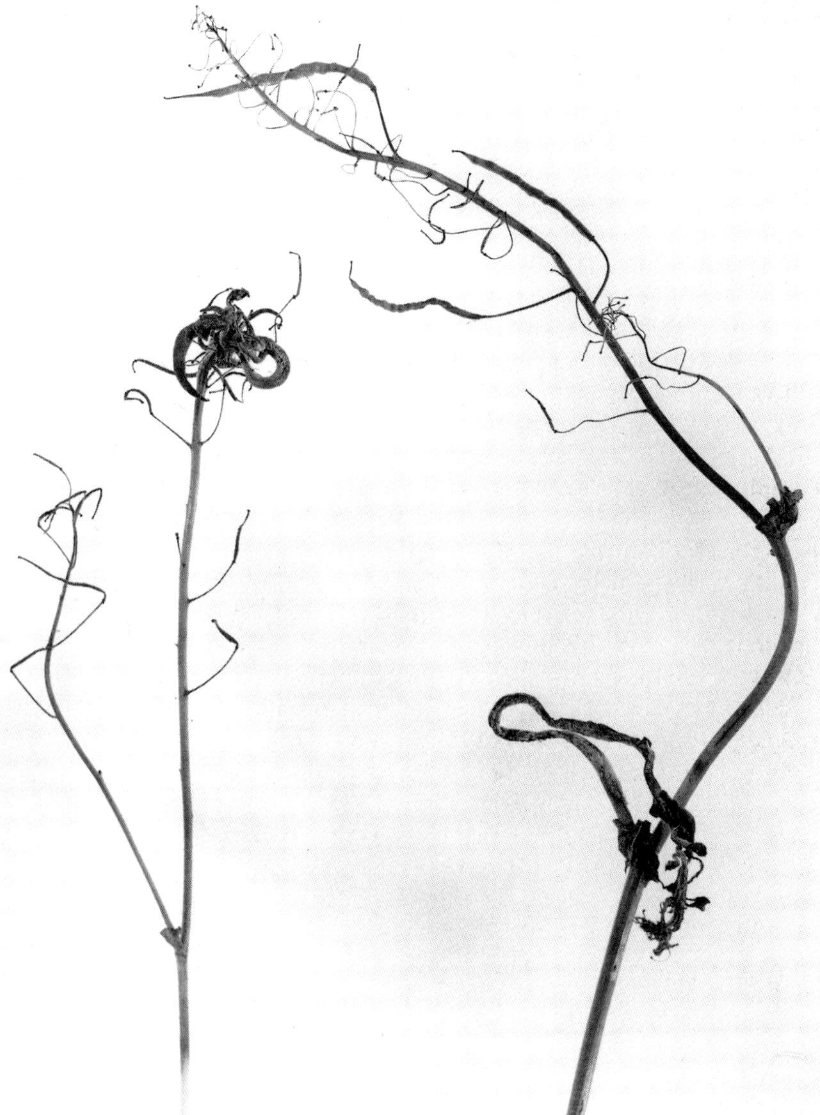

Fig. 7.14. White blister or white rust symptoms (*Albugo candida*).

oleracea and obtained resistance to *A. brassicae* equal to that in *S. alba*, although developing true breeding lines with high levels of resistance has been difficult. Resistance to *A. brassicicola* does not necessarily result in resistance to *A. brassicae* (Ryschka *et al.*, 1996). Variations in resistance in rape (*B. napus*) and field mustard (*B. rapa*) have been attributed to differences in the thickness of epicuticular waxes.

Limited fungicidal control is achieved with sulphur-based compounds and dithiocarbamates. The latter have been particularly effective when applied to cauliflower (*B. oleracea* var. *botrytis*). Sprays or seed dressings with synthetic molecules provide a more consistently reliable method of control. Hot water treatment of seed at 50°C for 25 min is a traditional technique for decontaminating infested seed.

Isolation of seed crops, especially of high-value horticultural crops, from farm oilseed rape is essential to avoid cross-infection during harvesting. Humpherson-Jones (1989) found that marrowstem kale (*B. oleracea* var. *acephala*) liberated 50 spores/mm^3 of air at cutting and 3200 spores/mm^3 of air when windrowed crops were harvested, and spores could be carried up to 100 m downwind during harvesting. Biological control may be of value in the future since saprophytic phylloplane fungi such as *Aureobasidium pulliulans* and *Epicoccum nigrum* are pathogenic to *A. brassicicola*.

Parasitism of *A. brassicae* by the *Verticillium* state of *Nectria inventa* occurs by either penetration or contact without penetration.

Erysiphe cruciferarum – powdery mildew (Fig. 7.15)

Considerable efforts are being made to control *E. cruciferarum* using resistant cultivars. Resistance in cabbage is attributed to a single dominant gene that is

Fig. 7.15. Powdery mildew (*Erysiphe cruciferarum*) symptoms on Brussels sprout leaf.

influenced by numbers of minor genes, since in parental generations resistance is incompletely dominant. Thus, under conditions of heavy inoculum, a heterozygotic host may support limited fungal growth. Multiple disease-resistant cabbages were produced in the northern USA for sauerkraut production and the fresh market trade in Florida during winter. Singh *et al.* (1997) performed inheritance studies involving interspecific crosses between *B. juncea* and *B. carinata* and found that resistance from *B. carinata* was dominant. The resistance gene was on the C genome, a diploid progenitor of *B. carinata*.

Powdery mildew has been recorded recently on broccoli raab (*B. rapa* subsp. *rapa*) in California, USA. This vegetable is also known in Europe as Rappini, a leafy vegetable recently popularized in America. Symptoms are characteristically seen as a white mycelium on leaves and stems. This strain will cross-inoculate on to *B. oleracea* var. *botrytis* and hence has the potential to cause substantial damage.

In the UK, powdery mildew on Brussels sprouts increased in importance following the intensification of production from the 1970s onwards (Dixon, 1981, 1984), and with the use of very susceptible F_1 hybrid cultivars that mature rapidly and are aimed at supplying the processing industry (Dixon, 1974). Originally, *E. cruciferarum* was found sporadically in Bedfordshire and the Vale of Evesham, and now it is a national problem. Eradicant fungicides based on sulphur compounds give some control of *E. cruciferarum*. Some newer chemicals provide considerable control, and are now regularly incorporated with aphicidal sprays applied routinely in August and early September when the primary lesions become visible.

Husbandry control by later sowing avoids the worst epidemics of *E. cruciferarum* and is used with swede and turnip crops. Although moderately successful in avoiding infection, there is a significant penalty as yield is reduced concomitantly with shortening of the growing season.

Leptosphaeria maculans (previously named *Phoma lingam*) – blackleg, stem and leaf canker (Fig. 7.16)

Hot water treatment can be used to free seed of *L. maculans* infestation, by soaking for 25 min at 50°C for cabbage and Brussels sprout (*B. oleracea* var. *capitata* and *gemmifera*) and 20°C minimum for cauliflower and broccoli (*B. oleracea* var. *botrytis* and *italica*). The technique is unreliable, however, often impairing subsequent germination. Treatments with fungicidal slurries are effective in disinfesting artificially infected cabbage (*B. oleracea* var. *capitata*) seed, and the use of zinc-based compounds has been successful in eradicating soil borne infection for cabbage (*B. oleracea* var. *capitata*) crops.

Breeding for resistance has been made difficult because of the wide variability in the pathogen populations. Crop types can be qualitatively categorized as *very susceptible* – red and green cabbage, savoy cabbage, Chinese cabbage, Brussels sprout, some radish cultivars, some swede

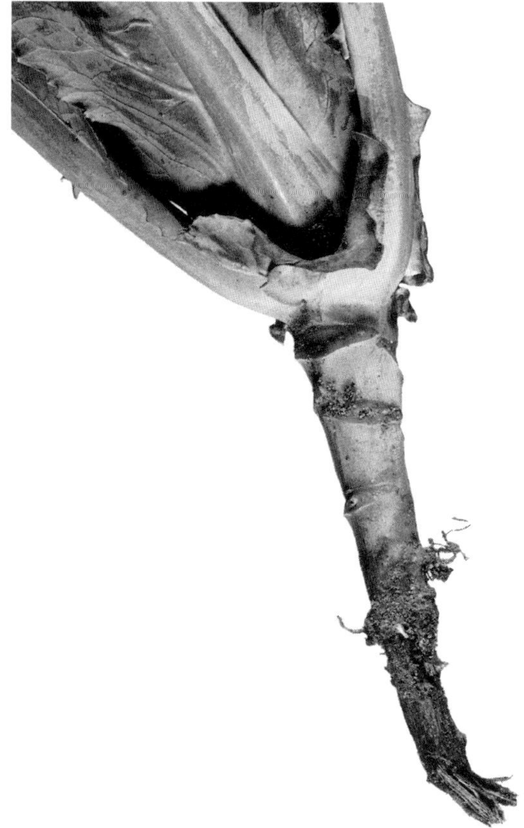

Fig. 7.16. Blackleg (*Leptosphaeria maculans,* previously *Phoma lingam*) symptoms on *Brassica.*

cultivars, white mustard and kohlrabi; *medium susceptible* – cauliflower, broccoli, rape, kale, collards, some turnip cultivars, wild radish and black mustard; *mildly susceptible* – some turnip and swede cultivars, Chinese mustard, garden cress and many strains of mustard; and *resistant* – horseradish, penny-cress, ball mustard, yellow rocket, shepherd's purse and pepper cress.

Swede cultivars often appear to be a mixture of susceptible and resistant populations, but the Wilhelmsburger types contain a high proportion of resistant material. *Carmelina sativa, S. alba, S. arvensis* and *B. nigra* are good sources of resistance, and attempts are being made to transfer this resistance to *B. napus* and *B. oleracea* through both embryo rescue and fusion and resistance tests applied to protoplasts using toxins. Some savoy cabbage, kale, turnip rape and white mustard express quantitative resistance and are being used to transfer resistance to white cabbage.

Husbandry controls include the use of rotations that ensure breaks of 4–5 years between *Brassica* crops. Legumes such as lucerne (*M. sativa*) or clover (*Trifolium* spp.) are particularly good break crops. Provision of adequate soil drainage and manipulation of plant stand density ensuring rapid air movement are essential, thereby discouraging the build up of a moist microclimate within the crop. Ploughing should be done deeply in the autumn to hasten destruction of infested debris.

Mycosphaerella brassicicola – ringspot (Fig. 7.17)

Cultural measures are of paramount significance for the control of *M. brassicicola*; seedbeds should be sited well clear of existing crops and be free of infected debris. Infection leads to premature leaf abscission, particularly in overwintered Anger-type cauliflower (*B. oleracea* var. *botrytis*); growers encourage rapid production of new foliage in the following spring by applications of nitrogenous top dressings. Autumnal applications of

Fig. 7.17. Ringspot (*Mycosphaerella brassicicola*) symptoms on cauliflower leaf.

potassium fertilizers (400 kg/ha) are considered as a further means of combating infection. Hot water treatment may be used to eradicate seed-borne infection by steeping the seed for 20 min at 45°C.

Chemical control has been achieved by using high volume sprays of manganese and zinc compounds applied every 14 days with between four and six times the recommended rate of wetter. The use of such compounds has now been largely discontinued. Resistance to *M. brassicicola* is claimed to exist in Roscoff-type cauliflower (*B. oleracea* var. *botrytis*), which have been selected in commerce for centuries against this pathogen in Brittany, France. In India, differences in disease severity have also been noted, with the cauliflower cv. Improved Japanese exhibiting greatest resistance.

Peronospora parasitica – downy mildew (Fig. 7.18)

Attempts to control *P. parasitica* by use of resistance appear to have been restricted to the use of a highly specific major gene, and this (as might be anticipated) has little practical value. Several studies suggest, however, that more generalized resistance is available which could be exploited. There are many physiological races of the pathogen and they tend to be specific to single species of *Brassica*. Resistance was considered to be due to a single dominant gene for seedling resistance. More recently, the seedling resistance has been

Fig. 7.18. Downy mildew (*Peronospora parasitica*) symptoms on *Brassica*.

shown to be quantitative, and there is a continuous variation in the level of resistance and susceptibility as well as variation in the seedling and mature plant resistance after seven leaves have been produced. Therefore, breeders must test for resistance in both seedling and mature plants. At the same time, there continue to be reports of forms of gene-for-gene resistance. As the number of physiological races identified expands, it appears that selection for horizontal resistance may be the only way to develop widespread seedling resistance. As with other downy mildews, the organism evolves new physiological races quite readily. Mapping for resistance to downy mildew is far more advanced in *A. thaliana* than in *B. oleracea*.

Coelho *et al.* (1997) reported that there were no accessions with higher resistance at the cotyledonary stage when compared with adult plants. There were, however, accessions where resistance at the adult plant stage exceeded that found in seedlings. Considerable mapping of different markers for various isolate and resistance genes has occurred and is continuing. Reports of new physiological races and sources of resistance have recently come from around the world. Evaluation of ever larger race collections and sources of resistance have been reported, along with efforts to find a universal and wide-ranging source of resistance. In most cases, resistance has been found to be dominant. 'Everest' broccoli has been reported to be resistant to up to nine physiological races of the pathogen, but not to all races, while Portuguese land races 'Couve Algarvia', 'Couve Murciana' and 'Couve Covacao de Bois' showed resistance to seven Portuguese isolates (Soursa *et al.*, 1997). Mitchell *et al.* (1995) using *B. rapa* showed a positive correlation for resistance to *P. parasitica* and *L. maculans*. This suggests that some resistance genes provide defence against these two very different pathogens. Perhaps such multiple disease resistance could contribute to durable resistance that might not be so easily circumvented by rapidly evolving microbial pathogens.

Jensen *et al.* (1999) found that resistance genes *br8* and *br9* reduced conidia production on cotyledons by 50–70% compared with some susceptible lines. The *Br9*, most resistant of the two lines tested, showed rather uniform resistance to 13 isolates from different geographical origins; *br8* showed some isolate specificity. Partial resistance in a half diallel set of six broccoli lines showed additive genetic effects explaining 45 and 38% of total variation of sporulation and conidial production, respectively. This suggests that recurrent selection for partial resistance might efficiently result in cotyledon resistance.

This pathogen was controlled chemically by use of 0.2% a.i. dichlofluanid, particularly with seedling crops. Addition of 0.1% non-ionic wetter greatly improves the efficacy of this compound, especially when infection spreads to the true leaves. Repeated use, however, may lead to phytotoxic damage. Good husbandry is of major importance in the control of downy mildew; excessive overhead irrigation will encourage pathogen spread. Every effort should be made to keep seedling foliage as dry as possible by use of abundant ventilation and preventing water from remaining on or around plants for prolonged periods. Where sprays are applied, use should be made of ultra-low volume techniques or dust formulations to prevent the accumulation of moisture on the foliage. Plant density should be regulated to prevent overcrowding in the seedbed. Deliberate 'checks-to-growth' often given to seedlings in order to prevent them becoming too large before transplanting are often associated with an attack by *P. parasitica*. With direct-drilled crops, wider spacing should be employed where downy mildew is known to be a hazard to crops. Growth stimulation by fertilizer application can be used to enable seedlings to outgrow infections. Crop debris must be removed from seedbeds since this fungus can perennate as oospores in old foliage. Where crops are grown intensively for the 'baby-leaf' production, as with turnips (*B. rapa*) in California, USA, or in Europe, *P. parasitica* is a major disease problem.

Soil-borne microbes

Thanetophorus cucumeris (also known as *Rhizoctonia solani, Corticium solani, Hypochnus cucumeris, H. solani* and *Pellicularia filamentosa*), *Pythium* spp. – damping-off and other diseases

Damping-off is a good example where the modification of cultural practices involving the avoidance of pathogen transmission with propagating material is of major importance as a control measure. Alteration of the sowing date will reduce the risk of infection from soil-borne inoculum for crops where the seed requires high temperatures to start germination. Shallow seeding encouraging rapid emergence is essential. Movement of soil around the hypocotyls of emerging seedlings should be avoided; this usually occurs during irrigation. Crop sequence can influence the extent of *T. cucumeris* infection by increasing or reducing inoculum levels in soils. Growing lucerne (*Medicago sativa*) or clover (*Trifolium* spp.) increased the level of infection in subsequent crops, whereas cropping with cereals or incorporation of straw into the soil reduced the *T. cucumeris* population. This effect possibly results from increased carbon dioxide concentrations, reduced nitrogen levels in the soil solution (*T. cucumeris* is favoured by soil nitrogen) and stimulation of antagonistic soil microorganisms. The antagonist *Trichoderma lignorum*, for instance, produces a toxin that destroys *T. cucumeris*. The search for biological forms of control is especially active in Japan and other parts of southeastern Asia where radish and Chinese cabbage are essential crops and seriously damaged by damping-off. An ingenious alternative control method is the stimulation of hypovirulent forms of *T. cucumeris* that compete for infection sites with pathogenic forms, rendering their recognition and occupation unavailable to virulent strains.

Fusarium oxysporum f.sp. *conglutinans* – cabbage yellows

Control of *F. oxysporum* f.sp. *conglutinans* is largely achieved by the use of resistant cultivars and is a classic example of the success of this approach, which has remained effective for decades in the USA. Although the pathogen is extremely variable in its growth characteristics *in vitro*, it is stable in pathogenicity towards its hosts. Initially, field selection for resistance within diseased crops resulted in the release of cv. Wisconsin Hollander. Subsequently, this cultivar was found to have incomplete resistance at high temperatures; when pathogen growth is most vigorous, some yellowing would develop. Further work, using inbred lines produced from selfing plants by bud pollination to overcome host self-incompatibility mechanisms, led to the development of cultivars possessing resistance unaffected by temperature below 26°C.

Fusarium yellows and turnip mosaic virus (TuMV) (Fjellstrom and Williams, 1997) are serious diseases of *B. oleracea* and *R. sativus*. Turnip mosaic virus causes substantial losses to *B. rapa* and *B. juncea* (Sako, 1981). Resistance to *Fusarium* is widespread in *Brassica* subspecies, but resistance to TuMV is scarce. Cultivars of Chinese cabbage are increasingly being developed with heat tolerance as a major attribute (traditionally it is grown in the cool seasons when *Fusarium* is not a problem), and hence wilt is developing as a constraint to production (Shattuck, 1992). Some chemicals are effective in controlling *F. oxysporum* f.sp. *conglutinans*. Dipping the roots of transplants in a solution of zinc dimethyldithiocarbamate is reportedly effective. Husbandry controls include use of high levels of potassium fertilizers. Soil solarization has been used successfully to reduce populations of this pathogen.

In Germany, soil treatment with steam, alone and in combination with calcium cyanamide, controlled *F. oxysporum* f.sp. *conglutinans*. Rotation is an effective control measure since *F. oxysporum* f.sp. *conglutinans* survives for only limited periods in the absence of host tissues. Biological control may be developed since cross-protection against *F. oxysporum* f.sp. *conglutinans* followed prior inoculation with other formae speciales of *F. oxysporum*. Some soils are suppressive to *F. oxysporum* f.sp. *conglutinans*. This characteristic may be related to the presence of siderophore-producing antagonists such as *Pseudomonas putida*. Other antagonistic organisms, for example *P. oxalicum* and *Trichoderma viride*, reduce the rate of hyphal development by the pathogen. Composted hardwood bark amended with *Trichoderma hamatum* and *Flavobacterium balustinum* restored soil-suppressive characteristics where they were lost.

Plasmodiophora brassicae – clubroot (Fig. 7.19)

Most resting spores of *P. brassicae* in field soils are concentrated in the upper profiles, decreasing gradually with depth, soil type, tillage practice and cropping history. Spore dissemination is via drainage water, organic material derived from infested roots, wind-borne soil particles and diseased transplanting material. Contaminated transplants are a major means for the long-range spread. Recently, *P. brassicae* has been identified in peat used in composts for container-grown *Brassica* transplants. This presents a significant new avenue for the widespread dissemination of clubroot. Irrigation water is also a regular means of spread in some countries.

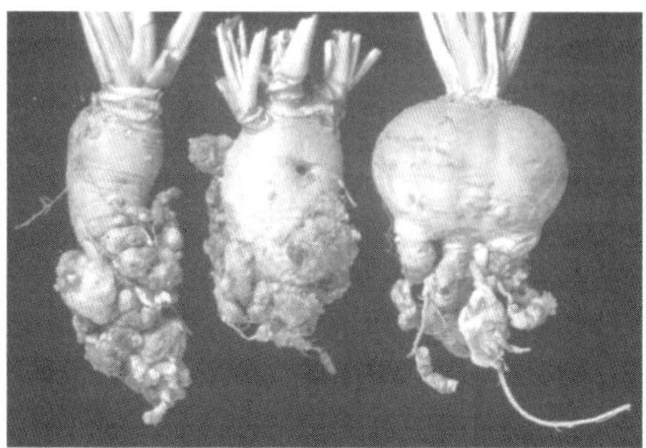

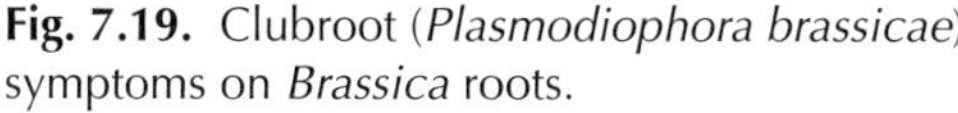

Fig. 7.19. Clubroot (*Plasmodiophora brassicae*) symptoms on *Brassica* roots.

Studies of wild *Brassica* revealed that there are few sources of resistance to *P. brassicae*, posing a substantial problem for plant breeders. Resistance in *B. rapa* (turnip) and *B. napus* (swede) is of a monogenic dominant type, while that of *B. oleracea* (cole crops) is polygenic and recessive. Several lines of *B. rapa* were selected in The Netherlands carrying strong resistance to *P. brassicae*. These were developed from turnip land races identified in Belgium and southern Holland. Resistance results from the action of three genes: *A*, *B* and *C*. In *B. napus*, there are four resistance factors which may be inherited as single independent genes, with resistance being dominant to susceptibility. Resistant swedes are mostly derived from either the green-skinned 'Wilhelmsburger' or bronze-skinned 'Bangholm' types. Clubroot resistance is rarely encountered in *B. oleracea*. The cabbage cv. Badger Shipper was developed from selfed progeny of a chance cross of kale × cabbage hybrid and was originally suggested to possess polygenic resistance. Subsequent research indicated that susceptibility is dominant to resistance and probably only results from a limited number of genes. Resistance to *P. brassicae* in the land races of the cabbage 'Bindsachsener' and kale 'Verheul' results from the interactive effects of several common genes of quantitative effect. Further resistance has been identified in the land race 'Bohmerwaldkohl', which shares common resistance genes with 'Bindsachsener'. In general, resistance in *B. oleracea* is quantitatively expressed, recessively inherited and dependent on the environment. For example, the cv. Resista developed from 'Bohmerwaldkohl' and land races from Shetland exhibits a high level of variation in reaction to *P. brassicae*. The combined characteristics of resistance and tolerance to *P. brassicae* are inherited recessively, with probably only a few genes involved. Similar characteristics are present in land race material from northern, eastern and central Europe. Incomplete dominance for tolerance to *P. brassicae* is linked to storage quality and in favour of extended growing periods for white cabbage types. Attempts have been made in the USA and Japan to transfer resistance to Brussels sprout, cauliflower, broccoli and calabrese from *B. oleracea* sources via interspecific hybrids with *B. napus*.

Interest centres particularly on transferring resistance from *B. rapa* to *B. napus* var. *oleifera* (one of the oilseed crops) in view of the increasing economic significance of these crops and their extreme vulnerability to *P. brassicae* which has already restricted production areas in France, Germany and Scandinavia. Development of resistance to *P. brassicae* in Chinese cabbage (*B. rapa* var. *pekinensis* and *B. rapa* var. *chinensis*) is of major economic significance in Asia. The cv. Michehili possesses one dominant gene while at low inoculum densities several other hosts express resistance. Numerous commercial breeding programmes in Japan and other parts of Asia aim to exploit these characteristics. Elsewhere, cytoplasmic male sterility has been utilized to develop F_1 possessing clubroot resistance. Use may also be made of resistance in *Raphanus* (radish) by forming the intergeneric hybrids *Raphanobrassica* and *Brassicoraphanus*, both of which are resistant to *P. brassicae*.

The kales are generally the most resistant of vegetable brassicas, together with some rutabagas (swedes) and turnips. Each has been used in resistance breeding programmes; an early example, as mentioned above, was the cabbage cv. Badger Shipper which was selected in Wisconsin (the Badger State), USA from kale in the 1930s. Tolerant broccoli, cauliflower and Chinese cabbage have been developed using resistance derived from *B. napus* and *B. rapa*. The Roscoff cauliflower and Nine-star broccoli are amongst the most resistant.

The Italian *B. oleracea* lines have displayed forms of tolerance possibly reflecting selection in crops grown for several human generations on pathogen-infested land and possibly reflecting close proximity to the centres of origin of these crops. Most cases of resistance especially in *B. oleracea* are partially recessive and partially additive traits governed by a few major genes. There is a debate as to the presence or absence of gene-for-gene relationships between *Brassica* species and this pathogen. Work is in progress to develop genetic maps for resistance to clubroot in various occidental and oriental host species and includes *A. thaliana* (Dixon, 2006).

Husbandry control using lime to increase soil pH above 7.0 has been a traditional method of clubroot control over several centuries, but this is far from a reliable technique. Liming will control the pathogen if the spore load is low, but even heavy applications are ineffective if soil is heavily contaminated by the pathogen. Sufficient lime must be used to raise the pH to 7.3–7.5 for several seasons in order to have substantial impact on inoculum potential. A more rapid effect is achieved with calcium oxide (hot lime). The fertilizer calcium cyanamide provides sources of nitrogen and calcium and is associated with a diminution in the severity of clubroot disease especially when used over several seasons. Calcium nitrate fertilizer has a similar effect and offers a readily soluble source of calcium. Boron also has an effect of diminishing the impact of *P. brassicae*, and may be used possibly in combination with non-ionic wetters as a treatment applied at transplanting.

Extended research demonstrated that part of the mode of action of boron, calcium, nitrogen and pH is to reduce the rate at which *P. brassicae* grows and reproduces following primary colonization of root hairs. There are also effects that diminish the abilities of primary zoospores to locate root hairs and which stimulate sources of generalized resistance in host plants.

Rotations with at least 5-year breaks between cruciferous crops are effective as control methods but frequently not economically feasible for horticultural brassicas. Breaks in cropping are equally important where resistant brassicas are grown. This is because the root hairs of resistant types are invaded, during the primary stages of pathogenesis, to a similar extent to those of susceptible forms.

Weeds such as grasses can also harbour the primary stages of *P. brassicae*, and should be used with caution in rotations. Even longer rotations are required for spring- and summer-growing brassicas that occupy the soil during optimal environmental conditions for the growth of *P. brassicae*. Some break crops such as maize and lucerne diminish the inoculum potential of *P. brassicae*. Green manuring following ploughing reduces infestation, as will trap cropping with summer rape for 5 weeks prior to planting cauliflower. Organic soil amendments have been demonstrated to reduce clubroot in Taiwan, where intensive *Brassica* production is common. Soil solarization using plastic sheeting is also an effective control, as reported from Australia, California and Japan. Direct soil heating reducing the viability of *P. brassicae* spores rather than any induced form of biological control is responsible for the effects of solarization. Soil cultivation can be used effectively for clubroot control; in southwestern Wales, for instance, continual rotary cultivation lowered the inoculum potential sufficiently for summer cauliflower production. Presumably loose, dry, friable soil allows the more rapid desiccation of resting spores. Suppressive soils have been identified which inhibit the development of *P. brassicae*. The suppressive potential remained unchanged after steam sterilization but was increased with alkaline pH and calcium content. Intensive research in Victoria, Australia has developed strategies for the control of *P. brassicae* related to the initial soil-borne inoculum potential involving the use of combinations of husbandry, fertilizers and agrochemicals (Donald, *et al.*).

Sclerotinia sclerotiorum – white rot (white mold in the USA)

Control of this pathogen is extremely difficult because of its wide host range and formation of persistent sclerotial resting bodies. Resistance is recessive and quantitative. The introduction line PI206942, a non-heading cabbage from Turkey, has shown superior levels of resistance (Dickson and Petzoldt, 1996). Resistance was transferred to cabbage, broccoli and cauliflower. Sharma *et al.* (1995) reported that cauliflower cv. Early Winter Adam's White Head and EC 162587 were highly resistant, and RSK 1301 and MRS1 were moderately resistant. Harindar and Kalda (1995) indicated that EC 103576,

EWAWH and EC 177283 were resistant. Some systemic pesticides offer protection from white rot, but tolerant strains can develop with remarkable speed. Cultural controls such as soil cultivation, aeration and flooding are helpful in combating the pathogen. Forms of nitrogen fertilizer such as calcium cyanamide are also associated with diminished pathogen activity. Control of storage rots incited by *S. sclerotiorum* is aided by retaining fresh produce in a turgid state and reducing the temperature to 5°C. Some encouraging results have been achieved using biological control, particularly with hyperparasites such as *Coniothyrium* species that attack the sclerotia of *S. sclerotiorum*.

White rot is a serious pathogen of Indian mustard (*B. juncea*) (Sharma and Sharma, 2001), attacking the stems before and after flowering, leading to substantial yield losses through reductions in height, fewer siliquae in the primary and secondary inflorescence branches, fewer pods and shrivelled seed. Even where seed matures it has reduced germination capacity with diminished radicle and plumule growth. Similar effects are reportedly caused by *Alternaria* spp. on brown sarson.

Sclerotial survival of *S. sclerotiorum* was evaluated in the field following amendment with whole crops of white mustard or oats incorporated by rotary tilling (Thaning and Gerhardson, 2001) (Table 7.4). Sclerotial survival was reduced significantly by plastic covering and amendment with brassica residues, and in one instance by mycoparasitism with *Coniothyrium minitans*. The controlling effects of brassica residues result from sulphur compounds – glucosinolates which by enzymic breakdown produce antimicrobial substances. Soil solarization is widely practised in Mediterranean countries where sufficiently high soil temperatures can be achieved. In more northerly areas, a combination of plastic covering and crop residues has proved a valuable form of control (Kirkegaard *et al.*, 1998). Use of plastic sheeting in cool climates, although sublethal, retards the dissipation of volatile derivatives of the glucosinolates and keeps oxygen concentrations low; these conditions add stress to the sclerotia and hence increase the control of this pathogen.

Bacterial pathogens

Xanthomonas campestris pv. *campestris* – black rot

This pathogen is one of the few afflicting *Brassica* crops where the pathogen is seed borne. Because this pathogen can cause devastating epidemics in seedlings being propagated under the intensive conditions used to raise transplants for eventual field crops, it is essential to increase the sensitivity and accuracy of test methods used by seed analysts. Application of prolonged extraction procedures and semi-selective agar media is recommended as a result of internationally coordinated tests (Koenraadt *et al.*, 2005). Once seed

Table 7.4. Survival of sclerotia of *Sclerotinia sclerotiorum* in the field following covering with a plastic sheet and amendment with white mustard residues.

Treatment	Viable sclerotia (%)
Control	93
Mustard residues	82
Plastic sheeting	66
Plastic sheeting + mustard residues	39

Mean of three field experiments (adapted from the data presented by Thaning and Gerhardson, 2001).

infestation is established, then hot water treatment of fresh plump seed at 50°C for 25–30 min for cabbage, plus dressing with organomercurial dusts can provide effective control. The use of organomercurial seed dressings is being phased out, however, for reasons related to environmental protection. Treatment of infested seed with calcium hypochlorite for 16 h in sealed containers reduces infection but only below the threshold established for detection in seed intended for direct field drilling. Hence this technique has only limited value for seed destined to be used by transplant propagators. Other primary control measures include: phytosanitary inspections of seed crops, accompanied by roguing and seed certification; crop rotation; and avoidance of excessive irrigation. Wide spatial separation of seed and ware crops is essential for the production of healthy seed.

Periodic epidemics of black rot disease follow the introduction of susceptible cultivars, careless use of contaminated seeds and seedlings, and weather conditions favourable to the disease. Research suggests there are new highly aggressive variants of the pathogen, and breeding previously has been carried through in the absence of recognition of their existence. The pathogen can survive in soil even on plant debris for only 1–2 growing seasons. Survival on contaminated seeds and on weed crucifers is considered to be essential for the cycle of disease. Plant morphology and life cycle play important roles in the degree of black rot development in the field.

The rate of guttation is important in determining the susceptibility of cultivars. The ability of a pathogen to multiply in the vascular system is also an important factor in determining the eventual level of black rot symptoms. Vein plugging is due to accumulation of fibrillar material in the vessels, which plugs the veins and prevents spread. Resistance reactions develop at the gateway to entry in the hydathodes and in the vascular system to stop pathogen spread. Hence leaf resistance and stem resistance are governed by different genes. Race-specific resistance is seen at the site of inoculation as a hypersentive response (HR). Resistance in *B. oleracea* that is severely damaged by black rot is very low, with no true resistance in many botanical varieties and land races tested.

By comparison, a resistance developed in Japan was related to the use of heading Mediterranean kale of the 'Penca de Mirandella' type. Since the early 1990s, diseases caused by *X. campestris* have been spreading on to new hosts and into new regions. *Brassica oleracea* appears to be the most susceptible host; resistance genes have been identified and the gene-for-gene studies indicate that there are different physiological races. The relationship between *X. campestris* pv. *campestris* and the *Brassica* described in the triangle of U (1932) is shown in Fig. 7.20. This identifies sources of monogenic and multigenic resistance to the pathogen correlated with the six known physiological races of *X. campestris* pv. *campestris*.

Unrelated resistance has been identified in southeast Asian cabbage and in Portuguese Penca kale (Ignatov *et al.*, 1998). The origin of the Asian cabbage was traced to the Flat Dutch group of varieties and heading Mediterranean kale. Some forms of resistance may be available in Chinese kale, broccoli and cabbage. Where pathogen-tolerant cabbage lines are available, the mechanism of resistance appears to be present in the hydathode. Co-inoculation of cabbages with *X. campestris* pv. *carotae* and *X. campestris* pv. *campestris* increased resistance (Cook and Robeson, 1986), hence some forms of cross-protective biological control might be developed. Determinants that trigger host resistance responses may reside on the bacterial cell surfaces (Roberts and Summerfield, 1987). Sources of resistance have been found in two accessions of *B. carinata* (PIs 199947 and 199949); these are thought to result from a single dominant gene. In *B. rapa*, resistance is quantitative and of moderate effect. Resistance to black rot was found initially in the Japanese cabbage cvs Early Fuji and Hugenot in the early 1950s. Later, PI 436606 from China, another cabbage, provided further sources of resistance. Williams *et al.* (1972) demonstrated that resistance in Early Fuji was due to one dominant major gene *f*, modified by one dominant and one recessive gene where *f* is in the heterozygous condition. There is, however, almost a continuous variation in the level of resistance from high resistance, but far from immune, to extreme susceptibility. There is also a form of seedling resistance regulated by an additional recessive gene (Vincente *et al.*, 2000).

Screening can best be done at temperatures of 25–30°C. The source of resistance in *B. carinata* PI 199947 and also 1999949 gives quasi-immunity. This has been transferred into broccoli by fusion (Hansen and Earle, 1995). Following fusion, several generations of rigorous selection were necessary to obtain a true breeding line with the immunity level of resistance found in the *B. carinata* parent. Broccoli line 9811B was identified as the best and, following further selection, a true breeding subselection with high seedling and mature or field resistance was obtained.

The *B. carinata* resistance is dominant, but back-crosses still segregated. Guo, Dickson and Hunter (1991) transferred resistance from *B. carinata* to *B. rapa* by classical breeding. Ignatov *et al.* (1998) collected isolates from the UK, Japan and Russia, and identified five races whose inheritance for resistance

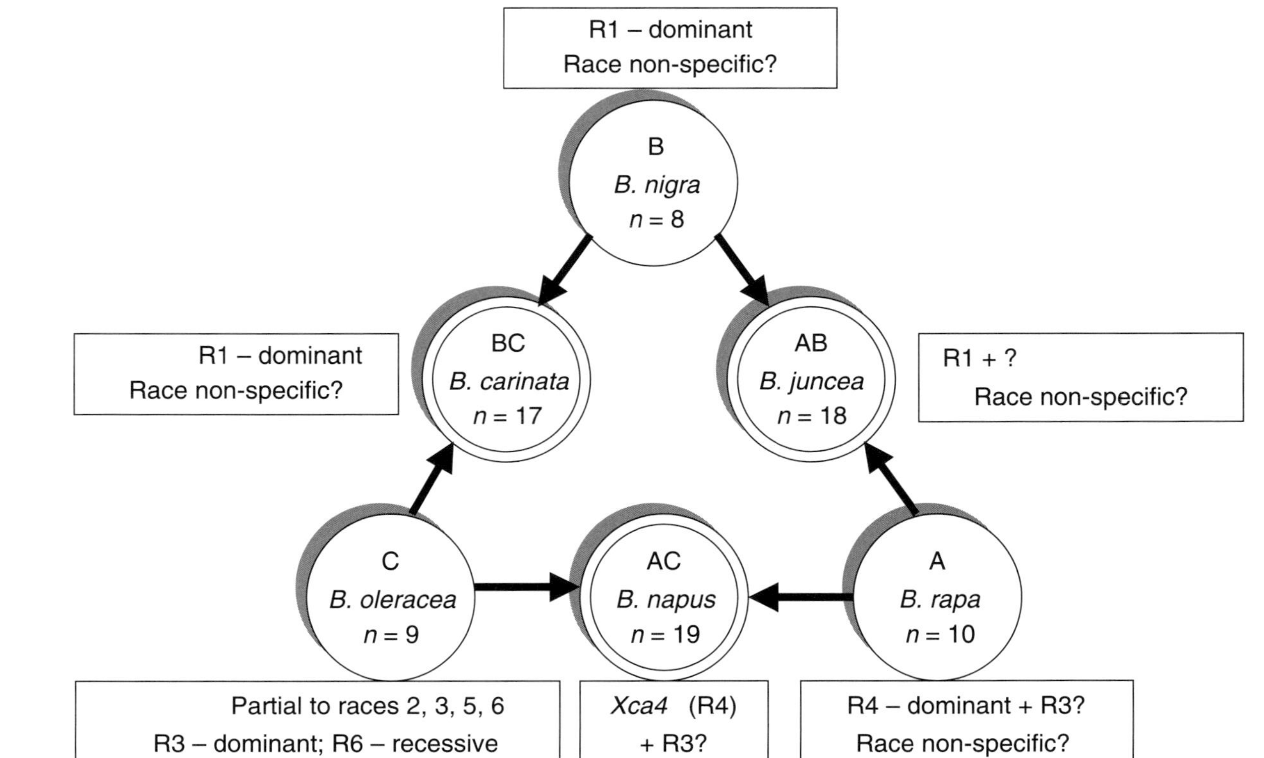

Fig. 7.20. Resistance to *Xanthomonas campestris* pv. *campestris* in *Brassica* spp., related to their position in the triangle of U (1935).

was controlled by dominant genes in races 1–4, and by a recessive gene *r5* in race 5. A gene-for-gene relationship best explained the data. Kamoun *et al.* (1992) identified five races (0–4). Vincente *et al.* (2000) modified the numbering of races. They withdrew race 3 and separated Kamoun's race 1 into three new races (1, 3 and 5), and also proposed a gene-for-gene model to explain interactions between *B. rapa* and *Brassica* differentials. In the UK, race 1 predominates, and worldwide races 1 and 4 are of most significance.

Recent intensive screening of the USA Department of Agriculture (USDA) *B. carinata* (Ethiopian mustard) collection identified several potentially important sources of resistance to *X. campestris* pv. *campestris.* These are thought to offer dominant single gene resistance for use in cabbage and cauliflower (Tonguç and Griffiths, 2004b). Interspecific hybrids of two *B. carinata* lines with *B. oleracea* are resistant to races 1 and 4 of *X. campestris* pv. *campestris* and were developed using embryo rescue. These offer a potentially valuable source of breeding material (Tonguç and Griffiths, 2004c).

No line, so far, has been found to be resistant to all six races. The cv. Wrosa (*B. oleracea*) was susceptible to all six races, cv. Cobra (*B. napus*) and cv. Just Right (*B. nigra*) were resistant to race 4, cv. Seven Top Turnip (*B. rapa*) to races 2 and 4, PI199947 (*B. carinata*) to races 1, 3 and 4, cv. Florida Broad Leaved Mustard (*B. juncea*) to races 1, 3 and 4, and cv. Miracle (*B. oleracea*) to races 2, 3 and 5. Resistance in cauliflower is reputedly governed by dominant polygenes. The US PI 436606 originating from China is resistant in both juvenile and adult stages, these effects apparently being controlled by a single recessive allele (Dickson and Hunter, 1987; Hunter *et al.*, 1987). Two ecotypes of *A. thaliana* have demonstrated a differential response to *X. campestris* pv. *campestris*, suggesting that ancestral resistance genes may be present outside the genus *Brassica* which could be of value in breeding programmes for crop brassicas.

Camargo *et al.* (1995) mapped quantitiative trait loci (QTLs) for juvenile and adult resistance in *B. oleracea*. Two regions on linkage groups 1 and 9 were associated with resistance at both juvenile and adult plant stages. Malvas *et al.* (1999) mapped 900 F_2 individuals from a line 'Badger 16' × 'LC201' cross, and six randomly amplified polymorphic DNA (RAPD) markers were associated with resistance linked to petal colour. Alleles from the susceptible 'LC201' contributed towards resistance. The research field of mapping markers as an aid to developing resistance to black rot is developing rapidly.

A novel molecular approach was made by Morra and Earle (2001) utilizing chitinase genes that have been implicated in plant defence against fungal pathogens because of their chitinolytic activity. Chitin is a linear homopolymer of β-1,4-linked *N*-acetylglucosamine. It is an important component of cell walls of fungi, constituting 3–60% of the wall. Chitinase genes cloned from plants and microorganisms have been inserted in the genome of a number of plant species including brassicas in order to achieve resistance against several fungal pathogens.

Erwinia caratovora subsp. *caratovora, Pseudomonas marginalis* and *P. fluorescens* – soft rots

Cultivars vary in susceptibility to soft rots; the broccoli (calabrese) cvs Marathon, Headline and Trixie appear to have some tolerance to *P. fluorescens*. Major efforts are being made to breed for resistance to soft rot in Chinese cabbage in Japan, Taiwan and other parts of Asia. In studies of >800 accessions, only low levels of tolerance were found, but three cycles of recurrent selection starting with the best 23 accessions improved this tolerance significantly (Ren *et al.*, 2001) (Fig. 7.21). Resistance appeared to be controlled by additive genetic effects with narrow sense heritability of 42%. From this research, resistance was suggested to be controlled by two major partially dominant genes and associated minor genes. It is also suggested that *Erwinia* was more virulent than *Pseudomonas*.

Broccoli (calabrese) cvs Shogun, Green Defender and related hybrids are quite tolerant, and some cauliflowers had higher levels of tolerance in both seedlings and mature plants. When cv. Shogun was crossed with the best Chinese cabbage lines using fusion, higher levels of resistance were obtained in the back-cross than in either fusion parent (Ren *et al.*, 2000). Improved resistance to bacterial soft rot following fusion between *B. rapa* and *B. oleracea* has been recorded. Field trials of Chinese cabbage (*B. rapa* var. *pekinensis*) in Ontario, Canada identified that the cvs Yuki, Manoko and Summer Top had

Fig. 7.21. Development of resistance to *Pseudomonas fluorescens*. The effects of three cycles of recurrent selection of broccoli (calabrese, *Brassica oleracea* var. *italica*) on resistance (M.H. Dickson).

the least wastage (<10% plant loss) from bacterial soft rot (*E. caratovora* subsp. *caratovora*) (Warner *et al.*, 2003).

Cultural controls involving the avoidance of excessive applications of nitrogenous fertilizers and encouraging aeration through the crop canopy by wider spacing can help minimize the impact of these pathogens. With broccoli (calabrese) in particular, crop spacing is used to determine head size according to market demands, and this leaves little option for change. Straw or plastic mulch placed around the base of plants minimizes the splash transference of bacteria from the soil on to foliage and helps reduce disease incidence.

Where permitted, copper oxychloride or similar compounds can provide some control, except under conditions of high inoculum potential and moist weather favouring the pathogen. Two applications are required for broccoli (calabrese), the first when spears are 10 mm diameter and the second after 7 days.

Virus pathogens (Fig. 7.22)

Elimination of the vector, where known, is the main route whereby virus-induced diseases in brassicas are controlled. Mostly the vectors are insects and their control is outlined earlier in this chapter. The principal viruses affecting *Brassica* crops include cauliflower mosaic virus (CaMV), broccoli necrotic yellows virus (BNYV), radish mosaic virus (RaMV), radish yellow edge virus (RaYEV), turnip crinkle virus (TuCV), turnip mosaic virus (TuMV), turnip yellow mosaic virus (TuYMV) and turnip rosette mosaic virus (TuRMV).

Losses caused by CaMV result from insufficient protection of curds by the foliage due to reduced leaf area and diminished size of the inflorescence that can make crops unmarketable. Although losses are generally limited, in severe cases 20–50% reductions in yield have been noted for some cultivars. Apparently CaMV is not seed transmitted so that protection of new *Brassica* crops is achieved by the destruction of the old season's infected plants prior to the emergence of new seedlings. There should be a break of at least 1 month between destruction of old crops and braiding of new ones. Seedbeds require geographical isolation from older *Brassica* crops or to be surrounded by trap crops of kale (*B. oleracea* var. *acephala*) or oilseed rape (*B. napus*) to produce either virus-free seed or transplants. The most satisfactory form of control is to raise transplants outside the areas of crop production. Virus-free (or more accurately virus-tested) plants may be raised by *in vitro* culture from inflorescence tissue.

Resistance to CaMV in Brussels sprout has been discussed recently by Walsh and Jenner (2002) as being controlled by at least one dominant gene and possibly also a recessive gene. Field studies of summer- and autumn-maturing cauliflowers showed that cultivars varied in their reaction to virus

Fig. 7.22. Cauliflower mosaic virus symptoms on a *Brassica* leaf.

infection, with significant differences in the effects of virus on yield, marketability and time of maturity. Genotype and environment interactions were established for these characteristics. The relative susceptibilities and symptom responses of several *Brassica* species infected with CaMV have been compared and related to the molecular events of the virus multiplication cycle.

A major component of the susceptibility of *Brassica* plants, and probably other hosts of CaMV, is at the level of the expression of viral minichromosome, and this is influenced by the host genotype.

Control of RaMV by the development of resistant cultivars may be possible since 'Tender Green Mustard' (*B. viridis*) is heterozygous for resistance that has also been identified in *B. rapa* cv. Purple Top White Globe, whereas cvs Early White Flat Dutch and Shogun are susceptible. Resistance is apparently present in the radish lines 5-white, 5–24–2a and 'L. R. Bombay'. Turnip mosaic virus is one of the most damaging pathogens of field-grown vegetables, particularly affecting brassicas in Asia, North America and Europe (Walsh, 1997). The biology, epidemiology and control of TuMV were reviewed by Shattuck (1992) and the first complete sequence published by Nicolas and Laliberté (1992; GenBank D10927). A further review by Walsh and Jenner (2002) concentrated on the use of host resistance.

Turnip mosaic virus incidence is usually greatest where seedbeds are sited near to old infected crops, consequently these should be placed well away from overwintered *Brassica* crops. If aphids are numerous, spraying the seedbeds with systemic insecticides may be useful to prevent TuMV spread within the seedbed. Use of resistant or tolerant cultivars is recommended as with white cabbage (*B. oleracea* var. *capitata alba*) (Walkey and Pink, 1988). Watercress crops should be replaced annually from seed-raised stock. Sources of resistance controlled by up to four genes in Brussels sprout have been identified. Rutabaga cvs Calder, Sensation, Vogesa, line 165 and a line developed by recurrent selection from the cross 'Laurentian' × 'Macomber' have shown a form of resistance to Canadian strains of TuMV which may be determined by a single dominant gene (Shattuck and Stobbs, 1987). White cabbage cvs Decema Extra, Vitala and Winter White III have high levels of resistance to the UK strain of TuMV. By comparison with other European virus isolates, those from the UK caused the highest level of external necrosis. While German isolates were associated with high levels of internal necrosis, both caused greater yield reductions than Danish or Greek strains. Differences in necrotic symptom expression may be utilized as markers in breeding for combined resistance to both German and UK strains. Virus infection had no effect on the incidence of pepper-spot necrosis.

Extracts from *Syzygium cumini*, *Acacia arabica* and *Callistemon lanceolatus* decrease production of local lesions by TuMV in *Chenopodium amaranticolor.* Grinding diseased leaves in a buffer produces an inoculum. The resultant suspension is rubbed on to the leaves to be tested that have been dusted previously with carborundum. Seedlings are inoculated and held for 7 days at 25°C to allow development of the virus and then grown at 15°C to encourage the expression of susceptibility. Provvidente (1980, 1981) found four isolates and appropriate resistance in Chinese cabbage from China; later a fifth isolate was identified in Taiwan.

The introduction PI 418957 is resistant or immune to races 1–4, PI 391560 to races 1–3 and other PIs and cultivars were sources of resistance to one or two races. Resistance to race 5 was found at the Asian Vegetable Research and Development Centre (AVRDC), Taiwan, and has been incorporated into some advanced breeding lines (Sako, 1981).

Resistance has been transferred to desirable types and is available. Turnip mosaic virus can be a problem in *B. oleracea*, and the cv. Globelle cabbage exhibits quantitative resistance to races 1 and 2 of TuMV. In China, 19 isolates were screened for resistance on 3000 *B. rapa* accessions and fell into seven strain groupings. Eight accessions were immune to all seven TuMV isolates (Lui *et al.*, 1996).

Mycoplasma-like organisms

Aster yellows complex

Yellows diseases are now recognized as being caused by mycoplasma-like organisms (MLOs) and associated forms of life. They are probably the simplest living organisms and are usually found in vascular and meristematic tissues. Aster yellows is a cause of excessive elongation of the internodes and generalized foliar yellowing. Antibiotics have provided effective controls for these pathogens in America. Their use is prohibited, however, in Europe because of fears that this will erode the usefulness of antibiotics in the treatment of human diseases. Hot water treatment is used with perennial planting material, but insecticidal control of the vectors is of limited efficacy (Lee *et al.*, 1998).

8

Postharvest Quality and Value

INTRODUCTION

Increasingly, research and development driven by the demands of the consumers of *Brassica* vegetables places ever greater emphasis on retaining quality after harvesting, in storage, during transport and while on display. This contrasts with previous objectives that were almost exclusively directed at improving the effectiveness and efficiency of operations in the field phase to achieve the greatest return on the units of resource employed from land, labour and capital. Recent changes to marketing systems have largely dictated and driven this revolution in producers' attitudes. The supermarkets now control most sales of greengrocery in developed countries and, increasingly, the methods of production in developing nations who supply the produce. Supermarkets establish criteria for quality as it will appear to customers in their stores at the end of the distribution chain. Growers and their supporting research services are required to ensure that quality is sustained in the postharvest phase, and this demand is in turn driving changing requirements for research.

The beginnings of this change can be detected in the 1950s as processing companies entered the *Brassica* vegetable market with 'quick-frozen fresh produce'. The processing industries formulated strict quality control and assurance procedures that were enforced at the point where crops were received into their factories. Producers obtained the comfort of agreed market pricing in advance of harvesting, but in return were required to provide produce adhering to rigorous standards of quality in terms of appearance, freedom from pest and pathogen damage, adherence to the regulations covering the use of agrochemicals and, increasingly, nutritional quality. This system turned sour when the processors themselves came under financial pressures and cancelled their contracts in the 1980s. The procedures of quality control and assurance developed in that 30-year period have been adopted and much refined by the supermarkets.

The delivery of high-quality fresh *Brassica* vegetables to the supermarket and onwards to the consumer is now a dominating marketing goal of growers. The quality of the raw product as it leaves the field is very largely determined by an interaction of plant growth, cultivar characteristics, the environment and husbandry practices. The extent to which quality is preserved after harvesting relates to the effectiveness of postharvesting handling, storage and transport systems. Attractive packaging will enhance the value of good quality produce, but does little to increase the visual appeal of poor-quality, blemished goods.

The vegetable brassicas are marketed mostly as fresh products, albeit that they may be packed in modified atmospheres with higher concentrations of carbon dioxide and lower quantities of oxygen compared with ambient conditions, and held at reduced temperatures. Significant quantities of cauliflower and broccoli particularly are blast-frozen, and these could be considered as semi-processed products.

WHAT IS QUALITY?

Quality is a subjective and somewhat nebulous attribute composed of components that vary with different *Brassica* crops and the attitudes of individual consumers. Basically these components may be grouped under four general headings:

- sight: colour, gloss, viscosity, size and shape, and obvious defects;
- touch or texture in hand, finger and mouth;
- smell and taste or flavour;
- hidden factors such as nutritional value, and the presence of either harmless adulterants or toxic elements.

Colour and gloss

Colour and gloss are products of light reflected from the *Brassica* in question. They result from the amount of light reflected in proportion to that absorbed by plant pigments. Thus, whiteness in cauliflower curds results from an almost total absence of pigmentation, leading to a very high proportion of light being reflected into the eye of the observer. Pigments such as the green chlorophyll that are present in high quantities in most *Brassica* crops contribute substantially to increased visual appeal. With cabbage heads, Brussels sprout buds and green broccoli (calabrese) spears, the consumer requires an appearance of freshness which is associated with a green and glossy appearance. Some crops such as red cabbage and coloured cauliflower curds are regarded as of good quality due to the presence of attractive red, orange or yellow anthocyanin pigments. In coloured cauliflower, there may be up to a 100-fold increase in vitamin A compared with white curds. Coloured cauliflower has become very popular worldwide since research started in the UK and North America about 15 years ago. By contrast, where there is yellowing of the foliage and flowers, it is associated with senescence and not acceptable to retail consumers. Consequently, colour is an important factor in the rapid evaluation of product quality and subsequent purchasing decisions. In red radish (*Raphanus sativus*), for example, changes to colour (Δ hue angle, Δ chroma) corresponded to postharvest changes of soluble and insoluble pectic substances, total glucosinolates and alkenyl glucosinolates (Schreiner *et al.*, 2003). In this study, a decrease in redness (hue angle increased, negative Δ hue angle) corresponded to increasing soluble pectin and decreasing insoluble pectin, indicating senescence and loss of crispness. A decrease in chroma (negative Δ chroma) revealed a loss in total glucosinolates and hence a reduction in flavour. Evaluation of changes in colour using colorimeters provides a rapid and non-destructive means of measuring postharvest quality of the product (Huyskens-Keil, 2003).

A clean, glossy appearance greatly enhances the apparent quality of *Brassica* vegetables. This appeal is increased by a film or droplets of moisture on the surface of leaves or buds. Moisture films or droplets increase the directional reflection of light, as compared with light being reflected evenly at all angles which conveys a dull, lifeless finish. Glossiness can be associated genetically with resistance to insects such as diamond back moth (*Plutella xylostella*), but is linked with the disadvantage of causing reductions in growth rate.

Viscosity

Viscosity describes high levels of internal friction in semi-fluid substances. It is not an attribute normally associated with fresh *Brassica* vegetables, only applying to crops such as swedes or turnip roots when they are processed into soups and purees. The degree of viscosity of such products is controlled in the factory during their manufacture and reflects the demands of consumers for 'thin' soup or very 'thick' broth. Such properties are well outside the control of growers who simply supply the raw materials for processing.

Size and shape

Size and shape are attributes of major significance for *Brassica* crops. Size requirements for fresh vegetables have diminished sharply in parallel with the reduction in the size of family units and rising dominance of single persons catering for themselves. Consumers demand produce that is easily and quickly prepared and can be consumed at one sitting. Few retail purchasers require products larger than 0.5 kg and frequently even smaller portions are demanded. Consequently, growers are required to divide cauliflower heads into florets before packaging or provide single heads and spears of cabbages or broccoli. The ability to present *Brassica* products at peak freshness and turgidity is increasing as their use in uncooked salad-style foods rises.

Shape is of particular importance in crops such as cauliflower and broccoli (calabrese). The market demands domed heads that differ markedly from the flat-headed types produced by the now obsolete open-pollinated cultivars. Well-formed spears of calabrese or the densely packed heads of Chinese cabbage, white cabbage or greens are essential quality characters. These qualities are frequently defined in considerable detail in the protocols issued by buyers working for the supermarkets.

Defects

Crop defects may have genetic, physiological, pathological or mechanical origins, or alternatively result from the presence of extraneous organic or inorganic items. Defects inherent in the crop resulting from interactions between genotype, environment and microbial pathogens are dealt with later in this chapter.

Surface blemishes, insect deposits, fungal growth, necrotic zones and virus-induced yellowing can all constitute defects of *Brassica* crops. Mechanical damage includes cuts, bruises and discoloration resulting from the harvesting processes. Cauliflower is probably the most easily damaged *Brassica* crop since rough handling will result in bruising and consequently the downgrading or rejection of heads. Where Brussels sprout buds are harvested mechanically, the failure to sharpen and adjust the stripper's blades correctly can ruin previously high-quality crops very quickly.

The effects of mechanical stress (dropping, compression and trimming) on Chinese cabbage cv. Yuki were tested by Porter *et al.* (2004). Dropping and compression did not affect marketable quality where the produce was sold immediately. Storage of damaged produce for 9 weeks (at 2°C), however, impaired quality. Heads that were repeatedly trimmed produced less ethylene at the end of the storage period compared with the start. This reflected the removal of outer senescing and rotting leaves. Marketable yield was not improved by trimming, emphasizing the point that treatments after harvest cannot improve on quality but only maintain what has already been established in the field.

The presence of extraneous items will also ruin crop quality. Pre-packaging by the grower for the supermarkets pre-supposes that there will be strict adherence to quality standards set as part of the contract. These are easily forfeited where, for example, leaves are packed with Brussels sprout buds or soil is allowed into the package along with cauliflower florets.

Texture

Texture is an interaction of those physical characteristics that are sensed by the feeling of touch. Purchasers still consider they are able to judge quality by the outer feel of vegetables, especially cabbage heads. Sensing firmness or softness by hand is used as a guide to maturity and quality by an experienced crop technologist for determining the start of harvesting. In *Brassica* crops, firmness is more important since softness indicates incipient enzyme-induced breakdown resulting from overmaturity or pathogen-related rotting. This attribute can be quantified by using standardized penetrometer tests that determine the rate at which needles pierce the produce when driven by a standard force. Compression can be determined numerically by exposing the produce to pressure imposed by a standard mass.

Texture in the mouth is generally sensed as chewiness, fibrousness, grittiness, mealiness, stickiness, oiliness or dryness. For *Brassica* crops which are cooked prior to consumption, the retention of flavour appeal and crispness is important. These are aspects of quality that cannot be controlled by the grower since they are very easily lost by overcooking. Increasingly, both leaf and root *Brassica* crops are being consumed in the raw state where the retention of crispness and flavour is of major importance. Here the application of modified atmosphere technology in packaging the produce to retain these qualities is of prime importance.

Flavour

Flavour is a composite of taste and odour. Taste is a four-dimensional characteristic distinguishing sweetness, sourness, saltiness and bitterness. Odour comes from the combination and interaction of a range of chemical constituents. Consumers would not expect odours from *Brassica* vegetables, and their presence reduces their value. The extreme forms of processing of *Brassica* as in the production of regional delicacies, such as the production of sauerkraut from head cabbage in Germany and North America and kimchi from Chinese cabbage in Korea, accentuate saltiness and bitterness as essential indicators of quality.

Taste is very much a combination of physical and chemical components. The chemical constituents can be identified in great detail by analytical techniques such as mass spectroscopy and liquid chromatography, which allow the isolation and quantification of molecules that are present in minute concentrations but which contribute very significantly to flavour. Equally, these analytical techniques allow plant breeders to select for the absence of molecules that contribute to bitterness or other undesirable characters.

HEALTH AND WELFARE BENEFITS

The hidden attributes of *Brassica* crops lie in their abilities to reduce the incidence of human cancer and coronary diseases when consumed over periods of years as part of a balanced diet. Medical evidence for these attributes has accumulated substantially in the past decade, and these properties are coming into prominence with the general public (Mazza, 2004) (see Table 8.1).

In this respect, the *Brassica* crops, such as broccoli (calabrese), have especial interest because of their sulphorathane content which is associated with the reduction of active oxygen in tissues and may provide protection from cancer and coronary diseases. Previously, *Brassica* crops lacked consumer appeal largely due to the presence of sulphur-containing amino acids that released unacceptable odours when overcooked and which can be mildly toxic. Now, broccoli (calabrese) is becoming a very popular convenience vegetable worldwide with significant benefits for long-term human health. It has been recommended as a dietary additive for cancer prevention since the early 1980s (Nestle, 1998).

The health benefits of *Brassica* vegetables are related to their content of glucosinolates. These are a group of sulphur-based secondary metabolites that are present in at least 16 families of the dicotyledonous plants (Fahey *et al.*, 2001). Wild and domesticated brassicas contain >100 different forms of the glucosinolate molecule; they all have a common basic structure composed of three parts: a β-D-thioglucose group, a sulphonated oxime moiety and a variable side chain. The latter can be a straight-chained alkyl or alkenyl structure, a ring-shaped aromatic group or an indolyl formation. Many common vegetable brassicas such as Brussels sprout or broccoli contain all three moieties. They remain inactive in intact cells, but damage releases myrosinase enzymes which breaks down the glucosinolate into several products, notably nitriles and isothiocyanates. These are the source of hot and bitter flavours in brassicas, especially condiments. Their original function

Table 8.1. Examples of the constituents of brassicas that may enhance human health.

Component	Potential benefit
Lutein	Contributes to healthy vision
Sulphoraphane	Neutralizes free radicals; may reduce cancer risk
Lignans	May protect against heart disease and some cancers; lowers low-density lipoprotein (LDL) cholesterol, total cholesterol and triglycerides
Allyl methyl trisulphide, dithiothiones	Lowers LDL cholesterol, maintains a healthy immune system

Adapted from Mazza (2004).

may have been to provide natural pesticides active against some insect and vertebrate pests. Some glucosinolate breakdown products have been thought to produce toxic, or anti-nutritional effects in grazing animals. Consequently, plant breeders have aimed to reduce their content in crops such as forage rape and fodder kale. They have also been ascribed goitrogenic effects in humans, but recent evidence offers a contrary view of their value.

There is a growing body of epidemiological and experimental evidence showing that the consumption of *Brassica* vegetables specifically reduces the risk of cancer in the human lung and alimentary tract (Chu *et al.*, 2002; Lester, 2006). Evidence suggests, for instance, that sulphorathane, an isothiocyanate present in broccoli (calabrese), aids in the detoxification of carcinogens, others such as those present in watercress help in the excretion of toxic agents from tobacco smoke while yet another group inhibits the proliferation of cancerous cells.

Since glucosinolates are highly labile, their health benefits depend on many variables relating to intake and metabolism following a range of factors which affect their concentrations in the crop and following its harvesting, storage and processing. Recent laboratory studies using rapid cycling *B. rapa* (see Chapter 2) suggest that zinc influences glucosinolate content and the resultant bitterness and medical properties of brassicas (Coolong *et al.*, 2004). In some crops such as red and white cabbage, concentrations of glucosinolates remain stable for several months after harvest. Where postharvest processing causes physical damage, then this is likely to lead to the activation of myrosinase and the release of glucosinolate breakdown products. Thus as the use by consumers of brassicas in chopped ready-to-use salads containing cabbage or broccoli (calabrese) increases, the concentration of anti-carcinogenic agents is likely to increase. Where the crops are subjected to heat treatment as in the blanching of chopped broccoli (calabrese), there will be reductions in the concentrations of glucosinolates. There is evidence that fermentation as in the production of sauerkraut or kimchi causes a complete loss of glucosinolates. These putative health-promoting effects of *Brassica* vegetables have substantial implications for producers, retailers and consumers. Breeding strategies are being developed aiming to produce new 'healthier' cultivars utilizing the abundant genetic variation present in *B. oleracea* (Johnson, 2000).

TRANSFERRING HIGH QUALITY FROM FIELD TO PLATE

Shelf life

Development, pre-maturation, maturation, ripening and senescence are the five phases that vegetables and fruit pass through during production, harvesting, storage and into the distribution chain to the ultimate retail consumer. *Brassica* crops are not required either to ripen or to senesce. These phases are mainly reserved for fruit crops. Indeed, senescent *Brassica* crops are generally unmarketable and worthless. Development in terms of product quality begins with the initiation and subsequent growth of the edible portion. In *Brassica* crops, this may be the flower curds and spears of cauliflower and broccoli (calabrese), buds of Brussels sprout, leaves of cabbage, including Chinese cabbage, or roots of swede, turnip and radish. This phase ends when there is a change in growth pattern or when natural enlargement ceases (see Chapter 4).

Maturation may be interpreted as the final development of size and quality in the field, and is the point where commercial crops are harvested. The maturation period for *Brassica* crops is short, providing only a brief period when the highest quality is available for harvest.

Genetic variation

The great variability inherent in *Brassica* species has led to the enormous spectrum of crop types (Chapter 1). This high variability ultimately resulted in a spread of maturity periods within individual crops that caused much difficulty for growers in achieving efficient harvesting of high-quality produce. The standard means of overcoming variation in the maturity of *Brassica* crops is the taking of multiple harvests in which the crop is picked over by hand. Field staff assess quality and maturity visually as they walk through a crop. This is very labour-intensive and inefficient, and makes scheduling for marketing extremely difficult. Minor changes in the weather, leading to a few abnormally hot days or a more prolonged cold period, will disrupt the most carefully prepared harvesting schedules. The development of F_1 hybrid *Brassica* crops offered a considerable advance in the control of their uniformity and reliability of maturation, and allowed the development of fully mechanized harvesting for Brussels sprouts and the use of gantry systems for crops such as cabbage, broccoli (calabrese) and cauliflower. Fully automated harvesting of fragile crops such as cauliflower is in the early phases of development. Here image analysis programs are used to assess the maturity of cauliflower curds and electronic signals are transmitted directly to automated cutting equipment when a head has been selected. These systems are relatively expensive, but this is set against the increasing problems that beset *Brassica* growers worldwide in locating and attracting field staff who are prepared to undertake harvesting in inclement weather at low rates of pay.

The move towards raising of transplants in modular containers by specialist propagators has considerably improved the uniformity of maturation gained from the genetic improvement pioneered by plant breeders. Propagation (Chapter 3) has been separated from crop growing, and the growing of seedlings prior to transplanting into the field is carefully controlled and regulated by specialist contractors. Standardized transplants are spaced mechanically in their field stations with accuracy and consistency. Provided soil fertility has been carefully adjusted to suit *Brassica* crops, then the subsequent transplants will establish quickly and form uniform root systems. Both the breeding of F_1 hybrids and production of modularized transplants have been major scientific advances reducing the variation of *Brassica* crops in their maturity phase and increasing their quality postharvest.

Senescence

Senescence applies to the physiological changes in flavour, composition and structure initiated by the cessation of growth at harvesting. The term deterioration includes senescence and the effects of pests, pathogens, disorders and mechanical damage before and after harvest. Deterioration in the *Brassica* crops can take on many forms, but is most widely characterized by chlorosis or yellowing resulting from a breakdown of chlorophyll pigments within the tissues. Taken to the ultimate stage, the tissues die and become necrotic. This final stage is characterized by dehydration, complete loss of colour and, finally, abscission.

Soil fertility and structure

Maintaining soil fertility and structure are cardinal to achieving high crop quality within the field and preventing its subsequent deterioration. Provided soil fertility has been adjusted to remove deficiencies in the macro- and microelements, then the availability of nitrogen during crop growth and development is of paramount significance in achieving consistent and even quality (Chapter 5). Luxury consumption of nitrogen or, alternatively, shortage of nutrients both contribute greatly towards the loss of crop quality (see Table 8.2).

As Babik *et al.* (1996) increased nitrogen applications to Brussels sprouts beyond 400 kg/ha, the buds lost sweetness and vitamin C content declined (see Table 8.2). The buds also lost firmness as indicated by the reduction in dry matter. This research supports the contention that excessive use of nitrogen fertilizer in Brussels sprouts, which was initially used to boost yield especially for crops destined for processing, resulted in increased acidity and reduced tastiness. Over a period of time, this loss of quality depressed the public liking for, and consequently the demand for, this crop.

There are no immediate remedies for the presence of excess nitrogen in the root zone. Incipient nitrogen deficiencies can be corrected, however, before visible stress symptoms appear. Diagnosis of incipient nutrient stress is

Table 8.2. The effect of increasing nitrogen fertilizer applications on quality characteristics of Brussels sprouts.

Nitrogen rate (kg/ha)	Dry matter (%)	Total sugars (%)	Vitamin C (mg %)	Chlorophyll (mg/l)
100 + 0[a]	18.4a	6.0b	148d	5.0
200 + 0	18.7a	6.4a	170a	5.6b
100 + 100	17.7b	6.0b	157b	5.8b
400 + 0	17.1b	6.0b	154c	5.9b
300 + 100	17.1b	5.8c	139e	6.6a
600 + 0	17.0b	5.4d	137e	6.9a
500 + 100	16.9b	5.2d	134f	7.1a

[a] Applications split between pre- and post-planting.
Data followed by different letters are significantly different using the Newman-Keuls test at $\alpha = 0.05$.
After Babik *et al.* (1996).

achieved by regular crop monitoring using foliage analyses. Traditionally, this is achieved by laboratory analysis of leaves and sap where several days may elapse between collecting the samples and the availability of results. Field test kits for such analyses are steadily coming on to the market. Automated and non-invasive systems for the measurement of incipient stress in crops are now readily available, however, mostly these utilize changes to chlorophyll fluorescence as the criterion (Wydrzynski *et al.*, 1995). This allows remedial action to be taken before obvious stress takes place. Once *Brassica* crops show visual yellowing, stress damage has become irreversible and variations in quality and maturity follow. Top dressings with highly soluble and readily available fertilizers such as calcium nitrate may prevent nutrient stress-induced variability from developing and maintain even growth towards maturity, thereby retaining the harvesting quality and scheduling.

Soil moisture

Moisture influences both nutrient uptake and the innate qualities of the *Brassica* product after harvest. Moisture stress increases the thickness of cell walls, relative dry weight and deposition of lignin, suberin and cellulose, resulting in a fibrous or woody texture. Fibrousness is unacceptable to the retail consumer in fresh vegetables, especially those purchased for their leaves, flowers or root and hypocotyl organs, as is the case with most *Brassica* crops. Quality decreases following erratic irrigation regimes that fail to sustain continuous development and impose stresses at all crop growth stages; these becomes critical as maturity approaches.

Temperature

Crop temperature is not normally controlled in open field crops. The advent of plastic film covers and mulching, however, permits some modest manipulation of temperature. Mulching in spring increases soil and air temperature. This advances the date of transplanting and accelerates growth, leading to earlier maturity. The timing of mulch removal becomes critical in late spring or early summer to sustain the quality in crops such as broccoli, leafy and hearted cabbage and some forms of early turnip production. Low field temperatures cause damage and reduce quality, especially when crops are approaching maturity. Rapid changes in temperature alter growth rates and induce stress, with resultant losses in quality.

Pesticides

Applications of pesticides aim to retain quality by preventing direct pest or pathogen damage or by eradicating weed competition (Chapters 6 and 7). Pesticides themselves can also adversely affect product quality. All pesticide applications are controlled by statutory regulations in terms of the last date for application to prevent toxic residues from remaining within the harvested portions of the crop at higher than acceptable levels (maximum residue limits (MRLs)). Very substantial margins of caution are used when framing MRLs such that they are pitched several orders of magnitude below concentrations that are likely to cause harm to humans, animals or the environment. These statutory regulations may not prevent the product from becoming unsightly because of the presence of spray residues that alter the colour of the product or cause burning to the foliage or flowers. Pesticide manufacturers normally foresee such eventualities when framing their label recommendations that are attached to the product's container. These directions indicate temperature conditions which are liable to be associated with spray-induced damage to *Brassica* vegetables.

Late application of pesticides can impart unpleasant flavours to a crop. Generally the deleterious effects of pesticides on quality result from ill-timed applications, miscalculation of application rates, failure to calibrate spraying equipment correctly or spraying when the ambient temperature is too low or too high in relation to the manufacturers' recommendations. Some crops are treated with growth-regulating chemicals prior to harvest to sustain postharvest shelf life, and these can enhance visual quality. The wholesale purchaser largely controls the use of growth regulators. Quality assurance schemes formulated by particular supermarkets frequently provide specifications that allow or forbid the use of such applications.

Integrated crop management (ICM)

The husbandry processes of soil cultivation, addition of fertilizers, transplanting, application of agrochemicals and irrigation are now regarded as a continuum. Collectively, they are aimed at achieving the highest possible quality in the product compatible with minimum use of resources, and are summarized as 'integrated crop management (ICM) (this is referred to in several previous chapters, notably Chapters 6 and 7). The formulation of standardized husbandry protocols well in advance of crop production and their review in the light of performance in preceding seasons has become known as 'smart scheduling'. This approach to crop husbandry has reached a high level of sophistication with all vegetable crops and is now often regulated by the wholesale purchasing system through the supermarkets, which collectively have established their own quality control systems as part of the purchasing contract. The UK Assured Produce scheme and more widely applied EUREPGAP scheme offer standardized protocols for crop production that are acceptable to the supermarket purchasers and should provide food safety and security for the consumer. These standards are becoming of widespread use throughout the European Union (EU). Suppliers in other countries will be required to apply them for fresh produce imported into the EU.

Postharvest treatment

The edible portions of the crop while attached to the growing plant derive their constituents of quality in relation to its rates of nutrient uptake, photosynthesis, respiration, transpiration and other metabolic processes. Once harvested, each portion becomes an independent entity where quality is controlled by its innate rates of respiration and transpiration. Excessive transpiration is the greatest source of postharvest damage to quality. All *Brassica* crops need to be maintained in a turgid state. Free surface water should be absent, however, since this provides conditions for the multiplication of bacteria and fungi that can rapidly cause degradation. At harvest, the water source of the crop product is severed. Respiration continues through the stomata and cut surfaces of detached organs.

The rate of water loss rises with increasing temperature and is exacerbated by decreasing relative humidity and atmospheric pressure. Loss rates are highest where there is a large area available for transpiration relative to unit weight, hence leafy crops lose water to the atmosphere more rapidly than densely packed produce such as cabbage heads, turnip or swede 'roots'. For most vegetables, a water loss of between 5 and 10% causes visual deterioration.

After harvest, changes to carbohydrate, organic acid and secondary metabolite content take place and these affect product quality. Probably the release of ethylene from tissues has the most critical effect on quality. Ethylene is involved in the acceleration of ripening and consequent senescence of tissues. For leafy or flower crops (cabbage, Chinese cabbage, cauliflower and broccoli), the presence of even small amounts of ethylene will accelerate senescence and increase deterioration. Respiration rates of vegetables vary widely and increase markedly in response to rising temperatures. High rates of respiration are characteristic of young and immature tissues, developing flowers, leaves and cauliflower heads, preventing their storage for more than a few days. The rates of heat loss in *Brassica* vegetables due to respiration are illustrated in Table 8.3.

Some crops such as cabbage heads may be stored for 6–8 months provided the temperature is reduced effectively. Use is made of controlled atmosphere storage, with increased concentrations of carbon dioxide and reduced oxygen levels minimizing postharvest deterioration. Uncontrolled accumulation of carbon dioxide concentrations and rising temperatures accelerate respiration in storage or transit, leading to lost quality. Detailed protocols for the storage of *Brassica* crops are given by Thompson (1998).

Green broccoli (calabrese) is an excellent example of a *Brassica* vegetable where the preservation of shelf life (i.e. high quality) is of paramount importance and most difficult to sustain. Fresh green broccoli is harvested when the flowering heads (spears or sprouts) are immature, and hence it is a highly perishable product, with a storage life of 2–3 days at 20°C and 3–4 weeks at 0°C (Makhlouf *et al.*, 1989). Hence, refrigeration is the primary means of slowing down the rate of senescence and maintaining product quality in broccoli. The major limitation in storage at ambient temperature is rapid yellowing of the flower buds due to chlorophyll breakdown, ethylene production and subsequent flower opening.

Table 8.3. Examples of heat loss caused by respiration in *Brassica* vegetables post-harvest.

	Heat loss (BTU/t/day)		
Cold type	0°C	10°C	20°C
Low rates			
Cabbage	n/a	4,000	9,500
Turnip	1,300	4,300	7,000
Moderate rates			
Brussels sprout	5,300	18,600	n/a
Cauliflower	n/a	7,400	17,600
High rates			
Broccoli (calabrese)	5,800	20,300	n/a
Watercress	5,800	20,300	n/a

n/a = not available.
BTU = British thermal unit.
After Ryall and Lipton (1972).

Modified atmosphere packaging (MAP) and controlled atmosphere storage will also extend the storage life of broccoli. Atmospheres of 6% carbon dioxide (CO_2) and 2% oxygen (O_2) delay yellowing, prolong chlorophyll retention, reduce microbial development and prevent the formation of odours.

At lower temperatures, loss of quality is due both to colour change from green to yellow and the onset of rotting induced by microorganisms. Practices that minimize the accumulation of ethylene in the storage environment or inhibit endogenous ethylene development can be effective in extending storage life and quality. Ethylene evolution is a key factor in the development of senescence in broccoli (Forney *et al.*, 2003).

Physiological changes preceding and accompanying visual deterioration of harvested broccoli floral tissue are discussed by Downs *et al.* (1997). There is a loss of sucrose in the head following harvesting; over the first 6 h, sucrose content declined by 50%. Treatment of florets with 6-benzylaminopurine (6-BA) delayed the associated large increases in the amides asparagine and glutamine that usually accompany sucrose loss in the first 48 h after harvesting. Also the normal decline in concentrations of amino acids and soluble proteins and increases in ammonia normally associated with postharvest deterioration were delayed by application of 6-BA. Breakdown of chlorophyll in harvested broccoli (*B. oleracea* L. var. *italica*) florets and the subsequent accumulation of degradation products are discussed by Yamauchi *et al.* (1997).

Floret maturity has the greatest effect on the rate of yellowing in broccoli, and Tian *et al.* (1995) showed that differing patterns of ethylene production, ethylene sensitivity, respiratory activity and chlorophyll loss occurred in florets of varying maturity. Floret maturity could be assessed by an examination of the developmental stage of the pollen. Dipping florets of cv. Shogun in 6-BA stimulated ethylene production, depressed respiration rate and delayed floret yellowing. The effect of this cytokinin analogue on broccoli was related to the concentration used and the postharvest age of the floret at the time of treatment.

Yellowing was further delayed where broccoli was treated at 20°C for 6 h with 1 μl/l of 1-methylcyclopropene (MCP), an ethylene inhibitor, at the same temperature; this doubled storage life where the atmosphere contained 0.1 μl/l ethylene. The use of MCP reduces the production of the volatiles dimethyl sulphide and trimethyl sulphide which are associated with accelerated senescence. Storage life was extended even further when the treatment temperature was retained at 20°C and the storage temperature reduced to 5°C. Treatment at a high temperature followed by storage at a lesser temperature provided the most satisfactory combination (Ku and Wills, 1999). Measurement of changes to chlorophyll fluorescence offers a technique for the assessment of quality in broccoli florets under commercial conditions (Toivonen and DeEll, 1998).

Green broccoli (calabrese) is a case where the product must be moved to the consumer as rapidly as possible. Head cabbage, on the other hand, offers opportunities for storage for periods of several months. Cabbage in store is affected by a range of physiological disorders and fungal pathogen problems, especially infection by *Botrytis cinerea* (grey mould). Prevention of grey mould was most effective at low temperatures as compared with controlled (modified) atmosphere conditions. Genetic resistance to grey mould would be a very valuable attribute, and it is suggested that this may be a character linked with resistance to white mould (*Sclerotinia sclerotiorum*).

Overwrapping cabbages in polyvinylchloride (PVC) and controlled atmosphere treatment reduced pepper spotting (internal black necrotic flecking) by 50% especially if the carbon dioxide concentration was raised to 10%. The combination of 3% oxygen and 5% carbon dioxide with PVC film delayed tissue yellowing compared with ambient air treatments. Controlled atmosphere treatments, however, induced the formation of off-odours and flavours when storage was extended beyond 74 days. Development of these contaminants was signalled by increased ethanol content in the stored cabbage heads. The most cost-effective storage conditions were 3% oxygen and 5% carbon dioxide; here storage life was limited by the onset of grey mould (*B. cinerea*)-induced rotting (Menniti *et al.*, 1997).

Packaging

Packaging is a standard requirement for *Brassica* crops in transit to the retail consumer. Effective packaging will enhance the attractiveness of the product and retain quality characters for longer periods. Defective packaging accelerates deterioration and destroys quality. Deterioration is especially rapid where packaging allows the accumulation of toxic compounds, such as ethylene, that contribute towards accelerated deterioration.

The formation of ethylene speeds up the processes of senescence and, in *Brassica* products such as cauliflower, broccoli (calabrese) and sprouting broccoli, leads to yellowing. Changing the concentrations of oxygen and carbon dioxide in small packs of *Brassica* vegetables retains their fresh quality and extends shelf life. This is an application of controlled atmosphere storage technology as used for bulk volumes of top fruit applied to small produce packs for the supermarket shelves. Lowering oxygen and increasing carbon dioxide concentrations around the produce blocks the synthesis of ethylene, thus slowing the ripening processes. The use of MAP started in the late 1940s when plastic film polymers became widely available for civilian use. It is defined as 'an alteration in the composition of the gases surrounding fresh produce when such commodities are sealed in plastic films'.

Plastic film technology and packaging techniques have advanced substantially in the intervening 50 years, and MAP is now a major marketing tool. Substantial problems remain, such as the development of off-flavours and fermentation in broccoli due to the innately high rates of respiration. Modified atmosphere packaging is now of great importance, however, for retaining freshness and quality. Few consumers realize that opening a pack of baby leaves releases a minute burst of carbon dioxide or nitrogen. The changed atmosphere is generated either by natural respiration or, more recently, by artificially adding nitrogen gas as the produce is sealed.

Reducing the oxygen content of atmospheres surrounding produce diminishes respiration and ethylene production and represses microbial and enzymic degradation. These processes start automatically with all fruit and vegetables from the instant they are harvested. Storage at around 5°C further reduces deterioration. The fresh turgid appearance of produce is helped by increasing the relative humidity in the pack and results from reducing the temperature. The danger is that excessive moisture from fruit and vegetables accumulates inside the pack, especially when temperatures fluctuate. The higher humidity inside the pack condenses into water droplets when the film surface is cooler than the air inside the pack. This can be overcome by using anti-fogging chemicals. A careful balance is required since vegetables such as broccoli (calabrese), cauliflower and Chinese cabbage need a moist atmosphere in order to avoid desiccation.

A key factor in MAP is finding films with a permeability that prevents an accumulation of damaging concentrations of either water vapour or carbon dioxide, or excessive depletion of oxygen. As an approximation, films are 4–6 times more permeable to carbon dioxide than they are to oxygen. Transmission of water vapour is made independently of gas diffusion by using multilayered plastics. One of the favoured newer products is 'linear low density polyethylene'. This has a better structure compared with simpler films, making it tougher and more suited to heat sealing with greater impact resistance and tear strength. Levels of carbon dioxide or oxygen take time to adjust inside the pack. Active adjustment by gas flushing speeds this up and uses mixtures of carbon dioxide, oxygen and nitrogen to purge the packages.

Alternatively, the package may be evacuated and refilled naturally by respiratory gases coming from the produce. Some plastic films are perforated with 'microperforations' that allow a greater degree of control of water vapour content but have lesser effect on gases. Plastic film manufacturers are now marketing products that are claimed to extend the life expectancy of fruit and vegetables. Some films have permeable windows made of silicon or polyvinyl polymers that control the diffusion process more efficiently. Early experiments in the USA with winter green cabbage showed that carbon dioxide and oxygen concentrations were lower than expected where plastic films incorporating silicon windows were used. The cabbage heads showed better colour retention, fresher appearance, firmer texture and smaller losses in weight.

The use of microperforated plastics and an ethylene retardant, such as MCP, reduced senescence-related yellowing in broccoli. A new generation of packaging materials uses environmentally responsive high permeability films equipped with molecular temperature switches. These could help reduce the fermentation risks in MAP packs. This approach can be extended further with packaging that has a built-in ethanol sensor that identifies the early stages of deterioration.

Perforation-mediated MAP relies on the use of perforations (tubes) of different sizes to control a package's gas content for both whole and freshly cut fruit and vegetables. Designing efficient packages demands knowledge of gas exchange rates through the perforations. In addition, circulating cooled air round the packages helps control the rate of gas exchange. A research aim for plastic manufacturers is to develop polymeric films with a selective barrier capable of matching the respiration of the produce; alternatives for such tasks are polypropylene, polyvinyl chloride and polyethylene films. For broccoli, packing in polypropylene film increased storage life. Broccoli stored in polyvinyl chloride deteriorated faster than when it was packaged in the other two materials. Broccoli has a high respiration rate, is very sensitive to ethylene and loses water rapidly, and consequently, is a difficult product to package. Using polypropylene extended the shelf life to 3–4 weeks in air at 0°C compared with only 2–3 days at 20°C. The standard storage recommendations for broccoli are 0°C with a relative humidity of 95%, 1–2% carbon dioxide and 5–10% oxygen.

New *Brassica* vegetable products are emerging suited to MAP packages, for example shredded Galega kale, which is very popular in Portugal where it is used as a companion to their traditional delicacy of salted, dried fish. Studies demonstrated that MAP with an atmosphere of 1–2% (v/v) oxygen and 15–20% (v/v) carbon dioxide will extend the shelf life of shredded Galega kale at 20°C to 4 days as compared with 2 days with ambient air storage.

PHYSIOLOGICAL DISORDERS

Storage losses of *Brassica* vegetables such as winter white cabbage and Chinese cabbage are regularly reported at 10% and frequently exceed this figure. The *Brassica* crops suffer from a range of physiological disorders that are major causes of these losses after harvesting. Physiological disorders have been attributed to a range of nutrient deficiencies interacting with environmental or climatic conditions and, more recently, the effects of virus pathogens. It appears that under specific conditions of nutrient stress, the viruses enhance the development of tissue chlorosis and necrosis.

Many of these conditions have been associated with a limitation of the availability of calcium within the tissues of rapidly growing organs that form the components of yield in brassicas. Calcium is an immobile element recognized as mainly being translocated from the roots through the xylem by mass flow. Mass flow results from transpiration, root pressure and diurnal change in water stress. There is some evidence that calcium ions in the vessels do not move primarily by mass flow but by exchange reactions along negatively charged sites on the walls of the xylem vessels. Once calcium has reached its destination in the plant, however, there is little or no subsequent redistribution. Most of the water flux is channelled to leaves exposed to the sun by transpiration that contributes to the overall cooling of the plant.

Developing tissues fed by the phloem are often disadvantaged by this competition from transpiring tissues. Competition between sinks such as buds and developing leaves and fruits is high when calcium in the xylem is low and transpiration is high. It is at this point that broccoli florets, cabbage heads or cauliflower curds are very susceptible to calcium deficiencies. Furthermore, the availability of excessive potassium and magnesium is likely to aggravate calcium deficiencies within the plant. Periods of excessive or fluctuating high and low temperatures also appear to contribute to the appearance of physiological disorders.

Tipburn

Internal tipburn is one of the most common physiological disorders affecting a wide range of vegetables and fruit. A necrotic breakdown of the marginal tissues of leaves is seen in cabbage heads; both Chinese and European cabbage are afflicted. Tipburn is usually attributed to localized calcium deficiency and related in incidence to genotype and prevailing weather conditions, and the availability of nitrogen fertilizer. Large applications of easily available nitrogen at transplanting increase the shoot to root ratio, unbalancing the plant metabolically and leading to physiological disorders such as tipburn. Magnusson (2002) suggested that rapidly increasing growth and high total nitrogen and nitrate concentrations at harvest increased the incidence of tipburn. He made comparisons of growth of Chinese cabbage with 'green mulch' grass leys used as intercrops (Chapter 6) and in combination with mineral fertilizers in which intercropping decreased the prevalence of tipburn.

An extensive review of tipburn in *Brassica* was published by Everaarts (2001), who subscribes to the view that a number of interacting factors lead to the expression of tipburn symptoms. This physiological problem is increasing with all forms of European and Oriental *Brassica*; possibly this is analogous to the upsurge in pathogenic and pest problems and reflects the increasingly concentrated genetic base from which commercial *Brassica* cultivars are drawn. Normally, there are no external symptoms, while internally they vary between genotypes and between storage and fresh market types.

In the fresh market types, the symptoms are desiccated papery thin leaf margins extending in zones from several millimetres to eventually covering the entire leaf. In genotypes intended for storage, the symptoms are dry, papery, thin, dark brown circular to oval spots with deep brown to black margins of several millimetres to several centimetres in size.

Localized calcium deficiency in the leaves is found in rapidly growing tissues with low transpiration rates and is exacerbated by constant high relative humidity, which stimulates calcium-related disorders. Since calcium is transported mainly in the xylem, the amount reaching the growing and meristematic tissues is closely related to the rate of transpiration; where this is low, then calcium deficiencies will develop.

Calcium accumulates in the outer leaves during the day by mass flow in the transpiration stream, and in heads at night when growth takes place and root pressure forces water and calcium into the head. Some genotypes exhibit levels of resistance or tolerance to tipburn. Since calcium is only taken up by the very young unsuberized roots, the number of young roots and the position in the soil of root apices relate to calcium uptake; consequently, root architecture is an important factor in the expression of this disorder. The large vigorous and deeply rooting genotypes are less likely to be prone to tipburn. Cultivars producing high yields and having rapid growth are prone to tipburn, but it is discouraged by the use of wider planting distances. Tipburn is most frequent where plants grow rapidly but fail to develop a sufficient root system, and in consequence there is a high leaf to root ratio.

The rate of biomass production in Brussels sprout is known to be proportional to the intercepted radiation. High radiation rates will increase growth rate and, thus, the amount of tipburn in this crop. The incidence of tipburn is associated with the use of high levels of nitrogen fertilizer. This is due to accelerated growth resulting from the availability of nitrogen, not the effect of the element *per se*. Ammonium as a nitrogen source reduces the uptake of calcium due to competition between the two cations. Husbandry factors such as planting date also affect the incidence of tipburn. Cabbages intended for long-term storage are normally planted between the end of April and mid-May, with head formation starting 60–75 days later. Early planting is associated with increased tipburn resulting from high growth rates during periods of limited duration of darkness. Delaying harvests increases the likelihood of tipburn developing; probable interacting factors are the age of tissue, and storage which increases tipburn with cool (low temperature) storage.

White cabbage roots can penetrate to 100 cm or more and the root shape is obconical reaching 150 cm deep, but the greatest intensity of rooting is in the top 20 cm of soil. Such size and vigour tend to minimize the incidence of tipburn. Soil characteristics can affect the incidence of tipburn, thus for example Dutch growers stop liming when the soil calcium content exceeds 2%. Normally they use well-cultivated and drained, moisture-retaining fertile soils for cabbage.

Soil where waterlogging is present leads to dysfunctional root growth, resulting in anaerobiosis. Even when anaerobiosis only lasts for short periods, it is sufficient to cause root death and, in consequence, reduces calcium uptake and initiates subsequent tipburn. Calcium applications made directly to the crop, however, are unlikely to be of benefit except possibly sprays of calcium chloride or nitrate. Tipburn is associated with inadequate calcium uptake by young, rapidly growing leaves. Several soil chemical and environmental factors that increase plant growth and decrease calcium mobility and transpiration have been implicated.

Soil cations (Ca^{2+}, K^{+}, Mg^{2+} and NH_4^{+}) play critical roles in the development and prevention of physiological disorders usually involving competition for plant uptake resulting in an excess or deficiency of a particular element or elements in the tissues (Cubeta *et al.*, 2000). Rapidity of growth in the presence of excessive nitrogen and minimal calcium appears to be the dominating factor in the development of this syndrome and affects all *Brassica* crops that form heads or inflorescences. The association with calcium deficiency is shown in Table 8.4

The association between tipburn and root size was demonstrated by Johnson (1991). Root systems of a tipburn-susceptible cultivar were smaller than those of a tipburn-resistant one, hence the plants were more susceptible to moisture stress, but *per se* this does not necessarily correlate with the development of tipburn. The calcium efficiency ratio (CaER = mg of dry matter produced per mg of calcium in tissue) in young leaves was greater, however, for tipburn-tolerant cultivars than for susceptible ones, and provides a valuable gauge to the likelihood of symptom expression (Table 8.5).

Internal browning

Internal browning of Brussels sprout buds was seen mainly as a problem for quick-freeze crops where even a small percentage of blemished buds would cause rejection since affected buds could not be identified and extracted from

Table 8.4. Nutrient composition of field-grown cauliflower leaves showing normal and tipburn growth.

	Nutrient composition (percent dry weight)		
Nutrient	Normal	Tipburn	Probability based on *t*-test
Nitrogen	4.70	4.80 (102%)	0.40
Phosphorus	0.64	0.52 (81%)	0.50
Potassium	1.31	1.58 (121%)	0.20
Calcium	0.50	0.18 (36%)	0.05
Magnesium	0.33	0.26 (79%)	0.30

From Maynard *et al.* (1981).

Table 8.5. Calcium efficiency ratio for collards (*Brassica oleracea* var. *acephala*) grown in controlled conditions.

		Calcium efficiency ratio[1]		
		Leaves[2]		
Cultival	Tipburn reaction	Young	Old	Total plant
Vates	Suceptible	107b	44a	69a
Blue Max	Tolerant	206a	43a	64a
Heavi Crop	Tolerant	183a	42a	66a

[1] Mean separation among cultivars within a calcium level by least significant difference at $P = 0.05$.
[2] Young leaves (blade and midrib) = terminal five leaves that were 2–6 cm long; old leaves (blade and midrib) = all leaves older than the fifth leaf from the terminal.
After Johnson (1991).

the processing line. Consequently, they would end up in the packs sold by supermarkets and found by the consumer. The latter would react badly to having purchased damaged goods to the disadvantage of the supermarket, processor and, ultimately, the grower.

Factors that have been implicated in the development of internal browning include: bud size (larger sprouts more prone to the syndrome); the density of leaf packing in the bud (the greater the density then the greater the likelihood of internal browning); seasonality (the incidence is most common in early maturing rapid, growing, and mid-season types (October to December)); and genotype (some cultivars are more susceptible than others). Symptoms are characterized by the death and subsequent brown discoloration of leaf tissue in the apical third of the sprout bud; symptoms are not seen frequently on either the older, outer or the youngest internal leaves. In severe cases, the browning can spread downwards along the petiole to the base of the sprout bud. It has been suggested that water condenses within the sprout bud, which may then restrict calcium transport and lead to marginal leaf necrosis in the bud.

Brown bead

Brown bead was first seen in broccoli (*B. oleracea* var. *italica*) in California in the 1970s as a physiological disorder and caused the abandonment of substantial areas of crop because of lost quality. This disorder was also found in Canada (Jenni *et al.*, 2001). The evidence used to explain its incidence is confused and inconclusive. There is no correlation with nitrogen fertilization, but low calcium content associated with rapid growth has been related to the incidence of brown bead; there is also some association with low potassium levels. In addition, there is tentative evidence for an association with excessive temperatures (22–38°C) in the 5 days prior to maturity. Some evidence also exists suggesting a relationship with elevated ethylene levels.

Experimental studies showed that less brown bead was expressed in the fastest growing crops but, where there is nutritional imbalance especially between calcium, potassium and magnesium, brown bead developed particularly when calcium was deficient. A likely additional contributory factor is high air temperature, whereas regular even water supplies to the root diminished brown bead. Adequate nitrogen supplies and moderate availability of potassium and magnesium help to prevent this syndrome. Whenever there is a restriction in water supply, calcium or nitrogen shortage associated with high temperatures, the condition is prone to occur. Applications of calcium chloride decreased brown bead.

Pepper spotting

Pepper spotting develops inside cabbage heads, especially the Dutch or white cabbage that are stored for several months (Cox, 1977). Pepper-spot is seen as clusters of small black spots <1 mm diameter. The causes of pepper spotting (synonymous with pepper-spot, grey speck and black speck) are unknown. The use of controlled atmosphere storage (2.5–3.0% oxygen and 5.0–6.0% carbon dioxide) will extend cabbage storage life for 5–6 months at 0°C, delay yellowing and maintain good quality characteristics. Differences in the expression of pepper spotting by several genotypes were reported by Shipway (1978).

Black speck

This syndrome is characterized by small sharply sunken brown or black specks and is similar to pepper-spot or grey speck disorder (Loughton and Riekels, 1988). Reduced oxygen content in controlled atmospheres will control the disorder, e.g. 2–2.5% oxygen and a temperature of 0°C (Geeson and Browne, 1980). It is a non-parasitic disorder of cabbage, most frequently seen about 1 week after cold storage starts. In many cases, black speck is found on the outer leaves and is accompanied by fungal and bacterial infections of leaves 3–10 towards the centre of the head necessitating substantial post-storage trimming to provide a marketable head. On occasion, black speck may extend to the core, making the entire head unmarketable. Black speck in stored broccoli is well recognized. Broccoli can be stored for 3 weeks at 0°C and 90–97% relative humidity. On this crop, black speck is characterized by small sunken black spots on the inflorescence stalk, which can coalesce into lesions 0.5–4 mm diameter. It is encouraged in broccoli by the use of high nitrogen fertilizers, rapid vigorous growth, inadequate mineral uptake and sometimes related to hollow stem or internal browning, and is also found in cauliflower which indicates boron deficiency and reduced availability of potassium. Black speck is related to genotype, soil type and the use of some postharvest fungicidal dips.

Broccoli scaring

Broccoli scaring is a relatively rare disorder developing where the leaves are removed or the stem is cut, and the exposed tissue is initially whitish but in store turns grey or blackish. Sodium hypochlorite dips up to 300 p.p.m. in concentration have reduced black speck and also scarring discoloration in both ambient and controlled atmosphere stores.

Cigar burn

Cigar burn is one of the major internal disorders of cabbage (*B. oleracea* var. *capitata*). It is seen as sunken necrotic spots 5–10 mm diameter on internal tissues. Walsh *et al.* (2004) added evidence that virus pathogens may also be implicated in these syndromes (Table 8.6). They proposed that cigar burn is caused by turnip mosiac virus (TuMV) on cvs Polinius and Impala. The condition reached its most severe level at 4 months in store and did not progress after that date. The presence of cauliflower mosaic virus (CaMV), while not directly causing cigar burn, enhanced the symptoms resulting from TuMV infection.

Cauliflower mosaic virus itself can cause raised pale lesions 3–8 mm diameter. Storage conditions may influence the progression of these physiological disorders. High definition control of temperature and atmospheric composition in a sealed store repressed the development of losses due to cigar burn and also rotting caused by microbial pathogens. In the field, these viruses reduced yield by between 16 and 76%.

The use of an enzyme-linked immunoassay (ELISA) test for the presence of TuMV in cabbage heads at harvest permits an estimate to be made of the risk of cigar burn developing. As a result, those heads at most risk may be marketed more quickly in advance of symptom expression (Walsh and Hunter, 2004).

Beet western yellows virus (BWYV) can be associated with increased severity of tipburn, but this is heavily affected by genotype, for example, cv. Impala appears to be very susceptible. All these viruses have aphid vectors and BWYV is likely to originate from adjacent oilseed rape crops. Walsh and Hunter (2004) suggest that BWYV spread by the peach-potato aphid (*Myzus persicae*) encourages the tipburn syndrome. Cigar burn is encouraged by TuMV and

CaMV spread by the aphids *M. persicae* and cabbage aphid, *Brevicoryne brassicae*, respectively. Pepper-spot has not as yet been associated with a virus vector, but is linked by these authors with the use of excessive nitrogen fertilizer. They also identify veinal streak, which is seen as large black lesions coalesced along the midrib and only visible when the head is split open. No association with virus vectors is proposed for this disorder. As control measures, they advocate cautious site selection that avoids proximity to oilseed rape crops.

It is considered that waterlogged fields exacerbate the problems of calcium uptake, and cultivars vary in susceptibility to the condition. In all areas where physiological problems are likely to occur, it is important to monitor crops and avoid using those showing symptoms of virus infection in the field for storage. The use of ELISA testing prior to storage can aid the identification of the presence of viruses, and the invasion of crops by aphids should, where feasible, be prevented (Walsh *et al.*, 2004).

Riceyness

Cauliflower curds suffer from a range of defects from discoloration and yellowing to riceyness and overmaturity where the florets begin to turn green. Broccoli of both the calabrese and sprouting (white or purple) types can easily become overmature and the florets start expanding and opening. This may be accelerated during display at the point of sale where the produce is held beyond its sell-by date.

Cold and heat damage

Brassica vegetables are evolutionarily suited to cool to warm temperate conditions. Some of the head cabbage, Brussels sprout and kale types can withstand low and freezing temperatures. Opportunities to improve frost tolerance in cauliflower have not been exploited sufficiently despite some strains being regularly used over winter (Deane *et al.*, 1996). Breeders have

Table 8.6. Viruses associated with physiological disorders in brassicas.

Disorder	Cause	Aphid transmission	Storage
Tipburn	BWYV	Slow	Continues to develop in store
Cigar burn	TuMV	Rapid	Does not get worse after 4 months in store
Cigar burn (severe)	TuMV + CaMV	Rapid	Does not get worse after 4 months in store

BWYV = beet western yellows virus; TuMV = turnip mosaic virus; CaMV = cauliflower mosaic virus. From Walsh *et al.* (2004).

developed forms of Chinese cabbage capable of withstanding low temperatures such that in countries such as Japan and Korea fresh *Brassica* vegetables are now available all year-round. Cold tolerance has been utilized from Japanese radish (Ogura, 1968; Heath *et al.*, 1994). There is a correlation between high dry matter content and cold tolerance. Kale has a dry matter content of >18% and is the most cold-tolerant *B. oleracea*. Savoy cabbage has dry matter of around 12%, while, in comparison, that for summer cabbage is about 6–7%.

Dry matter content is a useful indicator when selecting for frost tolerance. Winter hardy cabbage will survive –15 to –20°C. Crosses made between kale, broccoli and cabbage suggest there are two epistatic genes controlling tolerance to frost.

Heat tolerance in brassicas is a character of major value as production of these crops expands, especially for cauliflower and Chinese cabbage. Cauliflower curds tend to be initiated in response to temperature, but this varies according to their maturity type. In India, cauliflower has become a popular vegetable and moderate temperature tolerance is known. For example, Nowbuth and Pearson (1998) reported the development of lines that are capable of being grown in warmer conditions of Mauritius. While Chinese-cabbage breeders have long sought to develop high temperature-tolerant types for use throughout Asia. In broccoli, the critical time for heat response is the bud initiation at about 3–4 weeks prior to harvest. There are substantial differences in heat tolerance in broccoli. Some of the Japanese seed houses have marketed cultivars with extended heat tolerance for some time (Yang *et al.*, 1998).

PATHOGENS

Postharvest damage from diseases usually results from infections that happen either in the field or as a result of damage during harvesting and storage followed by the invasion of pathogenic microbes. Control of many storage rots was achieved with fungicidal dips applied postharvest, but this practice is becoming less acceptable since there is little opportunity with stored produce for the active ingredients to be dissipated by active metabolism. The ubiquitous grey mould fungus, *B. cinerea*, is probably responsible for the most widespread losses worldwide and can limit the storage period severely. This disease is characterized by causing a soft watery rot of cabbage head tissues with felt-like mats of spreading grey mycelium and spores. In mild infections, only the outer leaves are affected and these may be removed by trimming, but the pathogen soon colonizes cut stems and petioles, penetrating deeply into the head and resulting in total loss. Mechanical injury during harvesting has been associated with increased losses caused by this pathogen. The practice of 'mid-term trimming' during longer term storage has been associated with accelerated disease spread. Possibly this results from the wounding caused by the trimming or more probably as a consequence of the greater susceptibility of younger leaves to the pathogen compared with the older wrapper leaves. Freezing damage inflicted by frost in the field or poor control of the storage environment increases the risk of grey mould development. Good husbandry practice requires the early harvesting of crops destined for long-term storage before damage is inflicted by autumnal night frosts. The grey mould pathogen has weak powers of invasion requiring entry through wounds or as a secondary invader following damage caused by other organisms such as *Alternaria* spp. (dark leaf spot), *Mycosphaerella brassicicola* (ringspot) or TuMV. Susceptibility to grey mould disease varies with host genotype and previous husbandry practices.

There is some evidence that manipulation of fertilizer strategy (Chapter 5) can be used to minimize subsequent disease development. The excessive application of nitrogen (especially in ammonium forms) has been associated with enhanced disease development. Once infection foci are initiated, the rate of spread accelerates due to senescence induced by ethylene formation that may itself encourage symptoms.

Once in store, the major factors controlling disease development are temperature and relative humidity. Lowering the temperature below 4°C reduces disease severity, and in the USA environments of 0 to −1°C are routinely used to limit losses from grey mould. Caution is required, however, as reductions below −1°C may result in low temperature damage to the cabbage tissues. Results from studies of the effects of relative humidity are conflicting. Some reports suggest that maintaining the wrapper leaves in a turgid state diminished losses due to grey mould, whereas others have identified converse effects. It is likely that some dehydration of the wrapper leaves will slow the rate at which *B. cinerea* is able to colonize leaves and improve the longevity of produce in store. Hence losses are probably increased in ice-bank-cooled stores which maintain 98% relative humidity, whereas brine-cooled or direct expansion-cooled stores operating at 90–95% relative humidity are less conducive to fungal spoilage.

Manipulation of the storage atmosphere can be used to inhibit pathogen development. Reduced concentrations of oxygen (1.0%) at 0°C delayed yellowing in Chinese cabbage and decreased the incidence of grey mould-induced decay. This approach requires careful testing since higher levels of carbon dioxide (>6.0%) have been associated with the development of off-odours and off-flavours (Menniti *et al.*, 1997). Fungal inoculum will reside on the walls of stores, on containers and other surfaces ready to cause infection should conditions become favourable. It is essential to maintain a strict sanitary policy involving washing and cleaning stores and containers during the periods between crops.

Several species of *Alternaria* are responsible for serious losses of cabbage during storage and in the subsequent distribution and marketing chain; *A. brassicae*, *A. brassicicola* and *A. alternata* cause 'Alternaria spot' and 'dark leaf spot' symptoms. All three pathogens have been identified in northwestern Europe and North America, causing substantial crop losses. Symptoms are common to all these pathogens: invaded tissues are surrounded by chlorotic margins within which are discoloured areas that are dark brown or black in appearance with a dry or leathery texture which may show a superficial growth of mycelium and dark spore bodies (conidia). *Alternaria brassicicola* is characterized by causing large lesions bearing uniform, dark olive-black sporulation. In distinction, *A. alternata* produces mycelial growth that is characteristically dark grey to greyish-black and *A. brassicae* has brown to dark brown sporulation and distinct concentric zonation.

Infection initiated in the field continues to spread mainly on the outer wrapper leaves after harvest. Trimming can remove much infection, but areas of damage to leaves, petioles and stem butts offer portals for further invasion and the spread of existing infections. Although lesions caused by *Alternaria* spp. are typically limited in size (<5 cm diameter), damage may be greater than is initially apparent due to penetration below the wrapper leaves extending deeply into the head tissues. Damage may increase following secondary invasion by grey mould (*B. cinerea*) and soft-rotting bacteria.

Primary infection by *Alternaria* spp. can encourage secondary organisms by predisposing the tissues to infection through the release of ethylene. High relative humidity and temperatures above 5°C favour the growth of *Alternaria*. Disease risk increases as the length of the storage period increases, and *Alternaria* spp. are frequent damaging pathogens on cabbage stored beyond 6–9 months.

The field phase pathogen *M. brassicicola*, causing ringspot disease, incites dark grey to black lesions where, as with *Alternaria* spp., the tissues desiccate becoming dry, leathery and corky in texture. The randomly distributed lesions of *M. brassicicola* are best identified by the presence of small brown pycnidia that exude pale pink or whitish droplets containing pycnospores. Infection is initiated through bruised outer leaves or through cut stems and petioles. Lesions frequently extend deeply into the head tissues and may cause dark discoloration of the stem vascular tissues, but spread between heads is not frequent. In the field, ringspot disease is associated with cool moist growing periods hence, in store, high relative humidity and low temperatures (0–2°C) encourage infection.

Storage losses of cabbage in Scandinavia have been attributed to species of *Phytophthora* and, in one incident in Norway, the pathogen *P. porri* (the cause of white tip disease on leeks (*Allium porrum*) and salad onions (*Allium fistulosum*)) was isolated. Similar infections have now been identified in the UK and The Netherlands. This pathogen can cause substantial losses in store. Infection is characterized by dark brown or grey-brown lesions spreading upwards from the stem base. Diseased tissue remains firm but acquires a distinctly acidic, vinegary odour. No mycelium is evident on the outer surfaces of infected heads and hence the disease may be confused with bacterial rotting or frost injury. Small areas of white hyphal growth are found between the surfaces of heart leaves and within cavities in the stem medulla. It is possible that soil-borne oospores or chlamydospores of *P. porri* enter cuts made to cabbage stems during harvesting, especially in wet muddy conditions. Disease appears to be more frequent when harvesting takes place in very wet field conditions. Mid-term trimming in the store is associated with increased losses due to the spread of infection on knives. This pathogen can remain viable in soils for 3 years after *Allium* crops, consequently rotations should be chosen carefully to avoid their close cultivation with brassicas.

Several other fungi have been associated with losses to stored cabbage. *Fusarium avenaceum* is responsible for rapidly spreading, soft brown rots characterized by dense pink or whitish-pink woolly mycelial overgrowths. Infection by *S. sclerotiorum* (white rot) leads to watery soft rotting where the head collapses and the pathogen moves swiftly between heads, frequently affecting entire batches of cabbage. The pathogen is identified by black, sclerotial, resting bodies and dense cottony white mycelium found on diseased leaves and heads. Disease spread is retarded by storage close to 0°C.

Dark internal discoloration of cabbage and cauliflower heads can result from early infection at the propagation stage by downy mildew (*Peronospora parasitica*) or infection of growing crops. This pathogen becomes latent between seedling infection and harvest, only becoming obvious after cutting, illustrating the impact that early field-phase infection can have on subsequent marketable quality.

Black leg disease (*Leptosphaeria maculans* or *Phoma lingam*) invades cabbage heads causing black rotting of the stem and leaf bases following invasion in the field of root collars and stem bases. High temperatures allow the normally saprophytic fungus *Rhizopus stolonifer* (*R. nigricans*) to become parasitic and invade wounded tissues, leading to watery soft rotting.

The bacterium *Pseudomonas marginalis* is responsible for wet, slimy, soft rotting and the emission of foul, sour odours in cabbages held under refrigeration. At subzero temperatures, this organism spreads rapidly and is encouraged by poor ventilation of the store. Barn-stored crops are especially susceptible to this pathogen, other *Pseudomonas* spp. and *Erwinia carotovora*, particularly when crops have been harvested in a frosted condition.

Virus pathogens such as the aphid-transmitted TuMV are responsible for storage losses. The virus causes either large necrotic leaf spots (0.5–1 cm diameter) or small necrotic flecks. The common CaMV can similarly cause internal necrosis, particularly of stored white cabbage heads.

THE INTERACTION OF *BRASSICA* GENOTYPE, PATHOGENS AND QUALITY

Head cabbage

Dense-headed white or Dutch autumn-maturing cabbage (*B. oleracea* var. *capitata*) crops are stored for up to 10 months and ultimately sold all year-round on to the fresh market or for processing into coleslaw or as part of prepared salads. Such crops are grown extensively in north-west Europe (Germany, the UK and The Netherlands) and North America. Extended storage of cabbage is detrimental where it is going to be processed further, as in the production of coleslaw.

Increasingly, cabbage is used as a part of mixtures of vegetable products in what has become known in marketing jargon as the 'added-value chain' of 'minimally processed products'. Traditionally, crops such as cabbage have been held in low esteem because they demand considerable preparation for use in the home. This perception is changed by processing outside the home to a limited extent before sale to the retail customer. Naturally such processing increases the price charged to the consumer and is termed 'value-added'. The amount of processing is normally small ('minimal'), usually consisting of cutting into portions. Raw cabbage can be shredded and mixed with other vegetables such as diced carrot to provide a dry mixture that is stored in MAP for 10–14 days. Adding mayonnaise to the shredded mixture forms coleslaw. There are significant differences in the suitability of cultivars for this trade (Cliffe-Byrnes and Beirne, 2005).

Late-maturing, dense-headed, red leaf types (*B. oleracea* var. *capitata* f. *rubra*) may be similarly stored for subsequent processing, especially in The Netherlands. The stage when cabbages are suitable for cutting and storage is difficult to determine solely by visual estimation. Characteristics such as head firmness, density, size, days of growth from transplanting, estimates of heat units and summation of solar radiation from planting have all been tested as means of determining readiness for storage. The estimation of sucrose content has also been offered as a means of identifying maturity and has considerable attractions since this character is easily tested in the field with the type of portable, pocket refractometers used to test sugar content of maturing beet roots.

Maturation during the harvesting period is characterized by dry matter and sugar accumulation and increasing firmness, with the highest quality in white cabbage developing towards the end of the autumn harvesting period (Suojala, 2003) (see Fig. 8.1).

Quality during storage declines gradually as a result of disease incidence, dehydration and metabolic changes. Firmness diminishes during storage as a result of dehydration and dry matter losses caused by continued transpiration and respiration. As a result, juiciness and crispness are lost. Other sensory indicators of quality, however, such as chewiness, sweetness, lack of bitterness, intensity of flavour and lack of off-flavours seem to be retained. During storage, there appears to be a shift of energy reserves from the head leaves to the core. Sugars in particular are concentrated in the core towards the end of storage. This may be a physiological preparation for regrowth, demonstrating that the harvested cabbage head retains a seasonal rhythm despite being severed from the roots and placed in cool temperatures. These

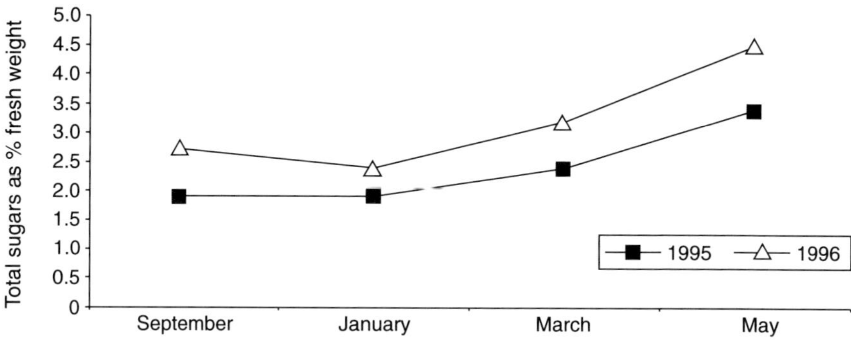

Fig. 8.1. White cabbage core sugar content.

processes will be under genetic control, and unravelling them (after Suojala, 2003) enables plant breeders to improve quality and longevity.

Storage technology has developed from traditional clamp methods as used for root vegetables. Use is made of barns that are equipped with forced night air ventilation for cooling and dispersal of field heat. This permits preservation for 3–4 months until late February or early March. More sophisticated methods are required for longer term storage involving purpose-built refrigerated stores in which cabbages are packed into palletized containers.

White cabbage has been stored for extended periods under refrigeration in nitrogen atmospheres modified with 2–6% carbon dioxide and 1–5% oxygen. These result in less storage and trimming losses, longer retention of fresh colour, flavour and texture, and lower pathogen-induced spoilage. There may be indirect benefits from modified atmosphere storage since reduced time is required in preparing the product for processing and marketing, and the storage period can be extended to 10 months.

Cauliflower

Cauliflower is normally only stored for short periods to fill gaps in the supply chain caused by periods of hot dry weather. There also appears to be interest in storing cauliflower florets for longer periods following blanching and dehydration for subsequent use in catering packs (Kadam *et al.*, 2005). Under optimal conditions (0°C and 95% relative humidity), cauliflower may be stored for up to 6 weeks. In practice, storage beyond 2–3 weeks is inadvisable. Cauliflower curd is very susceptible to damage during harvesting and even cryptic injuries will enlarge during storage into blemishes and discoloration associated with fungal and bacterial invasions. Spoilage can develop very quickly, leading to downgrading or outright rejection of the curds. Even during the normal time periods of distribution, curd discoloration develops and this is exacerbated where the heads are overwrapped with polyethylene film. *Alternaria* spp. cause leaf spots prior to harvest; these affect the wrapper leaves and can cause downgrading during cutting and grading.

Infection of the white curds is a source of much more serious wastage during storage and marketing. Small lesions (<5 mm) will spread and coalesce within 7 days of cutting, favoured by warm, moist conditions. Rapid cooling to remove field heat and cool chain marketing at 4°C are recommended to inhibit *Alternaria*-induced brown rotting of the curds. Since both *A. brassicae* and *A. brassicicola* are seed-borne pathogens also capable of survival on soil debris, there are several avenues by which cauliflower crops may become infected. Transfer to the developing curds leads to postharvest sporulation and spoilage during storage, distribution and marketing.

Infection of cauliflower heads by downy mildew (*P. parasitica*) causes pale greyish to brown discoloration on the curd surface. Inside the head, grey or black spotting and streaking extend through the bract tissues. Apparently healthy curds stored at 20°C and 70% relative humidity are rapidly spoiled by downy mildew sporulation. Storage at 4°C retards the expression of symptoms. Secondary invasion by soft rotting bacteria increases the rate of disintegration and deterioration of cauliflower heads. As with cabbage, ringspot (*M. brassicicola*), grey mould (*B. cinereae*) and in some instances sooty moulds (*Cladosporium* spp.) have been associated with postharvest spoilage of cauliflower.

Several *Pseudomonas* spp. and *E. carotovora* cause bacterial soft rots in cauliflower postharvest. Mechanical damage or bruising during harvesting and transit provide entry for soft rotting bacteria that may be spread between infected and healthy heads by the knives used for harvesting. Bacteria may also enter following primary infection by fungi such as *Alternaria*. Free water on cauliflower curds increases the rapidity of bacterial rotting, hence harvesting under wet conditions and overwrapping heads increases the likelihood of spoilage. Temperatures in excess of 10°C also encourage the rotting of cauliflower curds.

Brussels sprouts

Once sprout buds have been harvested, they are either sent immediately to processing factories or are packed into nets for the fresh market. The latter may be stored for a few days, but in ambient conditions the sprout butts rapidly become yellow or brown, reducing their acceptability to the consumer. Yellowing results from tissue senescence, and this is accelerated by ethylene released from the crop or from microbes colonizing the sprouts. Storage for longer periods is achieved at 1–2°C and 95% relative humidity; sprouts held in ice-bank coolers retain marketable quality for several weeks. A similar range of fungal and bacterial pathogens to those that reduce the postharvest quality of cabbage and cauliflower also spoil Brussels sprout. In particular, the normally saprophytic fungus *R. stolonifera* (*R. nigricans*) colonizes sprout buds, especially in warm moist conditions.

Green broccoli (calabrese) and sprouting broccoli

Strict quality criteria control the marketing of these *Brassica* vegetables. Factors such as bead size and uniformity, head shape, tightness and its evenness of green coloration significantly affect acceptability for sale. Physiological senescence and yellowing are frequently followed by secondary fungal and bacterial soft rotting, which is promoted by high relative humidity (>95%). Pathogens that damage broccoli heads include those affecting other brassicas. Each of these microbes is encouraged by moist, warm (>10°C) conditions.

As with cauliflower, modified and controlled atmosphere storage of these crops aids the retention of green coloration and delays senescence and the extension of flower buds. There may be changes, however, to texture and the development of off-odours. Reduced oxygen content (1–6%) failed to diminish losses caused by bacteria and, while some fungi (*Alternaria* spp. and *Cladosporium* spp.) were less apparent, the saprophytic *Mucor* spp. were encouraged. The physical condition of broccoli is also of great importance. The appearance of succulence is paramount. Coarse stems with shortened internodes resulting in increased numbers of leaf scars or their thickening will substantially reduce cash value (Sterrett *et al.*, 2004). In California, USA, broccoli is harvested at an immature stage that extends its shelf life. An export market has developed whereby such spears are shipped rapidly to Japan where they achieve a significant price that more than compensates for the sacrifice in yield and transport costs. Domestic American consumption of broccoli increased by 177% over the period 1986–2003 (http://www.ers.usda.gov/publications/vgs/tables/precap.pdf), and similar increased demand has been seen elsewhere. Consumption looks set for further increases as a consequence of the association of broccoli with reductions in the risks of contracting cancer and coronary diseases. Autumn production in Virginia USA has been encouraged, for example, by the development of heat-tolerant cultivars (O'Dell *et al.*, 1993) that help to satisfy increasing market demand.

Chinese cabbage

The storage life of Chinese cabbage is relatively short, extending to only a few weeks at 0–1°C; yellowing caused by tissue senescence is a major source of spoilage even under these conditions. Leaf spotting (*Alternaria* spp.) will develop in stores at 0–1°C and 95–97% relative humidity, and may be accelerated at higher temperatures. Soft rotting (*E. carotovora*) is a major source of postharvest damage during the transit and storage, and blackening of the leaf veins is caused by *Xanthomonas campestris*.

STORAGE ENVIRONMENT

Most research shows that the optimum controlled or modified atmospheres for brassicas are achieved by substantially increasing carbon dioxide concentration and decreasing oxygen. Detailed specifications depend on the field conditions under which the crop is grown, length of storage required and storage facilities to be used. Broadly, reduced temperatures combined with carbon dioxide between 2.5 and 7% and oxygen of 2.5–6% atmospheres are required. With white cabbage, Geeson (1983) found that at temperatures of 0–1°C for 39 weeks, 92% heads were marketable after trimming when held in an atmosphere of 5% CO_2 and 3% O_2.

By contrast, only 70% of those held in ambient air were marketable. Cabbage and other brassicas (such as Brussels sprouts, broccoli and Chinese cabbage) held in controlled atmospheres retain green coloration, fresh appearance and texture far longer than those held at ambient conditions.

References

Abercrombie, J.M., Farnham, M.W. and Rushing, J.W. (2005) Genetic combining ability of glucoraphanin level and other horticultural traits of broccoli. *Euphytica* 143, 145–151.

A'Brook, J. (1964) The effect of planting date and spacing on the incidence of groundnut rosette disease and of the vector *Aphis craceivora* Koch. at Mokwa northern Nigeria. *Annals of Applied Biology* 54, 199–208.

A'Brook, J. (1968) The effect of plant spacing on the numbers of aphids trapped over the groundnut crop. *Annals of Applied Biology* 61, 289–294.

Abuzeid, A.E. and Wilcocksen, S.J. (1989) Effects of sowing date, plant density and year on growth and yield of Brussels sprouts (*Brassica oleracea* var. *bullata* subvar. *gemmifera*). *Journal of Agricultural Science* 112, 359–375.

Adams, D. (2000) News Feature – Now the hard ones. *Nature* 408, 792–793.

Ahokas, A. (2004) On the evolution, spread and names of rutabaga. *Interdisciplinary Biology, Agriculture, Linguistics and Antiquities* 1, 1–32.

Ahuja, D.K., Badwal, S.S. and Labana, K.S. (1981) Qualitative and quantitiative changes in the seed and oil content of *Brassica juncea* mutants at different times of harvesting. *Quality of Plant Foods and Human Nutrition* 31, 61–66.

Alborn, H., Karlsson, H., Lundgren, L., Ruuth, P. and Stenhagen, G. (1985) Resistance in crop species of the genus *Brassica* to oviposition by the turnip root fly, *Hylemya floralis*. *Oikos* 44, 61–69.

Al-Khatib, K. and Libbey, C. (1992) Weed control in vegetable seed crops. *Proceedings of the Washington Horticultural Association* 83, 30–35.

Al-Khatib, K., Libbey, C. and Kadir, S. (1995) Broadleaf weed control and cabbage seed yield following herbicide application. *HortScience* 30, 1211–1214.

Altieri, M.A., Wilson, R.C. and Schmidt, L.L. (1985) The effects of living mulches and weed cover on the dynamics of foliage- and soil-arthropod communities in three cropping systems. *Crop Protection* 4, 201–213.

Andow, D.A., Nicholson, A.G., Wien, H.C. and Wilson, H.R. (1986) Insect populations on cabbage grown with living mulches. *Environmental Entomology* 15, 293–299.

Anon (1985) *Fertiliser Recommendations 1985–1986*. Reference Book No. 209, 15–17, Ministry of Agriculture, Fisheries and Food for England and Wales, London.

Anon (1997) *European Commission Regulation (EC) no 194/97*. Dated January 31,

1997. Published in the Official Journal of the European Community No. L 31/48–50.

Anon (2000) *Fertiliser Recommendations for Agricultural and Horticultural Crops.* Reference Book RB Number 209; Department for The Environment, Food and Rural Affairs (Defra) (previously, Ministry of Agriculture, Fisheries and Food), London.

Anon (2006) *The UK Pesticide Guide 2006*, Whitehead, R. (ed.). British Crop Protection Council, Berkshire and CAB International, Wallingford, UK.

Arabidopsis Genome Initiative (Multiple Authorship) (2000) Analysis of the genome sequence of the flowering plant, *Arabidopsis thaliana. Nature* 408, 796–815.

Astarini, I.A., Plummer, J.A., Lancaster, R.A. and Yan, G. (2004) Fingerprinting of cauliflower cultivars using RAPD markers. *Australian Journal of Agricultural Research* 55, 117–124.

Atherton, J.G., Hand, D.J. and Williams, C.A. (1987) Curd initiation in the cauliflower (*Brassica oleracea* var. *botrytis* L.). In: Atherton, J.G. (ed.) *Proceeding of Sutton Bonnington Easter School in Agricultural Science – Manipulation of Flowering.* Butterworth, London, pp. 133–145.

Babik, I., Rumpel, J. and Elkner, K. (1996) The influence of nitrogen fertilisation on yield, quality and senescence of Brussels sprouts. *Acta Horticulturae* 407, 353–359.

Bailey, L.H. (1922) The cultivated Brassicas. *Gentes Herbarum* 1, 53–108.

Bailey, L.H. (1930) The cultivated Brassicas, second paper. *Gentes Herbarum* 2, 211–267.

Bailey, L.H. (1940) Certain noteworthy Brassicas. *Gentes Herbarum* 4, 319–330.

Baker, H.G. (1965) Characteristics and modes of origin of weeds. In: Baker, H.G. and Stebbins, G.L. (eds) *The Genetics of Colonising Species – First International Union of Biological Sciences Symposium on General Biology.* Academic Press, New York, pp. 147–169.

Banga, O. (1976) Radish. *Raphanus sativus* (Cruciferae). In: Simmonds, N.W. (ed.) *Evolution of Crop Plants.* Longman, London, pp. 60–62.

Bannerot, H., Boulilard, L. and Chupeau, Y. (1974) Cytoplasmic male sterility transfer from *Raphanus* to *Brassica. Cruciferae Newsletter* 1, 52–54.

Barker, G.M. (ed.) (2002) *Molluscs as Crop Pests.* CAB International, Wallingford, UK.

Barnes, J.P. and Putnam, A.R. (1983) Rye residues contribute weed suppression in no-tillage cropping systems. *Journal of Chemical Ecology* 9, 1045–1057.

Baron, J.J., Holm, R.E. and Frank, J.R. (2002) The role of the IR-4 project in the registration of plant growth regulators in horticultural crops. *HorTechnology* 12, 59–63.

Baron, J., Kunkel, D. and Holm, R. (2004) The role of the IR-4 project in pest management for fruit, vegetable and other speciality crops in the USA. In: *Advances in applied biology: providing new opportunities for consumers and producers in the 21st century.* Association of Applied Biologists meeting December 15–17, 2004. Abstracts obtainable from Association of Applied Biologists, Warwick-Horticulture Research International, Wellesbourne, Warwickshire CV35 9EF, UK (unpaginated).

Baskin, C.C. and Baskin, J.M. (1998) *Seeds – Ecology, Biogeography and Evolution of Dormancy and Germination.* Academic Press, London.

Bello, D., Monaco, C.I. and Simon, M.R. (2002) Biological control of seedling blight of wheat caused by *Fusarium graminearum* with beneficial rhizosphere microorganisms. *World Journal of Microbiology and Biotechnology* 18, 627–636.

Benrey, B. and Denno, R.F. (1997) The slow-growth–high-mortality hypothesis: a test using the cabbage butterfly. *Ecology* 78, 987–999.

Bettey, M. and Finch-Savage, W.E. (1996) Respiratory enzyme activities during germination in *Brassica* seed lots of differing vigour. *Seed Science Research* 6, 165–173.

Bettey, M. and Finch-Savage, W.E. (1998) Stress protein content of mature *Brassica* seeds and their germination performance. *Seed Science Research* 8, 347–355.

Bewley, J.D. and Black, M. (1994) *Seeds, Physiology of Development and Germination.* Plenum Press, New York.

Bhattacharya, R.C., Maheswari, M., Dineshkumar, V., Kirti, P.B., Bhat, S.R. and Chopra, V.L. (2004) Transformation of *Brassica oleracea* var. *capitata* with bacterial *betA* gene enhances tolerance to salt stress. *Scientia Horticulturae* 100, 215–227.

Bhowmik, P.C. and McGlew, E.N. (1986) Effects of oxyfluorfen as a pretransplant treatment on weed control and cabbage yield. *Journal of the American Society for Horticultural Science* 11, 686–689.

Biemond, H., Vos, J. and Struik, P.C. (1995) Effects of nitrogen on accumulation and partitioning of dry matter and nitrogen of vegetables. 1. Brussels sprouts. *Netherlands Journal of Agricultural Science* 43, 419–433.

Bjorkman, T. and Pearson, K. (1997) High temperature arrest of inflorescence development in broccoli (*Brassica oleracea* var. *italica* L.). *Journal of Experimental Botany* 49, 101–106.

Blackman, R.L. and Eastop, V.F. (1984) *Aphids on the World's Crops. An Identification Guide.* John Wiley, Chichester, UK.

Bodson, M. and Remacle, B. (1987) Distribution of assimilates from various sources – leaves during floral transition of *Sinapis alba* L. In: Atherton, J.G. (ed.) *Proceeding of Sutton Bonnington Easter School in Agricultural Science – Manipulation of Flowering.* Butterworth, London, pp. 341–350.

Bond, J.M., Mogg, R.J., Squire, G.R. and Johnstone, C. (2004) Microsatellite amplification in *Brassica napus* cultivars: cultivar variability and relationships to a long term feral population. *Euphytica* 139, 173–178.

Booij, R. (1987) Environmental factors in curd initiation and curd growth of cauliflower in the field. *Netherlands Journal of Agricultural Science* 35, 435–445.

Booij, R. and Struik, P.C. (1990) Effects of temperature on leaf and curd initiation in relation to juvenility of cauliflower. *Scientia Horticulturae* 44, 201–214.

Booij, R., Enserink, C.T., Smit, A.L. and van der Werf, A. (1993) Effect of nitrogen availability on crop growth and nitrogen uptake of Brussels sprouts and leek. *Acta Horticulturae* 339, 53–65.

Booij, R., Kreuzer, A.D.H., Smit, A.L. and van der Werf, A. (1996) Effect of nitrogen availability on dry matter production, nitrogen uptake and light interception of Brussels sprouts and leeks. *Netherlands Journal of Agricultural Science* 42, 195–201.

Booij, R., Kreuzer, A.D.H., Smit, A.L. and van der Werf, A. (1997) Effects of nitrogen availability on the biomass and nitrogen partitioning in Brussels sprouts (*Brassica oleracea* var. *gemmifera*). *Journal of Horticultural Science and Biotechnology* 72, 285–297.

Booth, R.G., Cox, M.L. and Madge, R.B (1990) *Guides to Insects of Importance to Man.*

No 3. Coleoptera. International Institute of Entomology (An Institute of CAB International), The Natural History Museum, London. Published by CAB International, Wallingford, UK.

Boruagh, H.P.D. and Kumar, B.S.D. (2002a) Plant disease suppression and growth promotion by a fluorescent *Pseudomonas* strain. *Folia Microbiologica* 47, 137–143.

Boruagh, H.P.D. and Kumar, B.S.D. (2002b) Biological activity of secondary metabolites produced by a strain of *Pseudomonas fluorescens. Folia Microbiologica* 47, 359–363.

Bracy, R.P., Parish, R.L. and Bergeron, P.E. (1995) Sidedress N application methods for broccoli production. *Journal of Vegetable Crop Production* 1, 63–71.

Brainard, D.C. and Bellinder, R.R. (2004) Weed suppression in a broccoli–winter rye intercropping system. *Weed Science* 52, 281–290.

Brandsæter, L.O. (1996) Alternative Strategies for Weed Control: Plant Residues and Living Mulch for Weed Management in Vegetables. Doctor Scientiarum Thesis, 25, Norges Landbrukshøgskole (Agricultural University of Norway), Ås.

Brandsæter, L.O., Netland, J. and Meadow, R. (1998) Yields, weeds, pests and soil nitrogen in a white cabbage-living mulch system. *Biological Agriculture and Horticulture* 16, 291–309.

Bray, C.M., Davison, M., Ashraf, M. and Taylor, R.M. (1989) Biochemical changes during osmopriming of leek seeds. *Annals of Botany* 63, 185–193.

Brunel, D., Froger, N. and Pelletier, G. (1999) Development of amplified consensus genetic markers (ACGM) in *Brassica napus* from *Arabidopsis thaliana* sequences of known biological function. *Genome* 42, 387–402.

Bryne, D.N. and Bellows, T.S. (1991) Whitefly biology. *Annual Review of Entomology* 36, 431–457.

Burris, J.S., Edge, O.T. and Wahab, A.H. (1969) Evaluation of various indices of seed and seedling vigour in soybean (*Glycine max* (L) Merr.). *Proceedings of the Association of Seed Analysts* 59, 73–81.

Camargo, L.E.A., Williams, P.H. and Osbourn, T.C. (1995) Mapping of quantitative trait loci controlling resistance of *Brassica oleracea* to *Xanthomonas campestris* pv. *campestris* in field and greenhouse. *Phytopathology* 85, 1296–1300.

Cao, J., Tang, J.D., Strighov, N., Shelton, A.M. and Earle, E.D. (1999) Transgenic broccoli with high levels of *Bacillus thuringiensis* Cry1C protein control diamondback moth larvae resistant to Cry1A or Cry1C. *Molecular Breeding* 5, 131–141.

Cardi, T. and Earle, E.D. (1995) Transfer of cms 'Anand' cytoplasm from *Brassica rapa* through protoplast fusion. *Cruciferae Newsletter* 17, 44–45.

Carter, D.J. (1984) *Pest Lepidoptera of Europe with Especial Reference to the British Isles. Series Entomologia*, Vol. 31. Junk, Dordrecht, The Netherlands.

Cavell, A.C., Lydiate, D.J., Parkin, I.A.P., Dean, C. and Trick, M. (1998) Collinearity between a 30-centimorgan segment of *Arabidopsis thaliana* chromosome 4 and duplicated regions within the *Brassica napus* genome. *Genome* 41, 62–69.

Chancellor, R.J. (1982) Weed seed investigations. *Advances in Research and Technology of Seeds* 7, 9–29.

Charron, C.S. and Sams, C.E. (2004) Glucosinolate content and myrosinase activity in rapid-cycling *Brassica oleracea* grown in a controlled environment. *Journal of the American Society for Horticultural Science* 129, 321–330.

Chiang, M.S. (1972) Inheritance of head splitting in cabbage *Brassica oleracea* var *capitata. Euphytica* 21, 507–509.

Cho, H.S., Cao, J., Ren, J.P. and Earle, E.D. (2001) Control of lepidopteran insect pests in transgenic Chinese cabbage (*Brassica rapa* ssp *pekinensis*) transformed with a synthetic *Bacillus thuringiensis cry1C* gene. *Plant Cell Reports* 20, 1–7.

Chu, Y-F., Sun, J., Wu, X. and Liu, R.H. (2002) Antioxidant and antiproliferative activities of common vegetables. *Journal of Agriculture and Food Chemistry* 50, 6910–6916.

Chung, B. (1982) Effects of plant density on the maturity and once-over harvest yields of broccoli. *Journal of Horticultural Science and Biotechnology* 57, 365–372.

Cliffe-Byrnes, V. and O'Beirne, D. (2005) The effects of cultivar and physiological age on quality and shelf-life of coleslaw mix packaged in modified atmospheres. *International Journal of Food Science and Technology* 40, 165–175.

Coaker, T.H. (1980) Insect pest management in brassica crops by inter-cropping. *SROP/WPRS Bulletin* 111/1, 117–125.

Coaker, T.H. (1988) Insect pest management by intracrop diversity: potential and limitations. In: Cavalloro, R. and Pelerents, C. (eds) *Progress on Pest Management in Field Vegetables*. A.A. Balkema, Rotterdam, pp. 281–288.

Cobb, M. (1999) What and how do maggots smell? *Biological Review of the Cambridge Philosophical Society* 74, 425–459.

Coelho, P., Leckie, D., Bahcevandziev, K., Valerio, L., Astley, D., Boukema, L., Crute, I.R. and Monteiro, A.A. (1997) The relationship between cotyledon and adult plant resistance to downy mildew (*Peronospora parasitica*) in *Brassica oleracea*. *Acta Horticulturae* 459, 335–342.

Cook, D. and Robeson, D.J. (1986) Active resistance of cabbage (*Brassica oleracea*) to *Xanthomonas campestris* pv. *campestris* against the causal agent of black rot by coinoculation. *Physiological and Molecular Plant Pathology* 28, 41–52.

Coolong, T.W., Randle, W.M., Toler, H.D. and Sams, C.E. (2004) Zinc availability in hydroponic culture influences glucosinolate concentrations in *Brassica rapa*. *HortScience* 39, 84–86.

Costello, M.J. (1994) Interaction of leguminous living mulch, broccoli growth yield and aphid infestation. *Biological Agriculture and Horticulture* 10, 207–222.

Costigan, P.A. (1998) The placement of starter fertilisers to improve the early growth of drilled and transplanted vegetables. *Proceedings of the Fertiliser Society* 274, 1–24.

Cousens, R. (1985) A simple model relating yield loss to weed density. *Annals of Applied Biology* 107, 239–252.

Cox, E.F. (1977) Pepper spot in white cabbage – a literature review. *Agricultural Development and Advisory Service (ADAS) Quarterly Review* 25, 81–86.

Crisp, P. (1982) The use of an evolutionary scheme for cauliflowers in the screening of genetic resources. *Euphytica* 31, 725–734.

Crisp, P. and Gray, A. (1975a) Propagating cauliflowers vegetatively. *The Grower* 83, 860–861.

Crisp, P. and Gray, A. (1975b) Cauliflower breeding by grafting curd portions on to stock plants. *The Grower* 83, 915–916.

Crisp, P. and Gray, A. (1975c) Tissue culture in cauliflowers. *The Grower* 83, 966–967.

Crisp, P., Walkey, D.G.A., Bellaman, E. and Roberts, E. (1975) A mutation affecting curd colour in caulilfower (*Brassica oleracea* L. var. *botrytis* DC). *Euphytica* 24, 173–176.

Cubeta, M.A., Cody, B.R., Sugg, R.E. and Crozier, C.R. (2000) Influence of soil calcium, potassium and pH on development of leaf tipburn of cabbage in Eastern North Carolina. *Communications in Soil Science and Plant Analysis* 31, 259–275.

Dasgupta, S. and Mandal, R.K. (1993) Compositional changes and storage protein synthesis in developing seeds of *Brassica campestris*. *Seed Science and Technology* 21, 291–299.

Deane, C.R., Dix, P.J. and Fuller, M.J. (1996) Selection of hydroxyproline resistant proline accumulating mutants of cauliflower for improvement of frost resistance. *Acta Horticulturae* 407, 123–129.

Dell'aquilla, A. (2003) Image analysis as a tool to study deteriorated cabbage (*Brassica oleracea* L.) seed imbibition under salt stress conditions. *Seed Science and Technology* 31, 619–628.

Dell'aquilla, A., van der Schoor, R. and Jalink, H. (2002) Application of chlorophyll fluorescence in sorting controlled deteriorated white cabbage (*Brassica oleracea* L.) seeds. *Seed Science and Technology* 30, 689–695.

Dempster, J.P. and Coaker, T.H. (1974) Diversification of crop ecosystems as a means of controlling pests. In: Jones, D.P. and Soloman, M.E. (eds) *Biology in Pest and Disease Control; 13th Symposium of the British Ecological Society*. Blackwell Scientific Publications, Oxford, pp. 106–114.

de Tournefort, J.P. (1700) *Institutiones rei herbariae editio altera*. 1, 219–227, Paris.

Devi, L.C., Kant, K. and Dadlani, M. (2003) Effect of grading and ageing on sinapine leakage, electrical conductivity and germination percentage in the seed of mustard. *Seed Science and Technology* 31, 505–509.

Dickson, M.H. and Hunter, J.E. (1987) Inheritance of resistance in cabbage seedlings to black rot. *HortScience* 22, 108–109.

Dickson, M.H. and Lee, C.Y. (1980) Persistent white curd and other curd characters of cauliflower. *Journal of the American Society for Horticultural Science* 105, 533–535.

Dickson, M.H. and Peltzoldt, R. (1996) Breeding for resistance to *Sclerotinia sclerotiorum* in *Brassica oleracea*. *Acta Horticulturae* 407, 103–108.

Dickson, M.H. and Wallace, D.H. (1986) Cabbage breeding. In: Basset, M.J. (ed.) *Breeding Vegetable Crops*. AVI Publishing, Westport, Connecticut, pp. 395–422.

Dickson, M.H., Lee, C.Y. and Blamble, A.E. (1988) Orange-curd high carotene cauliflower inbreds, NY 156, NY 163 and NY 165. *HortScience* 23, 778–779.

Dixon, G.R. (1974) Field studies of powdery mildew (*Erysiphe cruciferarum*) on Brussels sprouts. *Plant Pathology* 23, 105–109.

Dixon, G.R. (1976) Disease assessment keys for powdery mildew (*Erysiphe cruciferarum*) on leaves and buds of Brussels sprout. In: *Manual of Plant Growth and Stages and Disease Assessment Keys*. Ministry of Agriculture, Fisheries and Food, London (now: The Department for Environment, Food and Rural Affairs, Defra).

Dixon, G.R. (1981) *Vegetable Crop Diseases*. MacMillan Publishers Ltd, London.

Dixon, G.R. (1984) *Plant Pathogens and their Control in Horticulture*. MacMillan Publishers Ltd, London in collaboration with the Horticultural Education Association and Royal Horticultural Society.

Dixon, G.R. (2006) Review of advances in understanding the biology of *Plasmodiophora brassicae* (clubroot). Proceedings of Brassica 2004 Symposium, Korea, October 2004. *Acta Horticulturae* 706, 271–282.

Dodsall, L.M., Good, A., Keddie, B.A., Ekvere, U. and Stringham, G. (2000) Identification and evaluation of root maggot (*Delia* sp) resistances within *Brassicae*. *Crop Protection* 19, 247–253.

Donald, C.M. (1958) The interaction of competition for light and for nutrients. *Australian Journal of Agricultural Research* 9, 421–435.

Donald, E.C., Porter, I.J., Faggian, R. and Lancaster, R.A. (2006) An integrated approach to the control of clubroot in vegetable brassica crops. *Acta Horticulturae* 706, 283–300.

Downs, C.G., Somerfield, S.D. and Davey, M.C. (1997) Cytokinin treatment delays senescence but not sucrose loss in harvested broccoli. *Post Harvest Biology and Technology* 11, 93–100.

Dufault, R.J. and Waters, L. (1985) Interaction of nitrogen fertility and plant populations on transplanted broccoli and cauliflower yields. *HortScience* 20, 127–128.

Earle, E.D. and Knauf, V. (1999) Genetic engineering in Brassica. In: Gomez-Campo, C. (ed.) *Biology of Brassica Coenospecies*. Elsevier, Amsterdam, pp. 287–313.

Edwards, M.D. and Williams, P.H. (1987) Selection for minor genes resistance to *Albugo candida* in a rapid-cycling population of *Brassica campestris. Phytopathology* 77, 527–532.

Egley, G.H. and Williams, R.D. (1990) Decline of weed seeds and seedlings over five years as affected by soil disturbances. *Weed Science* 38, 504–510.

Ellis, R.H. and Roberts, E.H. (1981) The quantification of ageing and survival in orthodox seeds. *Seed Science and Technology* 9, 373–409.

Ellis, R.H., Hong, T.D. and Roberts, E.H. (1987) The development of desiccation tolerance and maximum seed quality during seed maturation in six grain legumes. *Annals of Botany* 59, 23–29.

Ellis, P.R., Singh, R., Pink, D.A.C., Lynn, J.R. and Saw, P.L. (1996) Resistance to *Brevicoryne brassicae* in horticultural brassicas. *Euphytica* 88, 85–96.

Ellis, P.R., Pink, D.A.C., Phelps, K., Jukes, P.L., Breed, S.E. and Pinneghar, A.E. (1998) Evaluation of a core collection of *Brassica oleracea* accessions for resistance to *Brevicoryne brassicae*, the cabbage aphid. *Euphytica* 103, 149–150.

Ellis, P.R., Kift, N.B., Pink, D.A.C., Jukes, P.L., Lynn, J. and Tatchell, G.M. (2000) Variation in resistance to the cabbage aphid (*Brevicoryne brassicae*) between and within wild and cultivated *Brassica* species. *Genetic Resources and Crop Evolution* 47, 395–401.

Engelhard, A.W. (ed.) (1996) *Soilborne Plant Pathogens: Management of Diseases with Macro- and Micro-elements*. American Phytopathological Society, St Paul, Minnesota.

Ensminger, A.H., Ensminger, M.E., Konlande, J.E. and Robson, J.R.K. (1995) *The Concise Encyclopedia of Food*. CRC Press, Boca Raton, Florida.

Entwistle, P.F., Cory, J.S., Bailey, M.J. and Higgs, S. (1993) Bacillus thuringiensis *an Environmental Biopesticide*. John Wiley and Sons, New York.

Ester, A., van der Zande, J.C. and Frost, A.J.P. (1994) Crop covering to prevent pest damage to field vegetables and the feasibility of pesticide application through polyethylene nets. In: *Proceedings of the Brighton Crop Protection Conference (Pests and Diseases)*. British Crop Protection Council, Farnham, UK, pp. 761–766.

Evans, K.A. and Allen-Williams, L.J. (1998) Response of cabbage seed weevil (*Ceutorhynchus assimilis*) to baits of extracted and synthetic host-plant odor. *Journal of Chemical Ecology* 24, 2101–2114.

Everaarts, A.P. (1993) Strategies to improve the efficiency of nitrogen fertiliser use in the cultivation of Brassica vegetables. *Acta Horticulturae* 339, 161–173.

Everaarts, A.P. (2001) Review of tipburn in cabbage. Characterised tipburn by discoloration and desiccation of internal head leaves. *Journal of Horticultural Science and Biotechnology* 76, 515–521.

Everaarts, A.P. and de Moel, C.P. (1995) The effect of nitrogen and the method of application on the yield of cauliflower. *Netherlands Journal of Agricultural Science* 43, 409–418.

Everaarts, A.P. and de Willigen, P. (1999) The effect of nitrogen and method of application on yield and quality of broccoli. *Netherlands Journal of Agricultural Science* 47, 123–133.

Everaarts, A.P. and van Beusidem, M.L. (1998) The effect of planting date and plant density on nitrogen uptake and nitrogen harvest by Brussels sprouts. *Journal of Horticultural Science and Biotechnology* 73, 704–710.

Everaarts, A.P., Booij, R. and de Moel, C.P. (1998) Yield formation in Brussels sprouts. *Journal of Horticultural Science and Biotechnology* 73, 711–721.

Fahey, J.W., Zhang, Y. and Talalay, P (1997) Broccoli sprouts: an exceptionally rich source of inducers of enzymes that protect against chemical carcinogens. *Proceedings of the National Academy of Sciences of the USA* 94, 10367–10372.

Fahey, J.W., Haristoy, X., Dolan, P.M., Kensler, T.W., Scholtus, I., Stephenson, K.K., Talalay, P. and Lozniewski, A. (2002) Sulforathane inhibits extracellular, intracellular and antibiotic-resistant strains of *Helicobacter pylori* and prevents benzo[a]pyrene-induced stomach tumour. *Proceedings of the National Academy of Sciences of the USA* 99, 7610–7615.

Fahey, J.W., Zalcmann, A.T. and Talalay, P. (2001) The chemical diversity and distribution of glucosinolates and isothiocyanates among plants. *Phytochemistry* 56, 5–51.

Farnham, M.W., Stephenson, K.K. and Fahey, J.W. (2005) Glucoraphanin level in broccoli seed is largely determined by genotype. *HortScience* 100, 50–53.

Feller, C. and Fink, M. (2005) Growth and yield of broccoli as affected by the nitrogen content of transplants and the timing of nitrogen fertilization. *HortScience* 40, 1320–1323.

Feltwell, J. (1982) *Large White Butterfly: The Biology, Biochemistry and Physiology of* Pieris brassicae *(Linnaeus). Series Entomologica*, Vol. 18, Kluwer, The Netherlands.

Fenner, M. (1985) *Seed Ecology*. Chapman and Hall, London.

Finch, S. (1987) Horticultural crops. In: Burn, A.J., Coaker, T.H. and Jepson, P.C. (eds) *Integrated Pest Management*. Academic Press, London, pp. 257–293.

Finch, S. and Thompson, A.R. (1992) Pests of cruciferous crops. In: McKinley, R.G. (ed.) *Vegetable Crop Pests*. The Macmillan Press, London, pp. 87–138

Fischer, W., Bergfield, R., Plachy, C., Schäfer, R. and Schoper, P. (1988) Accumulation of storage material, precocious germination and development of desiccation tolerance during seed maturation in mustard (*Sinapis alba* L.). *Botanica Acta* 101, 344–354.

Fisher, N.M. and Milbourn, G.M. (1974) The effect of plant density, date of apical bud removal and leaf removal on the growth and yield of single-harvest Brussels sprouts (*Brassica oleracea* var. *gemmifera* D.C.) 1. Whole plant and axillary bud growth. *Journal of Agricultural Science (Cambridge)* 83, 479–487.

Fitter, A.H. and Hay, R.K.M. (1987) *Environmental Physiology of Plants*. Academic Press, London.

Fjellstrom, R.G. and Williams, P.H. (1997) Fusarium yellows and turnip mosaic virus resistance in *Brassica rapa* and *B. juncea*. *HortScience* 32, 927–930.

Fogg, J.M. (1975) The silent travellers. *Brooklyn Botanic Garden Records* 31, 12–15.

Forney, C.F., Song, J.L.F., Hildebrand, P.D. and Jordan, M.A. (2003) Ozone and 1-methylcyclopropene alter the postharvest quality of broccoli. *Journal of the American Society for Horticultural Science* 128, 403–408.

Fransden, K.J. (1943) The experimental formation of *Brassica juncea* Czern. Et Coss. (Preliminary report). *Dansk Botanisk Arkiv* 11, 1–17.

Freuler, P., Fischer, A.J., Gagnebin, S., Parko, F., Granges, J. and Mittaz, C. (2000)

Resistance of cauliflower against cabbage root fly. *Revue Suisse de Viticulture, d'Arboriculture et d'Horticulture* 32, 109–113.

Freyman, S., Hall, J.W. and Brookes, V.R. (1992) Effect of planting pattern on intra-row competition between cabbage and shepherd's purse (*Capsella bursa-pastoris*). *Canadian Journal of Plant Science* 72, 1393–1396.

Fujime, Y. (1983) Studies on thermal conditions of curd formation and development in cauliflower and broccoli, with especial reference to abnormal curd development. *Memoirs of the Faculty of Agriculture Kagawa University*, 40.

Fujime, Y. and Hirose, T. (1979) Studies on thermal conditions of curd formation and development in cauliflower and broccoli. *Journal of the Japanese Society of Horticultural Science* 114, 552–556.

Fujima, Y. and Okuda, N. (1996) The physiology of flowering in *Brassicas*, especially about cauliflower and broccoli. *Acta Horticulturae* 407, 247–254.

Gallagher, J.N. (1979) Field studies of cereal leaf growth. 1. Initiation and expansion in relation to temperature and ontogeny. *Journal of Experimental Botany* 30, 625–636.

Gardner, B.R. and Roth, R.L. (1989) Mid rib nitrate concentration as a means of determining nitrogen needs of cabbage. *Journal of Plant Nutrition* 12, 1073–1088.

Gates, R.R. (1953) Wild cabbages and the effects of cultivation. *Journal of Genetics* 51, 363–372.

Gavloski, J.E., Ekuere, U., Keddie, A., Dosdall, L., Kott, L. and Good, A.G. (2000) Identification and evaluation of flea beetle (*Phyllotreta cruciferae*) resistance within Brassicaceae. *Canadian Journal of Plant Science* 80, 881–887.

Geeson, J.D. and Browne, K.M. (1980) Controlled atmosphere storage of winter white cabbage. *Annals of Applied Biology* 95, 267–272.

Geeson, J.D. (1983) Brassicas. In: Dennis, C. (ed.) *Post-harvest Pathology of Fruits and Vegetables*. Academic Press, London, pp. 125–156.

Geraci, A., Divaret, I., Raimondo, F.M. and Cherve, A.M. (2001) Genetic relationships between Sicilian wild populations of *Brassica* analysed with RAPD markers. *Plant Breeding* 120, 193–196.

Gilbert, N. and Raworth, D.A. (1996) Insects and temperature – a general theory. *Canadian Entomology* 128, 1–13.

Giles, W.F. (1941) Cauliflower and broccoli. What they are and where they came from. *Journal of the Royal Horticultural Society (The Garden)* 66, 265–278.

Giles, W.F. (1944) Our vegetables: whence they came. *Journal of the Royal Horticultural Society (The Garden)* 69, 132–138, 167–173.

Gilmour, J.S.L. (chairman), Horne, F.R., Little, E.L., Jr, Stafleu, F.A. and Richards, R.H. (secretary) (1969) International Code of Nomenclature of Cultivated Plants-1969. *Regnum Vegetabile* 64. Published by the International Bureau for Plant Taxonomy and Nomenclature of the International Association for Plant Taxonomy, Utrecht, The Netherlands; obtainable from The Royal Horticultural Society, Vincent Square, London.

Goldberg, R.B., Barker, S.J. and Perez-Grau, L. (1989) Regulation of gene expression during plant embryogenesis. *Cell* 56, 149–160.

Gonzales, D. and Rawlins, W.A. (1968) Aphid sampling efficiency of Moericke traps affected by height and background. *Journal of Economic Entomology* 61, 109–144.

Górnik, K., deCastro, R.D., Liu, Y.Q., Bino, R.J. and Groot, S.P.C. (1997) Inhibition of cell division during cabbage (*Brassica oleracea* L.) seed germination. *Seed Science Research* 7, 333–340.

Graham, M.B. and Crabtree, G. (1987) Management of competition for water between cabbage (*Brassica olerace*) and perennial ryegrass (*Lolium perenne*) living mulch. *Proceedings of the Western Society for Weed Science* 40, 113–117.

Gray, A. (1989) Green curded cauliflowers. *Journal of the Royal Horticultural Society (The Garden)* 114, 31–33.

Gray, A.R. (1982) Taxonomy and evolution of broccoli (*Brassica oleracea* L var. *italica* Plenck.). *Economic Botany* 36, 397–410.

Gray, A.R. and Crisp, P. (1985) Breeding improved green-curded cauliflower. *Cruciferae Newsletter* 10, 66–67.

Greenler, J.McC. and Williams, P.H. (1990) Rapid-cycling *Brassica rapa* (Fast Plants) for hands-on teaching of plant science. In: McFerson, J.R., Kresovich, S. and Dwyer, S.G. (eds) *Proceedings of the 6th Crucifer Genetics Workshop*. Plant Genetic Resources Unit, Cornell University, Ithaca, New York, p. 24 (abstract).

Greenwood, D.J. (1990) Production or productivity: the nitrate problem? *Annals of Applied Biology* 117, 209–231.

Greenwood, D.J., Cleaver, T.J., Turner, M.K., Hunt, J., Niendorf, K.B. and Loquens, S.M.H. (1980) Comparison of the effects of nitrogen fertiliser on the yield, nitrogen content and quality of 21 different vegetable and agricultural crops. *Journal of Agricultural Science* 95, 471–485.

Greenwood, D.J., Neeteson, J.J. and Draycott, A. (1986) Quantitative relationships for the dependence of growth rate of arable crops on their nitrogen content, dry weight and aerial environment. *Plant and Soil* 91, 281–301.

Greenwood, D.J., Rahn, C., Draycott, A., Vaidyanathasn, L.V. and Paterson, C. (1996) Modelling and measurement of the effects of fertiliser-N and crop residue incorporation on N-dynamics in vegetable cropping. *Soil Use and Management* 12, 13–24.

Greuter, W., McNeill, J., Barrie, F.R., Burdet, H.M., Demoulin, V., Filgueras, T.S., Nicolson, D.H., Silva, P.C., Skog, J.E., Trehane, P., Turland, N.T. and Hawksworth, D.L. (2000) International Code of Botanical Nomenclature (Saint Louis Code). 16th International Botanical Congress, Missouri, July to August 1999. *Regnum Vegetabile* 138. Published by Koeltz Scientific Books, Königstein, Germany.

Grevsen, K. (1998) Effects of temperature on head growth of broccoli (*Brassica olerecea* L. var. *italica*): parameter estimates for a predictive model. *Journal of Horticultural Science and Biotechnology* 73, 235–244.

Grevsen, K. and Olesen, J.E. (1994) Modelling cauliflower development from transplanting to curd initiation. *Journal of Horticultural Science and Biotechnology* 69, 755–766.

Grieve, C.M., Shannon, M.C. and Poss, J.A. (2001) Mineral nutrition of leafy vegetable crops irrigated with saline drainage water. *Journal of Vegetable Crop Production* 7, 37–47.

Griffiths, B. and Scholl, R.L. (1991) Qualitative and quantitative genetic studies of *Arabidopsis thaliana*. *Genetics* 129, 605–609.

Griffiths, P.D. and Roe, C. (2005) Response of *Brassica oleracea* var. *capitata* to wound and spray inoculations with *Xanthomonas campestris* pv. *campestris*. *HortScience* 40, 47–49.

Grime, J.P. (1979) *Plant Strategies and Vegetative Processes*. John Wiley and Sons, Chichester.

Grundy, A.C., Mead, A. and Burston, S. (1999) Modellling the effect of cultivation on seed movement with application to the prediction of weed seedling emergence. *Journal of Applied Ecology* 36, 663–678.

Guo, H., Dickson, M.H. and Hunter, J.E. (1991) *Brassica napus* sources of resistance to black rot in crucifers and inheritance of resistance. *HortScience* 26, 1545–1547.

Gurusamy, C. and Thiagarajan, C.P. (1998) The pattern of seed development and maturation in cauliflower (*Brassica oleracea* L. var *botrytis*). *Phyton (Austria)* 38, 259–268.

Hailles, R., Rees, M., Kohn, D. and Crawley, M. (1997) Burial and seed survival in *Brassica napus* subsp. *oleifera* and *Sinapis arvensis* including comparison of transgenic and non-transgenic lines of the crop. *Proceedings of the Royal Society of London B* 264, 1–7.

Haine, K.E. (1959) Time of heading and quality of curd in winter cauliflower. *Journal of the National Institute of Agricultural Botany* 8, 667–674.

Hall, L., Topinka, K., Huffman, J., Davis, L. and Good, A. (2000) Pollen flow between herbicide-resistant *Brassica napus* is the cause of multiple-resistant *B. napus* volunteers. *Weed Science* 48, 688–694.

Hall, R. (ed.) (1996) *Principles and Practice of Managing Soilborne Plant Pathogens*. The American Phytopathological Society, St Paul, Minnesota.

Hand, D.J. (1988) Regulation of Curd Initiation in the Summer Cauliflower. PhD Thesis, University of Nottingham, UK.

Hand, D.J. and Atherton, J.G. (1987) Curd initiation in the cauliflower. 1. Juvenility. *Journal of Experimental Botany* 38, 2050–2058.

Hansen, L.N. and Earle, E.D. (1995) Transfer of resistance to *Xanthomomnas campestris* pv. *campestris* into *Brassica oleracea* L. by protoplast fusion. *Theoretical and Applied Genetics* 91, 1293–1300.

Hansen, L.N. and Earle, E.D. (1997) Somatic hybrids between *Brassica oleracea* and *Sinapis alba* L., with resistance to *Alternaria brassicae* (Berk.) Sacc. *Theoretical and Applied Genetics* 94, 1078–1085.

Hara, T. and Sonoda, Y. (1981) The role of micronutrients in cabbage head formation. IV. Effect of nitrogen supply or light intensity on the growth and $^{14}CO_2$ and $^{15}NO_3$-N assimilation of cabbage plants. *Soil Science and Plant Nutrition* 27, 185–194.

Haramoto, E.R. and Gallandt, E.R. (2005a) Brassica cover cropping: I. Effects on weed and crop establishment. *Weed Science* 53, 695–700.

Haramoto, E.R. and Gallandt, E.R. (2005b) Brassica cover cropping. II. Effects on growth and interference of green bean (*Phaseolus vulgaris*) and redroot pigweed (*Amaranthus retroflexus*). *Weed Science* 53, 702–708.

Harindar, S. and Kalda, T.S. (1995) Screening of cauliflower germplasm against *Sclerotinia* rot. *Indian Journal of Genetics and Plant Breeding* 55, 98–102.

Harrington, J.F. (1972) Seed storage and longevity. In: Kozlowski, T.T. (ed.) *Seed Biology*, Vol. 3. Academic Press, New York, pp. 145–243.

Harrison, H.F., Jackson, D.M., Keinath, A.P., Marino, P.C. and Pullaro, T. (2004) Broccoli production in cowpea, soybean and velvetbean cover crop mulches. *HorTechnology* 14, 484–487.

Heath, D.W., Earle, E.D. and Dickson, M.H. (1994) Introgressing cold tolerant Ogura cytoplasms from rapeseed into Pak Choi and Chinese cabbage. *HortScience* 29, 202–203.

Heather, D.W. and Sieczka, J.B. (1991) Effect of seed size and cultivar on emergence and stand establishment of broccoli in crusted soil. *Journal of the American Society for Horticultural Science* 116, 946–949.

Hegi, G. (1919) *Illustrierte Flora von Mittel-Europa.* NBd IV/1. Lehmanns Veralg, München.

Helm, J. (1963) Morphologische-taxonomische Gliederung der Kultursippen von *Brassica oleracea* L. *Die Kelturpflanze* 11, 92–210.

Henslow, G. (1908) History of the cabbage tribe. *Journal of the Royal Horticultural Society (The Garden)* 34, 15–23.

Herklots, G.A.C. (1972) *Vegetables in South-east Asia.* George Allen and Unwin, London, pp. 182–224.

Heydecker, W. (ed.) (1973) *Seed Ecology. Proceedings of the 19th Easter School in Agricultural Science.* University of Nottingham, 1972. Butterworths, London.

Hill, T.A. (1977) *The Biology of Weeds.* The Institute of Biology's Studies in Biology No. 79. Edward Arnold, London.

Hochmuth, G.J. (1992) Concepts and practices for improving nitrogen management for vegetables. *HortTechnology* 2, 121–125.

Hofsvang, T. (1991) The influence of intercropping and weeds on the oviposition of the brassica root flies (*Delia radicum* and *D. floralis*). *Norwegian Journal of Agricultural Sciences* 5, 349–356.

Hooks, C.R.R. and Johnson, M.W. (2001) Broccoli growth parameters and level of head infestation in simple and mixed plantings: impact of increased flora diversification. *Annals of Applied Biology* 138, 269–280.

Hopen, H.J. (1995) Herbicides available for commercial cabbage producers during 1965–94. *HortTechnology* 5, 25–26.

Hoyt, G.D., Bonnanno, A.R. and Parker, G.C. (1996) Influence of herbicides and tillage on weed control, yield and quality of cabbage (*Brassica oleracea* L. var. *capitata*). *Weed Technology* 10, 50–54.

Hulden, L. (1986) The whiteflies (Homoptera, Aleyrodidae) and their parasites in Finland. *Notulae Entomologicae* 66, 1–40.

Hulme, M., Jenkins, G.J., Lu, X., Turnpenny, J.R., Mitchell, T.D., Jones, R.G., Lowe, J., Murphy, J.M., Hassall, D., Boorman, F., McDonald, R. and Hill, S. (2002) *Climate Change Scenarios for the United Kingdom.* The UKCIP02 Scientific Report. Tyndall Centre for Climate Change Research, School of Environmental Sciences, University of East Anglia, Norwich, UK.

Humpherson-Jones, F.M. (1989) Survival of *Alternaria brassicae* and *Alternaria brassicicola* on crop debris of oilseed rape and cabbage. *Annals of Applied Biology* 115, 45–50.

Hunter, J.E., Dickson, M.H. and Ludwig, J. (1987) Source of resistance to black rot of cabbage expressed in seedlings and adult plants. *Plant Disease* 71, 263–266.

Huyskens-Keil, S. (2003) Quality determination of red radish by non-destructive root color measurement. *Journal of the American Society for Horticultural Science* 128, 397–402.

Ignatov, A., Hida, K. and Kuginuki, Y. (1998) Black rot of crucifers and sources of resistance in Brassica crops. *Japan Agricultural Research Quarterly* 32, 167–172.

Ignatov, A., Kuginuki, Y. and Hida, K. (1999) Vascular stem resistance to black rot in *Brassica oleracea. Canadian Journal of Botany* 77, 442–446.

Isenberg, F.M.R., Pendergress, A., Carroll, J.E., Howell, I. and Oyer, E.B. (1975) The use of weight, density heat units and solar radiation to predict maturity of cabbage for storage. *Journal of the American Society for Horticultural Science* 100, 313–316.

Jalink, H., van der Schoor, R., Birnbaum, Y.E. and Bino, R.J. (1999) Seed chlorophyll content as an indicator for seed maturity and seed quality. *Acta Horticulturae* 504, 219–227.

Jenni, S., Dutilleul, P., Yamasaki, S. and Tremblay, N. (2001) Brown bead of broccoli. II. Relationships of physiological disorder with nutritional and meteorological variables. *HortScience* 36, 1228–1234.

Jensen, B.D., Vaerbak, S., Munk, L. and Andersen, S.B. (1999) Characterization and inheritance of partial resistance to downy mildew *Peronospora parasitica*, in breeding material of broccoli *B. oleracea* convar. *botrytis* var. *italica*. *Plant Breeding* 118, 549–554.

Jett, W.L. and Welbaum, G.E. (1996) Changes in broccoli (*Brassica oleracea* L.) seed weight, viability and vigour during development and following drying and priming. *Seed Science and Technology* 24, 127–137.

Jett, W.L., Welbaum, G.E. and Morse, R.D. (1996) Effect of matric and osmotic priming treatments on broccoli seed germination. *Journal of the American Society for Horticultural Science* 121, 423–429.

Johnson, I.T. (2000) Brassica vegetables and human health: glucosinolates in the food chain. *Acta Horticulturae* 539, 39–44.

Johnson, J.R. (1991) Calcium accumulation, calcium distribution and biomass partitioning in collards. *Journal of the American Society for Horticultural Science* 116, 991–994.

Jones, F.G.W. and Jones, M.G. (1984) *Pests of Field Crops*. Edward Arnold, London.

Jourdan, P.S., Earle, E.D. and Mutschler, M.A. (1989) Synthesis of male-sterile, triazine-resistant *Brassica napus* by somatic hybridisation between cytoplasmic male sterile *B. oleracea* and atrazine-resistant *B. campestris*. *Theoretical and Applied Genetics* 78, 445–455.

Jyoti, J.L., Shelton, A.M. and Earle, E.D. (2001) Identifying sources and mechanisms of resistance in crucifers for control of cabbage maggot (Diptera: Anthomyiidae). *Journal of Economic Entomology* 94, 942–949.

Kadam, D.M., Lata, D.V.S. and Pandey, A.K. (2005) Influence of different treatments on dehydrated cauliflower quality. *International Journal of Food Science and Technology* 40, 849–856.

Kamoun, S., Kamdar, H.V., Tola, E. and Cado, C.I. (1992) Incompatible interaction between crucifers and *Xanthomonas campestris* involves a vascular hypersensitive response, role of the htpX locus. *Molecular Plant-Microbe Interactions* 5, 22–33.

Kaniszewski, S. and Rumpel, J. (1998) Effect of irrigation, nitrogen fertilisation and soil type on yield and quality of cauliflower. *Journal of Vegetable Crop Production* 4, 67–75.

Keen, N.T., Mayama, S., Leach, J.E. and Tsuyuma, S. (eds) (2001) *Delivery and Perception of Pathogen Signals in Plants*. The American Phytopathological Society, St Paul, Minnesota.

Keller, W.A., Rajhathy, T. and Lacapra, J. (1975) *In vitro* production of plants from pollen in *Brassica campestris*. *Canadian Journal of Genetics and Cytology* 17, 655–666.

Keng, H. (1974) Economic plants of ancient north China as mentioned in Shih Ching (Book of Poetry). *Economic Botany* 28, 391–410.

Kennedy, G.G.F., Gould, F., DePonti, O.M. and Stinner, R.E. (1987) Ecological, agricultural, genetic and commercial considerations in the development of insect resistant germplasm. *Environmental Entomology* 16, 327–338.

Khattra, S., Sharma, K. and Singh, G. (1993) The physiology of extremely desiccated *Brassica juncea* L. seeds. *Agrobotanica* 46, 5–13.

Kingsolver, J.G. (2000) Feeding, growth and the thermal environment of Cabbage White caterpillars, *Pieris rapae* L. *Physiological and Biochemical Zoology* 73, 621–628.

Kirkegaard, J.A., Sarwar, M. and Maatiessen, J.N. (1998) Assessing the biofumigant potential of crucifers. *Acta Horticulturae* 459, 105–111.

Kleinhenz, M.D. (2003) A proposed tool for preharvest estimation of cabbage yield. *HorTechnology* 13, 182–185.

Knavel, D.E. and Herron, J.W. (1981) Influence of tillage system, plant spacing and nitrogen on head weight, yield and nutrient concentration of spring cabbage. *Journal of the American Society for Horticultural Science* 106, 540–545.

Kocks, C.G. and Ruissen, M.A. (1996) Measuring field resistance of cabbage cultivars to black rot. *Euphytica* 91, 45–43.

Kocks, C.G., Ruissen, M.A., Zadoks, J.C. and Duijkers, M.G. (1998) Survival and extinction of *Xanthomonas campestris* pv. *campestris* in soil. *European Journal of Plant Pathology* 104, 911–923.

Koenraadt, H., van Bilsen, J.G.P.M. and Roberts, S.J. (2005) Comparative test of four semi-selective agar media for the detection of *Xanthomonas compestris* pv. *campestris* in brassica seeds. *Seed Science and Technology* 33, 115–125.

Kogan, M. and Ortman, E.F. (1978) Antixenosis – a new term proposed to replace Painter's 'non-preference' modality of resistance. *Bulletin of the Entomological Society of America* 24, 175–176.

Kole, C., Teutonico, R., Mengistu, A., Williams, P.H. and Osborn, T. (1996) Molecular mapping of a locus controlling resistance to *Albugo candida* in *Brassica rapa*. *Phytopathology* 86, 367–369.

Kole, C., Williams, P.H., Rimmer, S.R. and Osbourn, T.C. (2002) Linkage mapping of genes controlling resistance to white rust (*Albugo candida*) in *Brassica rapa* (syn. *campestris*) and comparative mapping to *Brassica napus* and *Arabidopsis thaliana*. *Genome* 45, 22–27.

Kopsell, D.A., Kopsell, D.E., Lefsrud, M.G., Curran-Celentano, J. and Dukach, L.E. (2004) Variation in lutein, β-carotene and chlorophyll concentrations among *Brassica oleracea* cultigens and seasons. *HortScience* 39, 361–364.

Krupinsky, J.M., Bailey, K.L., McMullen, M.P., Gossen, B.D. and Turkington, T.K. (2002) Managing plant disease risk in diversified cropping systems. *Agronomy Journal* 94, 198–209.

Ku, V.V.V. and Wills, R.B.H. (1999) Effect of 1-methylcyclopropene on the storage life of broccoli. *Post Harvest Biology and Technology* 17, 127–132.

Kumazawa, S. (1965) *Vegetable Gardening*. Yokendo, Tokyo, Japan.

Kumazawa, S. and Akiya, R. (1936) Study of vegetables in Taiwan and the south area of China. 2. Takana. *Agriculture and Horticulture* 11, 1741–1748.

Kurppa, S. and Ollula, A. (1993) Optimising insect pest control and nitrogen fertilizing of the summer turnip rape (*Brassica campestris sativa*). *Agricultural Science Finland* 2, 149–160.

Kurstjens, D.A.G. and Perdok, U.D. (2000) The selective soil covering mechanism of weed harrows on sandy soil. *Soil and Tillage Research* 55, 193–206.

Kushad, M.M., Cloyd, R. and Babadoost, M. (2004) Distribution of glucosinolates in ornamental cabbage and kale cultivars. *Scientia Horticulturae* 101, 215–221.

Laber, H. and Stützel, H. (2000) Nebenwirkungen mechanischer Unkrautregulationsmašnahmen im Gemüsebau. *Zeitscrift für Pflanzenkrankheiten und Pflanzenschutz* 17, 653–660.

Laibach, F. (1943) *Arabidopsis thaliana* (L.) Heynh. Als Objekt fürgenetische und entwicklungs-physiologische Untersuchungen. *Botanisches Archiv* 44, 439–455.

Lamb, R.J. (1989) Entomology of oilseed *Brassica* crops. *Annual Review of Entomology* 34, 211–229.

Lan, T. and Paterson, A. (2000) Comparative mapping of quantitative trait loci sculpting the curd of *Brassica oleracea*. *Genetics* 155, 1927–1954.

Lan, T., Delmonte, T.A., Reischmann, K.P., Hyman, J., Kowalski, S.P., McFerson, J., Kresovich, S. and Paterson, A.H. (2000) An EST-enriched comparative map of *Brassica oleracea* and *Arabidopsis thaliana*. *Genome Research* 10, 776–788.

Larkcom, J. (1991) *Oriental Vegetables. The Complete Guide for Garden and Kitchen*. John Murray (Publishers) Ltd, London.

Lawson, H.M. (1972) Weed competition in transplanted spring cabbage. *Weed Research* 12, 254–267.

Lee, I.-M., Gundersen-Rindal, D.E. and Bertaccinin, A. (1998) Phytoplasma: ecology and genomic diversity. *Phytopathology* 88, 1359–1366.

Lee, P.C., Taylor, A.G. and Paine, D.H. (1997) Sinapine leakage for detection of seed quality in *Brassica*. In: Ellis, R.H., Black, M., Murdoch, A.J. and Hond, T.D. (eds) *Basic and Applied Aspects of Seed Biology*. Kluwer Academic Publications, Boston, Massachusetts, pp. 537–546.

Leitch, I.J. and Bennett, M.D. (2003) Integrating genomic characters for a holistic approach to understanding plant genomes. *Biology International* 45, 18–29.

Leite, G.L.D., Picanco, M., Bastos, C.S., de Araujo, J.M. and Azevedo, A.A. (1996) Resistance of kale clones to the green peach aphid. *Horticultura Brasileira* 14, 178–181.

Leong, S.A., Allen, C. and Triplett, E.W. (eds) (2002) *Biology of Plant–Microbe Interactions*, Vol. 3. Proceedings of the 10th International Plant-Microbe Interactions Congress, Madison, Wisconsin, 10–14 July 2001. The International Society for Plant–Microbe Interactions, St Paul, Minnesota.

Lester, G.F. (2006) Environmental regulation of human health nutrients (ascorbic acid, B-carotene and folic acid) in fruits and vegetables. *Hort Science* 41, 59–64.

Li, B., Suzucki, J.-I. and Hara, T. (1999) Competitive ability of two *Brassica* varieties in relation to biomass allocation and morphological plasticity under varying nutrient availability. *Ecological Research* 14, 255–266.

Lin, J., Eckenrode, C.J. and Dickson, M.H. (1983) Variation in *Brassica oleracea* resistance to diamonback moth (Lepidoptera: Plutelidae). *Journal of Economic Entomology* 76, 1423–1427.

Linnaeus, C. (1735) *Systema Naturae*. Leyden, The Netherlands.

Liou, T.D. (1987) Studies on Germination and Vigour of Cabbage Seeds. PhD Dissertation, The Agricultural University, Wageningen, The Netherlands.

Lizgunova, T.V. (1959) The history of botanical studies of the cabbage *Brassica oleracea*. *Bulletin of Applied Botany, Genetics and Plant Breeding* 32, 37–70 (Russian with English summary).

Locascio, S.J. (2005) Management of irrigation for vegetables: past, present and future. *HorTechnology* 15, 482–485.

Lole, M. (2005) *Swede Midge Control in Brassica Crops.* Factsheet no: 28/05, Horticultural Development Council, East Malling, UK.

Lolle, S.J., Victor, J.L., Young, J.M. and Pruitt, R.E. (2005) Genome-wide non-Mendelian inheritance of extra-genomic information in *Arabidopsis. Nature* 434, 505–508.

Loughton, A. and Riekels, J.W. (1988) Black speck in cauliflower. *Canadian Journal of Plant Science* 68, 291–294.

Lui, X.P., Lu, W.C., Liu, Y.K., Wei, S.Q., Xu, J.B., Liu, Z.R., Zhang, H.J., Li, J.L., Ke, G.L., Yao, W.Y., Cai, Y.S., Wu, F.Y., Cao, S.C., Li, Y.H., Xie, S.D., Lin, B.X. and Zhang, C.L. (1996) Occurrence and strain differentiation on turnip mosaic potyvirus and sources of resistance in Chinese cabbage in China. *Acta Horticulturae* 407, 431–440.

Magnusson, M. (2002) Mineral fertilisers and green mulch in Chinese cabbage (*Brassica pekinsensis* (Lour.) Rupr.): effect on nutrient uptake, yield and internal tipburn. *Acta Agriculturae Scandinavica, Section B, Soil and Plant Science* 52, 25–35.

Makhlouf, J., Castaigne, F., Arul, J., Willemot, C. and Gosselin, A. (1989) Long-term storage of broccoli under controlled atmosphere. *HortScience* 24, 637–639.

Malvas, C.C., Coelho, R.M.S. and Camargo, L.E.A. (1999) Identification of resistance genes in *Brassica oleracea* to black rot by selective genotyping. *Fitopatologia Brasileira* 24, 143–148.

Mangan, F., DeGregorio, R., Schonbeck, M., Herbert, S., Guillard, K., Hazzard, R., Sideman, E. and Litchfield, G. (1995) Cover cropping systems for brassicas in the northeastern USA. *Journal of Sustainable Agriculture* 5, 15–36.

Marschner, H. (1995) *Mineral Nutrition in Higher Plants.* Academic Press, London.

Marshall, B. and Thompson, R. (1987a) A model of the influence of air temperature and solar radiation on the time to maturity of calabrese (*Brassica oleracea* var. *italica*). *Annals of Botany* 60, 513–519.

Marshall, B. and Thompson, R. (1987b) Application of a model to predict the time to maturity of calabrese (*Brassica oleracea* var. *italica*). *Annals of Botany* 60, 521–529.

Massie, I.H., Astley, D. and King, G.J. (1996) Patterns of genetic diversity and relationships between regional groups and populations of italian landrace cauliflower and broccoli (*Brassica oleracea* L. var. *botrytis* L. and var. *italica* Plenck). *Acta Horticulturae* 407, 45–54.

Matsumura, T. (1954) On the silique types of $n = 10$ group Brassicas of Japan. *Japanese Journal of Breeding* 4, 179–182.

Matthews, S. and Powell, A.A. (1987) Controlled deterioration test. In: Fiala, F. (ed.) *The Handbook of Vigour Test Methods.* International Seed Testing Association (ISTA), Zürich, Switzerland, pp. 49–56.

Maxwell, K. and Johnson, G.N. (2000) Chlorophyll fluorescence – a practical guide. *Journal of Experimental Botany* 51, 659–668.

Maynard, D.N., Warner, D.C. and Howell, J.C. (1981) Cauliflower leaf tipburn: a calcium deficiency disorder. *HortScience* 16, 193–195.

Mazza, G. (2004) Diet and human health: functional foods to reduce disease risks. *Acta Horticulturae* 642, 161–172.

McCall, D., Sorensen, L. and Jensen, B.D. (1996) *Broccoli Varieties.* S.P. Rapport-Status Planteaulsforsog No. 8.

McDonald, M.B. (1975) A review and evaluation of seed vigour tests. *Proceedings of the Association of Official Seed Analysts* 65, 109–139.

McGrady, J. (1996) Transplant nutrient conditioning improves cauliflower early growth. *Journal of Vegetable Crop Production* 2, 39–49.

McIlrath, W.J., Abrol, Y.P. and Heiligman, F. (1963) Dehydration of seeds in intact tomato fruits. *Science* 142, 1681.

Mehra, V., Tripathj, J. and Powell, A.A. (2003) Aerated hydration treatment improves the response of *Brassica juncea* and *Brassica campestris* seeds to stress during germination. *Seed Science and Technology* 31, 57–70.

Menniti, A.M., Maccaferri, M. and Folchi, A. (1997) Physio-pathological responses of cabbage stored under controlled atmospheres. *Postharvest Biology and Technology* 10, 207–212.

Metz, T.D., Roush, R.T., Tang, J.D., Shelton, A.M. and Earle, E.D. (1995) Transgenic broccoli expressing *Bacillus thuringuriensis* insecticidal crystal proteins: implications for pest resistance management. *Molecular Breeding* 4, 309–317.

Metzger, J. (1833) *Systematische Beschreibung der Kultivirten Kohlarten*. Heidelberg, 2, 1–68.

Meyerowitz, E.M. (1997) Plants and the logic of development. *Genetics* 145, 5–9.

Meyerowitz, E.M. and Somerville, C.R. (1994) *Arabidopsis*. Cold Spring Harbor Laboratory Press, Cold Spring Harbor, New York.

Milford, G.F.J., Pocock, T.O. and Riley, J. (1985) An analysis of leaf growth in sugar beet. 1. Leaf appearance and expansion in relation to temperature under controlled conditions. *Annals of Applied Biology* 106, 163–172.

Miller, A.B. and Hopen, H.J. (1991) Critical weed-control period in seeded cabbage (*Brassica oleracea* var. *capitata*). *Weed Technology* 5, 852–857.

Minks, A.K. and Harrewijn, P. (eds) (1987) *Aphids: Their Biology, Natural Enemies and Control*, Vols 2A, 2B and 2C. In: Helle, W. (editor-in-chief), *World Crop Pests*. Elsevier, Amsterdam.

Mitchell, P.A., Olds, T., James, R.V., Palmer, M.J. and Williams, P.H. (1995) Genetics of *Brassica rapa* (syn. *campestris*). 2. Multiple disease resistance to three fungal pathogens – *Peronospora parasitica*, *Albugo candida* and *Leptosphaeria maculans*. *Heredity* 75, 362–369.

Mithen, R., Faulkner, K., Magrath, R., Rose, P., Williamson, G. and Marquez, J. (2003) Development of *iso*-thiocyanate-enriched broccoli and its enhanced ability to induce phase 2 detoxification enzymes in mammalian cells. *Theoretical and Applied Genetics* 106, 727–734.

Mizushima, U. (1980) Genome analysis in *Brassica* and allied genera. In: Tsunoda, S., Hinata, K. and Gomez-Campo, C. (eds) *Brassica Crops and Wild Allies*. Japan Scientific Society Press, Tokyo, pp. 89–106.

Mizushima, U. and Tsunoda, S. (1967) A plant exploration in *Brassica* and allied genera. *Tohoku Journal of Agricultural Research* 17, 249–276.

Mohler, C.L. (1993) A model of the effects of tillage on emergence of seedlings. *Ecological Applications* 3, 53–73.

Molisch, H. (1922) *Der Einfluss einer Pflanze auf die andere-allelopathie*. Fischer (Jena), Jena, Germany.

Monreal, M.A., Derksen, D.A., Watson, P.R. and Monreal, C.M. (2000) Effect of crop management practices on soil microbial communities. In: *Proceedings of the Annual Manitoba Society of Soil Science Meeting*, 43rd Winnipeg, Manitoba, Canada, 25–26 January 2000, pp. 216–228.

Monteiro, A.A. and Williams, P.H. (1989) The exploration of genetic resources of

Portuguese cabbage and kale for resistance to several *Brassica* diseases. *Euphytica* 41, 215–225.

Monteith, J.L. (1981) Climatic variation and the growth of crops. *Quarterly Journal of the Royal Geographical Society* 107, 749–774.

Montgomery, J.A., Bressan, R.A. and Mitchell, C.A. (2004) Optimizing environmental conditions for mass aplication of mechano-dwarfing stimuli to *Arabidopsis*. *Journal of the American Society for Horticultural Science* 129, 339–343.

Moore, R.D. and Seward, D.L. (1986) No-tillage production of broccoli and cabbage. *Applied Agricultural Research* 1, 96–99.

Morra, A.A. and Earle, E.D. (2001) Resistance to *Alternaria brassicicola* in transgenic broccoli expressing a *Trichoderma harzianum* endochitinase gene. *Molecular Breeding* 8, 1–9.

Moss, S.R. (1985) The survival of *Alopecurus myosuroides* Huds. seeds in soil. *Weed Research* 25, 201–211.

Mound, L.A. and Halsey, S.H. (1978) *Whiteflies of the World. A Systematic Catalogue of the Aleyrodidae (Homoptera) with Host Plant and Natural Enemy Data.* British Museum (Natural History) and John Wiley, Chichester, UK.

Myers, J. (1985) Effect of physiological condition of the host plant on the ovipositional choice of cabbage white butterfly (*Pieris rapae*). *Journal of Animal Ecology* 54, 193–204.

Mylavarapu, R.S., Smith, J.P. and Munoz, F. (2005) Influence of soil and nutrient management on growth and quality of collards. *HorTechnology* 15, 163–168.

Narain, A. (1974) Rape and mustard. In: Hutchinson, J. (ed.) *Evolutionary Studies in World Crops*. Cambridge University Press, Cambridge, pp. 67–70.

Nestle, M. (1998) Broccoli sprouts in cancer prevention. *Nutrition Review* 56, 127–130.

Neutrofal, F. (1927) Zytologische Studien über die Kulturrassen von *Brassica oleracea*. *Österreichische Botanische Zeitschrift* 76, 105–115 (101–113).

Neuvel, J.J. (1990) *Nitrogen Fertilisation of Brussels Sprouts* (in Dutch). Report 102, Research Station for Arable Farming and Field Production of Vegetables, Lelystad, The Netherlands.

Nicholson, A.G. and Wien, H.C. (1983) Screening turfgrass and clovers for use as living mulch cropping system. MS Thesis, Cornell University, Ithaca, New York.

Nicolas, O. and Laliberté, J.-F. (1992) The complete nucleotide sequence of turnip mosaic virus RNA. *Journal of General Virology* 73, 2785–2793.

Nieto, J.H., Brondo, M.A. and Gonzales, J.T. (1968) Critical periods of the crop growth period for competition from weeds. *PANS (Pest Articles and News Summaries) section C* 14, 159–166.

Nieuwhof, M. (1969) *Cole Crops: Botany, Cultivation and Utilization.* Leonard Hill, London.

Nilsson, T. (1980) The influence of soil type and irrigation on yield, quality and chemical composition of cauliflower. *Swedish Journal of Agricultural Research* 10, 65–75.

Norton, G. and Harris, J.F. (1975) Compositional changes in developing rape seed (*Brassica napus* L.). *Planta* 123, 162–174.

Nowbuth, R.D. and Pearson, S. (1998) The effect of temperature and shade on curd initiation in temperate and tropical cauliflower. *Acta Horticulturae* 459, 79–88.

O'Dell, C.R., Morse, R. and Ramsey, P. (1993) *Field Seeded Fall Bunching Broccoli Production; a Guide for Virginia Farmers.* Virginia Cooperative Extension Service Publication No. 438–011.

O'Donnell, M.S. and Coaker, T.H. (1975) Potential of intra-crop diversity for the control of brassica pests. *Proceedings of the 8th British Insecticide and Fungicide Conference* 1, 101–107.

Ogura, H. (1968) Studies on the new male sterility in Japanese radish, with special reference to the utilization of this sterility towards practical raising of hybrid seeds and subsequently transferred to *B. oleracea* var *botrytis* and *B. napus. Memoirs of the Faculty of Agriculture, Kagoshima University* 6, 39–78.

Ohsawa, K. and Ohsawa, D. (2001) Efficacy of plant extracts for reducing larval populations of the diamondback moth, *Plutella xylostella* L. (Lepidoptera: Yponomeutidae) and cabbage webworm, *Crocidolomia binotalis* Zeller (Lepidoptera: Pyralidae) and evaluation of cabbage damage. *Applied Entomology and Zoology* 36, 143–149.

Okuda, N. and Fujime, Y. (1996) Plant growth characters of Chinese kale (*Brassica oleracea* L. var. *alboglabra*). *Acta Horticulturae* 407, 55–60.

Opeña, R.T., Kuo, G.C. and Yoon, J.Y. (1988) *Breeding and Seed Production of Chinese Cabbage in the Tropics and Sub-tropics.* Bulletin 17. Asian Vegetable Research and Development Centre (AVRDC), Shanhua, Taiwan.

Osborne, D.J. (1983) Biochemical control systems operating in the early hours of germination. *Canadian Journal of Botany* 61, 3568–3577.

Palaniswamy, P. and Lamb, R.J. (1992) Host preferences of flea beetles *Phyllotreta cruciferae* and *P. striolata* (Coleoptera: Chrysomelidae) for crucifer seedlings. *Journal of Economic Entomology* 85, 743–752.

Panetsos, C. and Baker, H.G. (1986) The origin of variation in 'wild' *Raphanus sativus* (Cuciferae) in California. *Genetica* 34, 243–274.

Parish, R.L. (2000) Stand of cabbage and broccoli in single- and double-drill plantings on beds subject to erosion. *Journal of Vegetable Crop Production* 6, 87–96.

Pearson, O.H. (1972) Cytoplasmically inherited male sterile characters and flavor components from the species *Brassica nigra* (L) Koch × *B. oleracea* L. *Journal of the American Society for Horticultural Science and Biotechnology* 97, 397–402.

Pearson, S. and Hadley, P. (1988) Planning calabrese production. *The Grower* 100, 21–22.

Pearson, S., Hadley, P. and Wheldon, A.E. (1994) A model of the effects of temperature on the growth and development of cauliflower (*Brassica oleracea* var. *bortytis* L.). *Scientia Horticulturae* 59, 91–106.

Pellan-Delourne, R. and Renard, R. (1987) Cytoplasmic male sterility in rapeseed (*Brassica napus* L.): female fertility of restored rapeseed with 'Ogura' and cybrid cytoplasm. *Genome* 30, 234–238.

Peltier, G., Primard, C., Vedel, F., Chetrit, P., Remy, R., Rouselle, P. and Renard, M. (1983) Intergeneric cytoplasmic hybridization in *Cruciferae* by protoplast fusion. *Molecular and General Genetics* 191, 244–250.

Perry, D.A. (1978) Report of the vigour test committee. 1974–1977. *Seed Science and Technology* 6, 159–181.

Pichard, B. and Thouvenot, D. (1999) Effect of *Bacillus polymyxa* seed treatments on control of black-rot and damping-off of cauliflower. *Seed Science and Technology* 27, 455–465.

Porter, K., Collins, G. and Klieber, A. (2004) Effect of post harvest mechanical stress on quality and storage life of Chinese cabbage cv. Yuki. *Australian Journal of Experimental Agriculture* 44, 629–633.

Powell, A.A. and Matthews, S. (1984) Application of the controlled deterioration vigour test to detect seed lots of Brussels sprouts with low potential for storage under commercial conditions. *Seed Science Technology* 12, 649–657.

Prakash, S. and Hinata, K. (1980) Taxonomy, cytogenetics and origin of crop *Brassica*, a review. *Opera Botanica* 55, 1–57.

Pratt, P.F. (1984) Nitrogen use and nitrate leaching in irrigated agriculture. In: Hauck, R.D. (ed.) *Nitrogen in Crop Production*. American Society of Agronomists, Madison, Wisconsin, pp. 319–334.

Provvidente, R. (1980) Evaluation of Chinese cabbage cultivars from Japan and the People's Republic of China for resistance to turnip mosaic and cauliflower mosaic vuirus. *HortScience* 105, 571–573.

Provvidente, R. (1981) Sources of resistance to turnip mosaic virus in Chinese cabbage. In: Talekar, N.S. and Griggs, T.D. (eds) *Chinese Cabbage*. The Asain Vegetable Research and Development Centre, Taiwan, pp. 423–430.

Pruitt, R.E., Chang, C., Pang, P.P.-Y. and Meyerowitz, E.M. (1987) Molecular genetics and the development of *Arabidopsis*. In: Loomis, W. (ed.) *Genetic Regulation of Development. 45th Symposium of the Society for Developmental Biology*. Wiley Liss, New York, pp. 327–338.

Qasem, J.R. and Hill, T.A. (1993) A comparison of the competitive effects and nutrient accumulation of fat-hen and groundsel. *Journal of Plant Nutrition* 16, 679–698.

Quiros, C.F. (1999) Genome structure and mapping. In: Gomez-Campo, C. (ed.) *Biology of Brassica Coenospecies*. Elsevier, Amsterdam, The Netherlands, pp. 217–245.

Radovich, T.J.K. and Kleinhenz, M.D. (2004) Rapid estimation of cabbage head volume across a population varying in head shape: a test of two geometric formulae. *HorTechnology* 14, 388–391.

Ramirez-Villapudua, J. and Munnecke, D.E. (1987) Control of cabbage yellows (*Fusarium oxysporum* f.sp. *conglutinans*) by solar heating of field soils amended by dry cabbage residues. *Plant Disease* 71, 217–221.

Rather, K., Schenk, M.K., Everaarts, A.P. and Vethman, S. (1999) Response of yield and quality of cauliflower varieties (*Brassica oleracea* var. *botrytis*) to nitrogen supply. *Journal of Horticultural Science and Biotechnology* 74, 658–664.

Reddy, G.V.P. and Guerrero, A. (2001) Optimum timing of insecticide applications against diamondback moth *Plutella xylostella* in cole crops using threshold catches in sex pheromone traps. *Pest Management Science* 57, 90–94.

Reeves, J., Fellows, J.R., Phelps, K. and Wurr, D.C.E. (2001) Development and validation of a model describing the curd induction of winter cauliflower. *Journal of Horticultural Science and Biotechnology* 76, 714–720.

Ren, J.P., Dickson, M.H. and Earle, E.D. (2000) Improved resistance to bacterial soft rot by protoplast fusion between *B. rapa* and *B. oleracea*. *Theoretical and Applied Genetics* 100, 810–819.

Ren, J.P., Petzoldt, R. and Dickson, M.H. (2001) Genetics and population improvement of resistance to bacterial soft rot in Chinese cabbage. *Euphytica* 117, 197–207.

Rice, E.L. (1984) *Allelopathy*. Academic Press, New York.

Ridgway, R.L., Silverstein, R.M. and Inscoe, M.N. (1990) *Behaviour-modifying Chemicals for Insect Management*. Marcel Dekker, New York.

Riggins, B.T.M. and Gould, F. (1997) Impact of intraplot mixtures of toxic and non-toxic plants on population dynamics of diamondback moth and its natural enemies. *Journal of Economic Entomology* 90, 241–251.

Riley, H., Løes, A.-K., Hansen, S. and Dragland, S. (2003) Yield responses and nitrogen utilisation with the use of chopped grass and clover material as surface mulches in an organic growing system. *Biological Agriculture and Horticulture* 21, 63–90.

Risch, S.J. (1981) Insect herbivore abundance in tropical monocultures and polycultures: an experimental test of two hypotheses. *Ecology* 62, 1325–1340.

Roberts, B.W. and Cartwright, B. (1991) Alternative soil and pest management practices for sustainable production of fresh-market cabbage. *Journal of Sustainable Agriculture* 1, 21–35.

Roberts, E.H. and Summerfield, R.J. (1987) Measurement and prediction of flowering in annual crops. In: Atherton, J.G. (ed.) *Proceedings of Sutton Bonnington Easter School in Agricultural Science – Manipulation of Flowering*. Butterworths, London, pp. 17–50.

Roberts, H.A. and Bond, W. (1975) Combined treatments of propachlor and trifluralin for weed control in cabbage. *Weed Research* 15, 195–198.

Roberts, H.A., Bond, W. and Hewson, R.T. (1976) Weed competition in drilled summer cabbage. *Annals of Applied Biology* 84, 91–95.

Robinson, D. (1991) Strategies for optimising growth in response to nutrient supply. In: Porter, J.R. and Lawlor, D.W. (eds) *Plant Growth: Interactions with Nutrition and Environment. Vol. 43, Society for Experimental Biology Seminar Series*. Cambridge University Press, Cambridge, pp. 177–205.

Robinson, R.A. (1969) Disease resistance terminology. *Review of Applied Mycology* 48, 593–605.

Röhrig, M. and Stützel, H. (2001) A model for light competition between vegetable crops and weeds. *European Journal of Agronomy* 14, 13–29.

Root, R.B. (1973) Organisation of a plant–arthropod association in simple and diverse habitats: the fauna of collards (*Brassica oleracea*). *Ecological Monographs* 43, 95–124.

Root, T.L., Price, J.T., Hall, K.R., Schneider, S.H., Rosenzweig, C. and Pounds, J.A. (2003) Fingerprints of global warming on wild animals and plants. *Nature* 421, 57–60.

Rubatsky, V.E. and Yamaguchi, M. (1997) *World Vegetables, Principles, Production and Nutritive Values*. Chapman and Hall, New York.

Ryall, A.L. and Lipton, W.J. (1972) *Handling, Transportation and Storage of Fruits and Vegetables; Volume 1, Vegetables and Melons*. AVI Publishing, Westport, Connecticut.

Ryan, J., Ryan, M.F. and McNaeidhe, F. (1980) The effect of interrow plant cover on populations of the cabbage root fly, *Delia brassicae* (Weidemann). *Journal of Applied Ecology* 17, 31–40.

Ryschka, U., Schumann, G., Klock, E., Scholze, P. and Neumann, M. (1996) Somatic hybridisation in Brassiceae. *Acta Horticulturae* 407, 201–208.

Sako, N. (1981) Virus diseases of Chinese cabbage in Japan. In: Talekar, N.S. and Griggs, T.D. (eds) *Chinese Cabbage*. The Asian Vegetable Research and Development Centre, Taiwan, pp. 129–142.

Salisbury, E. (1961) *Weeds and Aliens*. New Naturalist Series, Collins, London.

Salo, T. (1999) Effect of band placement and nitrogen rate on dry matter accumulation, yield and nitrogen uptake of cabbage, carrot and onion. *Agricultural and Food Sciences in Finland* 8, 159–231.

Salter, P.J. (1961) The irrigation of early summer cauliflower in relation to stage of growth, plant spacing and nitrogen level. *Journal of Horticultural Science and Biotechnology* 36, 242–253.

Salter, P.J. (1969) Studies on crop maturity in cauliflower. I. Relationship between times of curd initiation and curd maturity of plants within a cauliflower crop. *Journal of Horticultural Science and Biotechnology* 44, 129–140.

Salter, P.J. and Fradgley, J.R. (1969) Studies on crop maturity in cauliflower, II. Effects of cultural factors on the maturity characteristics of a cauliflower crop. *Journal of Horticultural Science and Biotechnology* 44, 141–154.

Salter, P.J. and James, J.M. (1975) The effect of plant density on the initiation, growth and maturity of curds of two cauliflower varieties. *Journal of Horticultural Science and Biotechnology* 50, 239–248.

Salter, P.J., Andrews, D.J. and Akehurst, J.M. (1984) The effects of plant density, spatial arrangement and sowing date on yield and head characteristics of a new form of broccoli. *Journal of Horticultural Science and Biotechnology* 59, 79–85.

Sanderfoot, A.A. and Raikel, N.V. (2001) *Arabidopsis* could shed light on human genome. *Nature* 410, 299.

Sangers, W.J. (1952) *De ontwikkeling van de Nederlandse tuinbouw (tot het jaar 1930).* Zwolle.

Sangers, W.J. (1953) *Gegevens betreffende de ontwikkeling van de Neherlandse tuinbouw (tot het jaar 1800).* Zwolle.

Santos, M.P., Dias, J.S. and Monteiro, A.A. (1996) Screening Portuguese cole landraces for white rust (*Albugo candida* (Pers.) Kuntze). *Acta Horticulturae* 407, 453–459.

Santos, R.H.S., Gliessmann, S.R. and Cecon, P.R. (2002) Crop interactions in broccoli intercropping. *Biological Agriculture and Horticulture* 20, 51–75.

Sato, F., Yoshioka, H., Fujiwara, T., Higashio, H., Uragami, A. and Tokuda, S. (2004) Physiological responses of cabbage plug seedlings to water stress during low-temperature storage in darkness. *Scientia Horticulturae* 101, 349–357.

Scaife, A. (1988) Derivation of critical nutrient concentrations for growth rate from field experiments. *Plant and Soil* 109, 159–169.

Scaife, A., Cox, E.F. and Morris, G.E.L. (1987) The relationship between shoot weight, plant density and time during propagation of four vegetable species. *Annals of Botany* 59, 325–334.

Schellhorn, N.A. and Sork, V.L. (1997) The impact of weed diversity on insect population dynamics and crop yield in collards, *Brassica oleracea* (Brassicaceae). *Oecologia* 111, 233–240.

Schiemann, E. (1932) Entstehung der Kulturpflanzen. *Handbuch Vererbwissenschaften* 3, 1–377.

Schonbeck, M., Herbert, S., DeGregorio, R., Mangan, F., Guillard, K., Sideman, E., Herbst, J. and Jaye, R. (1993) Cover cropping systems for Brassicas in the Northeastern USA. 1. Cover crop and vegetable yields, nutrients and soil conditions. *Journal of Sustainable Agriculture* 3, 105–132.

Schreiner, M., Krumbein, A., Schonhof, I., Widell, S. and Huyskens-Keil, S. (2003) Quality determination of red radish by nondestructive root color measurement. *Journal of the American Society for Horticultural Science* 128, 397–402.

Schultz, O.E. (1919, 1936) Cruciferae. In: Engler, A. and Harms, H. (eds) *Die Natürlichen Pflanzenfamilien*. Bd. 17b, Leipzig, pp. 227–658.

Sharma, S. and Sharma, G.R. (2001) Influence of white rot (*Sclerotinia sclerotiorum*) on growth and yield parameters of Indian mustard (*Brassica juncea*) varieties. *Indian Journal of Agricultural Science* 71, 273–274.

Sharma, S.R., Kapoor, K.S. and Gill, H.S. (1995) Screening against *Sclerotinia* rot in cauliflower. *Indian Journal of Agricultural Sciences* 65, 916–918.

Shattuck, V.I. (1992) The biology, epidemiology and control of turnip mosaic virus. *Horticultural Reviews* 14, 199–238.

Shattuck, V.I. and Stobbs, L.W. (1987) Evaluation of rutabaga cultivars for turnip mosaic virus resistance and the inheritance of resistance. *HortScience* 22, 935–937.

Shaw, R.H. and Loomis, W.E. (1950) Bases for the prediction of corn yields. *Plant Physiology* 25, 225–244.

Shebalina, M.A. (1968) The history of the botanical investigation and classification of turnip. *Bulletin of Applied Botany, Genetics and Plant Breeding* 38, 44–87.

Shelton, A.M. (1995) Temporal and spatial dynamics of thrips populations in a diverse ecosystem: theory and management. *International Conference on Thysanoptera (1993)*, pp. 425–432.

Shelton, A.M., Tang, J.D., Rousch, R.J., Metz, T.D. and Earle, E.D. (2000) Field tests on management of resistance to BT engineered plants. *Nature Biotechnology* 18, 339–342.

Shibutani, S. and Okamura, T. (1954) On the classification of turnips in Japan with regard to the epidermal layer of the seed. *Journal of the Japanese Society for Horticultural Science* 22, 235–238.

Shigaki, T., Nelson, S.C. and Alvarez, A.M. (2000) Symptomless spread of blight inducing strains of *Xanthomonas campestris* pv *campestris* on cabbage seedlings in misted beds. *European Journal of Plant Pathology* 106, 339–346.

Shipway, M.R. (1978) *Winter White Cabbage: Evaluation of 'Pepper-spot' Resistant Varieties*. Kirton Experimental Horticulture Station Annual Report for 1978, pp. 62–63.

Siemonsa, J.S. and Piluek, K. (eds) (1993) *Plant Resources of South-east Asia No 8 Vegetables*. Pudoc Scientific Publishers, Wageningen, The Netherlands.

Sigareva, M. and Earle, E.D. (1997a) Intertribal somatic hybrids between *Camalina sativa* and rapid cycling *Brassica oleracea*. *Cruciferae Newsletter* 19, 49–50.

Sigareva, M. and Earle, E.D. (1997b) *Capsella bursa pastoris*: regeneration of plants from protplasts and somatic hybridization with rapid cycling *Brassica oleracea*. *Cruciferae Newsletter* 19, 57–58.

Singh, C.B., Asthana, A.N. and Mehra, K.L. (1974) Evolution of *Brassica juncea* under domestication and natural selection in India. *Genetica Agraria* 28, 111–135.

Singh, D., Naveen, C. and Gupta, P.P. (1997) Inheritance of powdery mildew resistance in interspecific crosses of Indian and Ethiopean mustard. *Annals of Biology* 13, 73–77.

Singh, R.V. and Naik, L.B. (1991) Response of cauliflower (cv. Early Kunwari) to plant density, nitrogen and phosphorus level. *Progressive Horticulture, India* 23, 51–54.

Sinskaia, E.N. (1928) The oleiferous plants and root crops of the family Cruciferae. *Bulletin of Applied Botany, Genetics and Plant Breeding* 19, 555–626.

Siomos, A.S. (1999) Planting date and within-row plant spacing effects on Pak Choi yield and quality characteristics. *Journal of Vegetable Crop Production* 4, 65–73.

Smith, J.G. (1976a) Influence of crop background on aphids and other phytophagous insects on Brussels sprouts. *Annals of Applied Biology* 83, 1–13.

Smith, J.G. (1976b) Influence of crop background on natural enemies of aphids on Brussels sprout. *Annals of Applied Biology* 83, 15–29.

Smittle, D.A. (1994) Irrigation regimes affect cabbage water use and yield. *Journal of the American Society for Horticultural Science* 119, 20–23.

Somerville, C. (1989) *Arabidopsis* blooms. *Plant Cell* 1, 1131–1135.

Song, K.M., Osbourn, T.C. and Williams, P.H. (1988) *Brassica* taxonomy based on nuclear restriction fragment length polymorphisms. 1. Genome evolution of diploid and amphidiploid species. *Theoretical and Applied Genetics* 75, 784–794.

Song, K., Tang, K. and Osbourn, T.C. (1993) Development of synthetic *Brassica* amphidiploids by reciprocal hybridisation and comparison to natural amphidiploids. *Theoretical and Applied Genetics* 86, 811–821.

Song, K., Tang, K., Osbourn, T.C. and Lu, P. (1996) Genome variation and evolution of *Brassica* amphidiploids. *Acta Horticulturae* 407, 35–44.

Sorenson, J.N. (2000) Ontogenetic changes in macro nutrient composition of leaf-vegetable crops in relation to plant nitrogen status: a review. *Journal of Vegetable Crop Production* 6, 75–96.

Soursa, M.S., Dias, J.S. and Monteiro, A.A. (1997) Screening Portuguese cole landraces for resistance to seven indigenous downy mildew isolates. *Scientia Horticulturae* 68, 49–58.

Spitters, C.J.T. (1990) Weeds: population dynamics, germination and competition. In: Rabbinge, R., Ward, S.A. and van Laar, H.H. (eds) *Theoretical Production Ecology: Reflections and Prospects*. Pudoc, Wageningen, The Netherlands, pp. 182–216.

Sreeramulu, N., Tesha, A.J. and Kapuya, J.A. (1992) Some biochemical changes in developing seeds of bambarra groundnut (*Voandzeia subterranea* Thouars). *Indian Journal of Plant Physiology* 35, 191–194.

Stace, C. (2001) *New Flora of the British Isles*. Cambridge University Press, Cambridge.

Sterrett, S.B., Haynes, K.C. and Savage, C.P. (2004) Cluster analysis on quality attributes identify broccoli cultivars suitable for early and main-season harvest on the Eastern shore of Virginia. *HortTechnology* 14, 376–380.

Still, D.W. and Bradford, K.J. (1994) Development of seed quality in red cabbage. *HortScience* 29, 552 (abstract)

Still, D.W. and Bradford, K.J. (1998) Using hydrotime and ABA-time models to quantify seed quality of Brassicas during development. *Journal of the American Society for Horticultural Science* 123, 692–699.

Stirling, K. and Lancaster, R. (2005) Alternative planting configurations influence cauliflower development. *Acta Horticulturae* 694, 301–305.

Stivers-Young, L. (1998) Growth, nitrogen accumulation and weed suppression by fall cover crops following early harvest of vegetables. *HortScience* 33, 60–63.

Stokes, P. and Verkerk, K. (1951) Flower formation in Brussels sprouts. *Mededelingen van de Landbouw Hogeschool te Wageningen* 50, 143–160.

Stone, D.A. (1998) The effects of 'starter' fertilizer injection on the growth and yield of drilled vegetable crops in relation to soil nutrient status. *Journal of Horticultural Science and Biotechnology* 73, 441–451.

Stoner, K.A., Dickson, M.H. and Shelton, A.M. (1989) Inheritance of resistance to damage by *Thrips tabaci* Lindemann (Thysanoptera: Thripidae) in cabbage. *Euphytica* 40, 233–239.

Suojala, T. (2003) Compositional and quality changes in white cabbage during harvest period and storage. *Journal of Horticultural Science and Biotechnology* 78, 821–827.

Sureshgouda, R.S. and Kalidhar, S.B. (2005) Effects of karanj (*Pongamia pinnata* Vent.) methanolic seed extracts and fractions on growth and development of *Plutella xylostella* L., (Lep., Yponomeutidae). *Biological Agriculture and Horticulture* 23, 1–14.

Sutherland, R.A., Crisp, P. and Angell, S.M. (1989) The effect of spatial arrangement on the yield and quality of two cultivars of autumn cauliflower and their mixture. *Journal of Horticultural Science and Biotechnology* 64, 35–40.

Swarup, V. and Chaterjee, S.S. (1972) Origin and genetic improvement of Indian cauliflower. *Economic Botany* 26, 381–393.

Tahvanainen, J.O. and Root, R.B. (1972) The influence of vegetational diversity on the population ecology of a specialised herbivore, *Phyllotreta cruciferae* (Coleoptera: Chrysomelidae). *Oecologia* 10, 321–346.

Talekar, N.S. and Griggs, T.D. (eds) (1981) *Chinese Cabbage*. Asian Vegetable Research and Development Centre, Shanhau, Taiwan.

Talekar, N.S. (1986) *Diamond Back Moth Management*. Proceedings of the First International Workshop, Tainan, Taiwan, 11–15 March 1985. Asian Vegetable Research and Development Center (1986), Shanhua, Tainan, Taiwan.

Talekar, N.S. and Shelton, A.M. (1993) Biology, ecology and management of diamondback moth. *Annual Review of Entomology* 38, 275–301.

Talekar, N.S., Yong, H.C., Lee, S.T., Chen, B.S. and Sun, L.Y. (1985) *Annotated Bibliography of Diamond Back Moth*. Asian Vegetable Research and Development Center, Publication, Shanhua, Tainan, Taiwan, pp. 85–229.

Taylor, A.G., Allen, P.S., Bennett, M.A., Bradford, K.J., Burris, J.S. and Misra, M.K. (1998) Seed enhancements. *Seed Science Research* 8, 245–256.

Tesnier, K., Strookman-Donkers, H.M., van Pulen, J.G., van der Geest, A.H.M., Bino, R.J. and Groot, S.P.C. (2002) A controlled deterioration test for *Arabidopsis thaliana* reveals genetic variation in seed quality. *Seed Science and Technology* 30, 149–165.

Thaning, C. and Gerhardson, B. (2001) Reduced sclerotial soil-longevity by whole crop amendment and plastic covering. *Journal of Plant Diseases and Protection* 108, 143–151.

Theunissen, J. (1984) Supervised pest control in cabbage crops: theory and practice. *Mittelungen Biologische Bundesanstalt für Land-und Forstwirtschaft* 218, 76–84.

Theunissen, J. and Den Ouden, H. (1980) Effects of intercropping with *Spergula arvensis* on pests of Brussels sprouts. *Entomologia Experimentalis et Applicata* 27, 260–268.

Theunissen, J., Booij, C.J.H. and Lotz, L.A.P. (1995) Effects of intercropping white cabbage with clovers on pest infestation and yield. *Entomologia Experimentalis et Applicata* 74, 7–16.

Thompson, A.K. (1998) *Controlled Atmosphere Storage of Fruits and Vegetables*. CAB International, Wallingford, UK.

Tian, M.S., Davies, L., Downs, C.G., Liu, X.F. and Lill, R.E. (1995) Effects of floret maturity, cytokinin and ethylene on broccoli yellowing after harvest. *Post Harvest Biology and Technology* 6, 29–40.

Tilman, D. (1988) *Plant Strategies and the Dynamics and Structure of Plant Communities*. Princeton Monographs. Princeton University Press, Princeton, New Jersey.

Toivonen, P.M.A. and DeEll, J.R. (1998) Differences in chlorophyll fluorescence and chlorophyll content of broccoli associated with maturity and sampling section. *Post Harvest Biology and Technology* 14, 61–64.

Tokumasu, S., Kato, S. and Yano, F. (1975) The dormancy of seed as affected by different humidities during storage in *Brassica*. *Japanese Journal of Breeding* 25, 197–202.

Ton, J., Davison, S., van Loon, L.C. and Pieterse, C.M.J. (2001) Heritability of rhizobacteria-mediated induced systemic resistance and basal resistance in *Arabidopsis*. *European Journal of Plant Pathology* 107, 63–68.

Tonguç, M. and Griffiths, P.D. (2004a) Genetic relationships of *Brassica* vegetables determined using database derived simple sequence repeats. *Euphytica* 137, 193–201.

Tonguç, M. and Griffiths, P.D. (2004b) Evaluation of *Brassica carinata* accessions for resistance to Black Rot (*Xanthomonas campestris* pv. *campestris*). *HortScience* 39, 952–954.

Tonguç, M. and Griffiths, P.D. (2004c) Development of black rot resistant interspecific hybrids between *Brassica oleracea* L., cultivars and *Brassica* accession A 19182 using embryo rescue. *Euphytica* 136, 313–318.

Toxopeus, H. (1974) Outline of the evolution of turnips and coles in Europe and the origin of winter rape, swede–turnips and rape kales. *Proceedings of Eucarpia 'Cruciferae 1974' Meeting*, Dundee (September 25–27), pp. 1–7.

Toxopeus, H. (1979) The domestication of Brassica crops in Europe – evidence from the herbal books of the 16th and 17th centuries. *Proceedings of the Eucarpia 'Cruciferae 1979' Conference*, Agricultural University, Wageningen, The Netherlands (October 1–3), pp. 29–37.

Toxopeus, H. (1993) *Brassica rapa* L. In: Siemonsma, J.S. and Piluek, K. (eds) *Plant Resources of South-east Asia*. No. 8. *Vegetables*. Bogor, Wageningen, pp. 121–123.

Toxopeus, H., Oost, E.H. and Reuling, G. (1984) Current aspects of the taxonomy of cultivated brassica species. *Cruciferae Newsletter* 9, 55–58.

Trdan, S., Valic, N., Znidarcic, D., Vidrih, M., Bergant, K., Zlatic, E. and Milevoj, L. (2005) The role of Chinese cabbage as a trap crop for flea beetles (Coleptera: Chrysomelidae) in production of white cabbage. *Scientia Horticulturae* 106, 12–24.

Tsunoda, S., Hinata, K. and Gomez-Campo, C. (eds) (1984) *Brassica Crops and Wild Allies: Biology and Breeding*. Japanese Societies Press, Tokyo, Japan.

Tukahirwa, E.M. and Coaker, T.H. (1982) Effect of mixed cropping on some insect pests of Brassicas: reduced *Brevicoryne brassicae* infestations and influences on epigeal predators and the disturbance of oviposition behaviour in *Delia brassicae*. *Entomologia Experimentalis et Applicata* 32, 129–140.

U, N. (1935) Genome analysis in *Brassica* with special reference to the experimental formation of *B. napus* and peculiar mode of fertilization. *Japanese Journal of Botany* 7, 389–452.

Unger, P.W. and McCalla, T.M. (1980) Conservation tillage systems. *Advances in Agronomy* 33, 1–58.

Vågen, I.M. (2003) Nitrogen uptake in a broccoli crop. 1: Nitrogen dynamics on a relative time scale. *Acta Horticulturae* 627, 195–202.

Vågen, I.M., Skelvåg, A.O. and Bonesmo, H. (2004) Growth analysis of broccoli in relation to fertiliser nitrogen application. *Journal of Horticultural Science and Biotechnology* 79, 484–492.

Van der Meer, J. (1989) *The Ecology of Intercropping*. Cambridge University Press, London.

Vane-Wright, R.I. and Ackery, P.R. (eds) (1984) *The Biology of Butterflies. Symposium of the Royal Entomological Society of London.* Published for the Royal Entomological Society of London by Academic Press, London.

Vincente, J.G., Conway, J., King, G.J. and Taylor, J.D. (2000) Resistance to *Xanthomonas campestris* pv. *campestris* in *Brassica* spp. *Acta Horticulturae* 539, 61–68.

Viswakarma, N., Bhattacharya, R.C., Chakrabarty, R., Dargan, S., Bhat, S.R., Kirti, P.B., Shastri, N.V. and Chopra, V.L. (2004) Insect resistance to transgenic broccoli ('Pusa Broccoli KTS 1') expressing a synthetic *cry1A(b)* gene. *Journal of Horticultural Science and Biotechnology* 79, 182–188.

Walkey, D.G.A. and Pink, D.A.C. (1988) Reactions of white cabbage (*Brassica oleracea* var. *capitata*) to four different strains of turnip mosaic virus. *Annals of Applied Biology* 112, 273–284.

Walsh, J.A. (1997) Turnip mosaic virus. Data sheet for Commonwealth Agricultural Bureau International Global Crop Protection Compendium. CAB International, Wallingford, UK.

Walsh, J. and Hunter, P. (2004) Pest and disease management: sunken and disorderly. *The Grower*, pp. 14–15.

Walsh, J.A. and Jenner, C.E. (2002) Turnip mosaic virus and the quest for durable resistance. *Molecular Plant Pathology* 3, 289–300.

Walsh, J., Hunter, P. and MacDonald, N. (2004) *Internal Disorders of Stored White Cabbage.* Factsheet No. 11/04 Horticultural Development Council, East Malling, Kent.

Walters, T., Mutschler, M.A. and Earle, E.D. (1992) Protoplast fusion-derived Ogura male sterile cauliflower with cold tolerance. *Plant Cell Reports* 10, 624–628.

Wan-Zhi, Y.E., Jia-Shu, C., Xiang, X. and Zeng, G.-W. (2003) Molecular cloning and characterisation of the genic male sterility related gene *CYP86MF* in Chinese cabbage (*Brassica campestris* L. ssp. *chinensis* Makino var. *comunis* Tsen et Lee). *Journal of Horticultural Science and Biotechnology* 78, 319–323.

Warman, P.R. (2005) Soil fertility, yield and nutrient contents of vegetable crops after 12 years of compost or fertilizer amendments. *Biological Agriculture and Horticulture* 23, 85–96.

Warner, J., Cerkauskas, R., Zhang, T. and Hao, X. (2003) Response of Chinese cabbage cultivars to petiole spotting and bacterial soft rot. *HorTechnology* 13, 190–195.

Watanabe, S. (1953) Studies on the dormancy of seed in Cruciferous vegetables. *Journal of the Japanese Society for Horticultural Science* 10, abstract.

Watson, D.J. (1947) Comparative physiological studies on the growth of field crops. 1. Variation in net assimilation rate and leaf area between species and varieties and within and between years. *Annals of Botany* 11, 41–76.

Weaver, S.E. (1984) Critical periods for weed competition. *Weed Research* 24, 317–325.

Wellington, P.S. (1954) The heading of broccoli. Factors affecting quality and time. *Agriculture* 61, 431–434.

Wellington, P.S. and Quartley, C.E. (1972) A practical system for classifying, naming and identifying some cultivated Brassicas. *Journal of the National Institute of Agricultural Botany* 12, 413–432.

Wheeler, T.R., Ellis, R.H., Hadley, P. and Morison, J.I.L. (1995) Effects of CO_2, temperature and their interaction on growth, development and yield of cauliflower (*Brassica oleracea* var. *botrytis*). *Scientia Horticulturae* 60, 181–197.

Whitmore, A.P. and Groot, J.J.R. (1994) The mineralisation of N from finely or coarsely chopped crop residues. *European Journal of Agronomy* 4, 367–373.

Whyte, R.O. (1960) *Crop Production and Environment*. Faber and Faber, London.

Wiebe, H.J. (1972a) Wirkung von Temperatur und Licht auf Wachsum und Entwicklung von Blumenkohl. I. Dauer der Jungendphase fur Vernalisation. *Gartenbauwissenschaft* 37, 165–178.

Wiebe, H.J. (1972b) Wirkung von Temperatur und Licht auf Wachstum und Entwicklung von Blumenkohl. II. Optimale Vernalisationstemperature unf Vernalisationdauer. *Gartenbauwissenschaft* 37, 293–303.

Wiebe, H.J. (1972c) Wirkung von Temperatur und Licht auf Wachtsum und Entwiklung von Blumenkohl. III. Vegitative Phase. *Gartenbauwissenschaft* 37, 455–469.

Wiebe, H.J. (1975) Effect of temperature on the variability and maturity date of cauliflower. *Acta Horticulturae* 52, 69–75.

Wiebe, H.J. (1981) Influence of soil water potential during different growth periods on yield of cauliflower. *Acta Horticulturae* 119, 299–300.

Wien, H.C. (1997) Transplanting. In: Wien, H.C. (ed.), *The Physiology of Vegetable Crops*. CAB International, Wallingford, UK, pp. 37–67.

Wilcocksen, S.J. and Abuzeid, A.E. (1991) Growth of axillary buds of Brussels sprouts (*Brassica oleracea* var. *bullata* subvar. *gemmifera*). *Journal of Agricultural Science (Cambridge)* 117, 207–212.

Williams, P.H. (1980) Bee-sticks, an aid to pollinating *Cruciferae*. *HortScience* 15, 802–803.

Williams, P.H. (1981) *Screening Crucifers for Multiple Disease Resistance*. Department of Plant Pathology, University of Wisconsin, Madison, Wisconsin.

Williams, P.H. and Heyn, F.W. (1980) The origins and development of cytoplasmic male sterility in Chinese cabbage. In: *Chinese Cabbage*. Proceedings of the First International Symposium. Asian Vegetable Research Development Centre, Taiwan, pp. 293–300.

Williams, P.H. and Hill, C.B. (1986) Rapid cycling populations of *Brassica*. *Science* 232, 1385–1389.

Williams, P.H., Staub, T. and Sutton, J.C. (1972) Inheritance of resistance in cabbage to black rot. *Phytopathology* 62, 247–252.

Wills, A.B., Fyfes, K. and Wiseman, E.M. (1979) Testing F_1 hybrids of *Brassica oleracea* for sibs by seed isoenzyme analysis. *Annals of Applied Biology* 91, 263–270.

Wu Geng Min (1957) *Vegetable Gardening in China*. Scientific Publisher, Peking.

Wurr, D.C.E. and Fellows, J.R. (1998) Leaf production and curd initiation of winter cauliflower in response to temperature. *Journal of Horticultural Science and Biotechnology* 73, 691–697.

Wurr, D.C.E., Kay, R.H. and Allen, E.J. (1981a) Studies on the growth and development of winter heading cauliflower. *Journal of Agricultural Science* 97, 409–419.

Wurr, D.C.E., Kay, R.H. and Allen, E.J. (1981b) The effects of cold treatments on the curd maturity of winter-heading cauliflowers. *Journal of Agricultural Science* 97, 421–425.

Wurr, D.C.E., Elphinstone, E.D. and Fellows, J.R. (1988) The effect of plant raising and cultural factors on the curd initiation and maturity characteristics of summer/autumn cauliflower crops. *Journal of Agricultural Science* 111, 427–434.

Wurr, D.C.E., Fellows, J.R. and Hiron, R.W.P. (1990a) The influence of field conditions on the growth and development of four cauliflower cultivars. *Journal of Horticultural Science and Biotechnology* 65, 565–572.

Wurr, D.C.E., Fellows, J.R. and Hiron, R.W.P. (1990b) Relationships between the times of transplanting, curd initiation and maturity in cauliflower. *Journal of Agricultural Science* 114, 193–199.

Wurr, D.C.E., Fellows, J.R. and Hambridge, A.J. (1991) The influence of field environmental conditions on calabrese growth and development. *Journal of Horticultural Science and Biotechnology* 66, 495–504.

Wurr, D.C.E., Fellows, J.R. and Hambridge, A.J. (1992) The effect of plant density on calabrese head growth and its use in a predictive model. *Journal of Horticultural Science and Biotechnology* 67, 77–85.

Wurr, D.C.E., Fellows, J.R., Phelps, K. and Reader, R.J. (1993) Vernalisation in summer/autumn cauliflower (*Brassica oleracea* var. *botrytis* L.). *Journal of Experimental Botany* 44, 1507–1514.

Wurr, D.C.E., Fellows, J.R., Phelps, K. and Reader, R.J. (1994) Testing a vernalisation model on field grown crops of four cauliflower cultivars. *Journal of Horticultural Science and Biotechnology* 69, 251–255.

Wurr, D.C.E., Fellows, J.R. and Phelps, K. (1996) Growth and development of heads and flower stalk extension in field-grown Chinese cabbage in the UK. *Journal of Horticultural Science and Biotechnology* 71, 273–286.

Wurr, D.C.E., Fellows, J.R. and Fuller, M.P. (2004) Simulated effects of climate change on the production pattern of winter cauliflower in the UK. *Scientia Horticulturae* 1001, 359–372.

Wurr, D.C.E., Fellows, J.R., Sutherland, R.A. and Elphinestone, E.D. (1990) A model of cauliflower curd growth to predict when curds reach a specified size. *Journal of Horticultural Science and Biotechnology* 65, 555–564.

Wydrzynski, T.J., Chow, W.S. and Badger, M.R. (eds) (1995) Chlorophyll fluorescence: origins, measurements, interpretations and applications. *Australian Journal of Plant Physiology* 22, 1–355.

Yamamura, K. and Yano, E. (1999) Effects of plant density on the survival rate of cabbage pests. *Research into Population Ecology* 41, 183–188.

Yamauchi, N., Harada, K. and Watada, A.E. (1997) *In vitro* chlorophyll degradation in stored broccoli (*Brassica oleracea* L. var. *italica* Plen.) florets. *Postharvest Biology and Technology* 12, 239–245.

Yang, Y.W., Tsai, C.C. and Wang, T.T. (1998) Heat tolerant broccoli F1 hybrid, 'Ching-Long 45'. *HortScience* 33, 1090–1091.

Yarrow, S.A., Wu, S.C., Barnsby, T.L., Kemble, R.J. and Shepard, J.F. (1986) The introduction of CMS mitochondria to triazine tolerant *Brassica rapa* L., var. 'Regent', by micromanipulation of individual heterokaryons. *Plant Cell Reports* 5, 415–418.

Zeven, A.C. (1996) Sixteenth to Eighteenth century depictions of cole crops, (*Brassica oleracea* L.), turnips (*B. rapa* L. cultivar group vegetable turnip) and radish (*Raphanus sativus* L.) from Flanders and the present-day Netherlands. *Acta Horticulturae* 407, 29–33.

Zeven, A.C. and Brandenburg, W.A. (1986) Use of paintings from the 16th to 19th centuries to study the history of domesticated plants. *Economic Botany* 40, 397–408.

Zhang, G.Q., Tang, G.X., Song, W.J. and Zhou, W.J. (2004) Resynthesising *Brassica napus* between *Brassica rapa* and *B. oleracea* through ovary culture. *Euphytica* 140, 181–187.

Index

Use of this index: cross references are provided between the Latin binomial names of major crops and their colloquial common names. For the sake of brevity, minor crops are listed under their Latin binomials with colloquial names solely listed. The taxonomy of *Brassica* is complex and confused and some arbitrary decisions have been taken as to the nomenclature used, especially in the relationship between binomial and colloquial names.